Lecture Notes in Mathematics

A collection of informal reports and seminars
Edited by A. Dold, Heidelberg and B. Eckmann, Zürich

315

Klaus Bichteler

The University of Texas at Austin, Austin, TX/USA

Integration Theory
(with Special Attention to Vector Measures)

Springer-Verlag
Berlin · Heidelberg · New York 1973

AMS Subject Classifications (1970): 28 A 30, 28 A 45, 46 G 15

ISBN 978-3-540-06158-8 Springer-Verlag Berlin · Heidelberg · New York
ISBN 978-0-387-06158-0 Springer-Verlag New York · Heidelberg · Berlin

Offsetdruck: Julius Beltz, Hemsbach /Bergstr.

PREFACE

These notes reflect a course on vector measures given at The University of Texas at Austin in the summer term of 1970. They contain as much material as is usual for a course on this topic.

The presentation is unusual, though, and might possibly be of some interest even to the expert. It is motivated by the desire to carry M. H. Stone's unified treatment of set functions and Radon measures as far as possible.

This has been achieved by the use of a new and, I believe, natural definition of measurability, far-reaching enough to embrace the definitions both of Bourbaki and Caratheodory-Halmos and combining the flexibility and intuitive appeal of the former with the applicability to probability theory of the latter. It can also be applied to more general situations than either of these.

Another unorthodox feature is the axiomatic treatment of upper integrals and their generalizations, the upper gauges, leading to the Daniell-integral also for measures and linear maps that do not have finite variation.

I should like to express my gratitude to Professor Roy P. Kerr, who eradicated a frightful number of mistakes and Germanisms from the original manuscript, and to NSF for their support of these notes; and last but not least to Ms. Linda S. White who typed these notes with extreme skill and diligence and with inexhaustible patience.

Klaus Bichteler *

---

*) Supported by NSF Grant GP-20541

TABLE OF CONTENTS

CHAPTER IV. ELEMENTARY OPERATIONS ON MEASURE SPACES

§1. FIELDS OF MEASURES AND OF UPPER GAUGES

§2. IMAGES OF MEASURES

§3. PROJECTIVE LIMITS

CHAPTER V. LIFTINGS AND DERIVATIVES

§1. LIFTINGS

§2. INTEGRAL REPRESENTATION OF LINEAR MAPS

1. Introduction

A. Sketch of the Daniell integral

A Riemann step function on the line is a function of the form

$$s = \sum \{r_i \cdot \chi_{A_i} : 1 \leq i \leq n\},$$

where the A_i are disjoint finite intervals on the line, the χ_{A_i} are their characteristic functions, and the r_i are real numbers. The Riemann step functions form a vector space, $\mathcal{S}(\mathbb{R})$, which is closed under pointwise infima and suprema. The integral of the step function s is defined by

$$\int s(x)dx = \sum \{r_i \cdot |A_i| : 1 \leq i \leq n\},$$

where $|A_i|$ is the length of the interval A_i. In other words, $\int s(x)dx$ is the sum of the products

measure of step × height above ground,

taken over all steps. The map $s \to \int s(x)dx$ is a linear map from the vector space $\mathcal{S}(\mathbb{R})$ to $\mathbb{R}$, and is positive on positive step functions.

Following intuitive arguments, one says that an arbitrary function f is Riemann integrable if, for every $\varepsilon > 0$, there are two step functions ℓ and u in $\mathcal{S}$ such that

$$\ell \leq f \leq u \quad \text{and} \quad \int (u(x) - \ell(x))dx < \varepsilon.$$

By defining $s = u - \ell$, this can be rewritten as

$$(1) \qquad |f - \ell| \leq s \quad \text{and} \quad \int s(x)dx < \varepsilon.$$

1A

For any positive real-valued function h on $\mathbb{R}$, define

$$m^J(h) = \inf\{\int s(x)dx: h \le s \in \mathscr{S}\}$$

whenever the right hand side exists, and set $m^J(h) = \infty$ if h fails to be majorized by a Riemann step function. The "Jordan gauge" m^J has the following properties.

(a) It is subadditive: $m^J(f+g) \le m^J(f) + m^J(g)$ for $f,g: \mathbb{R} \to \mathbb{R}_+$.

(b) It is positively homogeneous: $m^J(rf) = rm^J(f)$ for $r \in \mathbb{R}_+$, $f: \mathbb{R} \to \mathbb{R}_+$

(c) It is increasing: $f \le g$ implies $m^J(f) \le m^J(g)$.

(d) $|\int s(x)dx| \le \int |s(x)|dx = m^J(|s|)$ for $s \in \mathscr{S}$.

Inspection of (1) shows that a function $f: \mathbb{R} \to \mathbb{R}$ is Riemann integrable if and only if for every $\varepsilon > 0$ there is a Riemann step function ℓ such that

(2) $$m^J(|f - \ell|) < \varepsilon.$$

In other words, f is Riemann integrable if its m^J-distance from $\mathscr{S}(\mathbb{R})$,

(3) $$\inf\{m^J(|f - \ell|): \ell \in \mathscr{S}(\mathbb{R})\},$$

is zero, i.e., if it belongs to the closure $\mathscr{L}^1(\mathscr{S}(\mathbb{R}), m^J)$ of $\mathscr{S}(\mathbb{R})$ under the seminorm

$$m_1^J(f) = m^J(|f|) \qquad f: \mathbb{R} \to \mathbb{R}.$$

From (d), the Riemann integral is continuous on $\mathscr{S}(\mathbb{R})$ in the seminorm m_1^J, and there is a unique continuous extension of the integral to all of $\mathscr{L}^1(\mathscr{S}(\mathbb{R}), m^J)$. This is the Riemann integral extension.

Unfortunately, it has serious shortcomings, all connected with the fact that it does not produce enough integrable functions. Reformulating Lebesgue's ideas, Daniell has shown how to improve on this by replacing the Jordan gauge m^J by a smaller gauge M having the same properties (a) - (d). Since it is easier for a function to be close to $\mathcal{S}(\mathbb{R})$ with respect to M than with respect to m^J, there will be more functions in the closure $\mathcal{L}^1(\mathcal{S}(\mathbb{R}), M)$ of $\mathcal{S}(\mathbb{R})$ with respect to M than there are in $\mathcal{L}^1(\mathcal{S}(\mathbb{R}), m^J)$; and the integral can be extended to all of them by continuity as before.

In fact, it is possible to find a gauge M obeying (a) - (d) and small enough to satisfy the following condition:

(e) If (f_n) is any increasing sequence of positive functions, then

$$\sup M(f_n) = M(\sup(f_n)).$$

Gauges satisfying (a) - (e) are called upper integrals, and most results in Lebesgue-Daniell integration theory are statements about these. Specifically, property (e) has a vast number of consequences, unmatched in Riemann's integration theory. Accordingly, the theory of upper integrals and their generalizations, the upper gauges, will occupy a large part of these notes.

B. Generalizations

More general situations than that of functions on the line occur frequently and can be handled along the same lines.

Definitions. Let X be a set. A vector space $\mathfrak{R}$ of real-valued functions on X is an integration lattice, or a Stone lattice, if for every pair of elements φ, ψ of $\mathfrak{R}$ it contains their pointwise infimum $\varphi \wedge \psi$, their pointwise supremum $\varphi \vee \psi$, and $\varphi \wedge 1$.

An integral or measure on $\mathfrak{R}$ is a linear map m from $\mathfrak{R}$ to some

Banach space E, and the pair $(\mathcal{R}, m)$ is called an E-valued <u>elementary integral</u> on X.#

Note that the Riemann step functions $\mathcal{S}(\mathbb{R})$ form an integration lattice and that $(\mathcal{S}(\mathbb{R}), \int \cdot dx)$ is an elementary integral. Linear maps on an integration lattice with values in an arbitrary topological vector space can also be considered, but this will not be done here. The following two examples describe whole classes of elementary integrals, and most elementary integrals appearing in nature belong to one of them.

<u>Example 1</u>. Let X be a locally compact Hausdorff space. A positive ($\varphi \geq 0$ implies $m(\varphi) \geq 0$) real-valued linear form m on the space $C_{00}(X)$ of continuous real-valued functions of compact support is called a <u>positive Radon measure</u> on X. The couple $(C_{00}(X), m)$ is evidently an elementary integral, and, conversely, most elementary integrals that play a role in functional analysis occur as Radon measures.

<u>1.1. Example 2</u>. The Riemann integral on the line can be generalized in another direction, seemingly different from the one that leads to the notion of an elementary integral: The integral of a Riemann step function, $\sum r_i \chi_{A_i}$, is known because the steps A_i can be measured: Let $\mathcal{C}(\mathbb{R})$ denote the family of subsets of $\mathbb{R}$ that are finite unions of bounded intervals, either open, half-open, or closed. It is closed under finite unions, finite intersections, and differences. The measure, or <u>Riemannian content</u>, of a set A in $\mathcal{C}(\mathbb{R})$ is the sum of the lengths of the disjoint intervals which make up A. A generalization of this situation is reflected in the following definition.

Definition. Let X be a set. A clan or ring on X is a family $\mathscr{C}$ of subsets of X that is closed under differences, finite unions and finite intersections. An elementary content on X is a couple $(\mathscr{C},m)$ consisting of a ring $\mathscr{C}$ on X and a map m from $\mathscr{C}$ to some Banach space, such that

$$m(A \cup B) = m(A) + m(B) \quad \text{for } A,B \in \mathscr{C} \text{ and } A \cap B = \emptyset. \#$$

Elementary contents occur frequently in probability theory.

An elementary content gives rise to an elementary integral in much the same way as the Riemann content gives rise to the integral of step functions. We shall go into the details presently, but first make a useful convention.

Convention. Henceforth the subsets of a set are identified with their characteristic functions. For example, $A \cap B = A \wedge B = AB$ is the intersection of A and B and equals $\chi_{A \cap B}$; and $A \Delta B = |A - B| = A \setminus B \cup B \setminus A$ is the symmetric difference of A and B. Furthermore, if f is any function then $[f = a]$, $[f \leq a]$, etc., denote the sets of points in the domain of definition of f in which the equality, or inequality, etc., holds. For instance, $f = f[f \neq o]$ is true for all Banach-valued functions f on X.#

Suppose we are given an elementary content m on a clan $\mathscr{C}$ of subsets of X, and let us construct the associated elementary integral: its integration lattice is $\mathscr{S}(\mathscr{C})$, set of all finite linear combinations of the (idempotent) functions in $\mathscr{C}$; on $\mathscr{S}(\mathscr{C})$, we define a linear form $\tilde{m}$ by

$$\tilde{m}(\sum r_i A_i) = \sum r_i m(A_i), \qquad \sum r_i A_i \in \mathscr{S}(\mathscr{C}).$$

The couple $(\mathscr{S}(\mathscr{C}), \tilde{m})$ is called the elementary integral associated with

$\mathcal{C}(m)$. To see that this definition makes sense we have to show that $\mathcal{S}(\mathcal{C})$ is, indeed, an integration lattice and that $\tilde{m}$ is well-defined. In preparation, a lemma on finitely generated clans is needed:

The intersection of any number of clans on X is clearly also a clan; and therefore the clan generated by any family F of subsets of X is well-defined as the intersection of all clans containing F.

1.2. Lemma. Let $A_1,\dots,A_n$ be subsets of X. There are $2^n - 1$ mutually disjoint (and quite possibly void) sets K_i in the clan generated by $A_1,\dots,A_n$ such that every A_k is the union of those of the K_i that are entirely contained in A_k.

Proof. The statement is clear for $n = 1$. Suppose it is true for $n - 1$ and let K_i $(i = 1,\dots,2^{n-1} - 1)$ be mutually disjoint sets in the clan generated by $A_1,\dots,A_{n-1}$, having the property in question. The $2^n - 1$ sets $K_i' = A_n \cap K_i$, $K_i'' = K_i - K_i'$ $(i = 1,\dots,2^{n-1} - 1)$ and $K = A_n \setminus \bigcup\{K_i : 1 \leq i \leq 2^{n-1} - 1\}$ meet the requirements of the lemma, for n.#

To see that $\mathcal{S}(\mathcal{C})$ is an integration lattice, let $\varphi = \sum r_k A_k \in \mathcal{S}(\mathcal{C})$. From the lemma, φ can be written as a linear combination of disjoint elements of $\mathcal{C}$:

$$\varphi = \sum_{i=1}^{2^n-1} K_i \cdot \sum\{r_k : K_i \subset A_k\}.$$

$\mathcal{S}(\mathcal{C})$ consists therefore exactly of all real-valued functions on X that take only finitely many values and that take their non-zero values in sets belonging to $\mathcal{C}$. This description easily shows that $\mathcal{S}(\mathcal{C})$ is closed under taking finite infima and infima with 1, and so that it is an integration lattice. To see that $\tilde{m}$ is well-defined, we will suppose that $\sum_{k=1}^{n} r_k A_k = 0$ and show that then $\sum r_k m(A_k) = 0$: if we choose the K_i $(1 \leq i \leq 2^n - 1)$ as in the lemma, then

$$\sum r_k m(A_k) = \sum_k r_k \sum \{m(K_i): K_i \subset A_k\}$$

$$= \sum_i \sum \{r_k m(K_i): K_i \subset A_k\} = 0,$$

since

$$\sum \{r_k : K_i \subset A_k\} = 0 \quad \text{for} \quad 1 \leq i \leq 2^n - 1.$$

To summarize example 1.1: Every Banach-valued elementary content $(\mathscr{K}, m)$ gives rise to an elementary integral $(\mathscr{S}(\mathscr{K}), \tilde{m})$, by the simple expedient of linear extension.#

The notion of an elementary integral covers the two forms in which the elementary data for integration theory appear, namely those represented by the two examples above. However, the elementary contents are usually taken to be basic, for predominantly (but not exclusively) historical reasons. Example 1 is subsumed via the celebrated Riesz representation theorem, which asserts that with a Radon measure m there is naturally associated an elementary content having the same integration theory as m. The proof of Riesz's theorem, though, is much more complicated than the simple considerations in 1.1, its counterpart, and uses large parts of the integration theory of elementary integrals. From that point of view it is therefore preferable to develop the integration theory of elementary integrals rather than of elementary contents.

Fortunately, a compromise is possible. Many results on elementary integrals have their complete and easily identifiable analogues in the theory of elementary contents, and vice versa. This is due to the fact that often only the structure common to both integration lattices and clans is used to establish a result. This common structure is identified in the following definition.

Definition. An integration domain on a set X is a collection $\mathscr{I}$ of real-valued functions on X such that whenever φ and ψ belong to $\mathscr{I}$ then so do $\varphi \wedge \psi$, $\varphi \vee \psi$, $\varphi - \varphi \wedge \psi$, and $\varphi \wedge 1$. A measure on $\mathscr{I}$ is a map m from $\mathscr{I}$ to some Banach space, satisfying

$$m(\varphi) + m(\psi) = m(\varphi+\psi) \quad \text{whenever } \varphi, \psi, \text{ and } \varphi+\psi \text{ are in } \mathscr{I}.$$

A real-valued measure is positive if it is positive on the positive functions in $\mathscr{I}$. A couple $(\mathscr{I},m)$ as above is an elementary measure on X, and the triple $(X,\mathscr{I},m)$ is an elementary measure space. #

Both a clan $\mathscr{C}$ and an integration lattice $\mathfrak{R}$ are evidently integration domains, and both an integral on $\mathfrak{R}$ and a content on $\mathscr{C}$ are measures. Note that an integration domain is a clan if it consists of sets, and an integration lattice if it is a vector space.

We shall compromise, and develop the theory of elementary measures as far as possible. Nearly every result of Chapter I has two interpretations, one for elementary integrals and one for elementary contents. Since one usually visualizes integration lattices and clans in a rather different way, it is advisable for the reader to formulate these two interpretations separately. From time to time the theories of elementary integrals and elementary contents differ naturally. In these instances we shall deal with the former in the text and with the latter in the exercises.

C. Layout of the material

Chapter I contains the study of elementary measures. The measures on a fixed integration domain are analyzed in the terms of Riesz space theory, and the measures accessible to Lebesgue-Daniell integration are singled out.

Chapter II contains the Daniell extension theory of elementary measures, as sketched in 1A. The treatment is slightly unorthodox in that upper gauges--the analogues of the usual N_p-norms--are defined axiomatically, and treated first. Next, it is shown how they are used to extend an elementary measure, and only in the following section are the standard upper integrals constructed. The reason for doing so is that there are many non-standard upper gauges which can be profitably used, notably in connection with integrals of measure fields (Ch. IV).

The integration theory of measures that do not have finite variation, such as some stochastic integrals and weakly compact linear maps is somewhat new: It is shown that they are amenable to Daniell integration as well and without additional effort. The last section exhibits a number of different upper gauges, among them the N_p-norms.

Chapter III treates measurability in Bourbaki's fashion, i.e., by taking Lusin's theorem for the definition. The topology of the locally compact spaces considered by Bourbaki is replaced by the uniformity generated by the integration lattices considered here. Our definition generalizes the one of Bourbaki and Carathéodory-Halmos, and is applicable to arbitrary upper gauges and to functions having values in uniform spaces, or in measure spaces, as well. It adds considerably to the unification of the integration of Radon measures and set functions initiated by Stone [46].

The particularities of tight measures are easily expressed in terms of the general theory.

Chapter IV on elementary operations treats fields of measures, products, images, and projective limits. Some martingale theorems are discussed from the point of view of projective limits.

Chapter V starts with some lifting theorems. They are later applied to give the general theorem of Radon-Nikodym, the disintegration of measures, and the structure of certain spaces of linear maps from L_E^p to other Banach spaces: e.g., the dual of L_E^p .

The sections are numbered consecutively. At the end of each there is a subsection "Supplements." The results of the text are numbered n.m , where n refers to the section and m to the particular result in question. The unproved results in the "Supplements" may be used as exercises; those that will be used later in the text are indicated by an asterisk.

D. Supplement. The integration lattice generated by a family of functions

1.3*. The intersection of any collection of integration lattices on a set X is itself an integration lattice. Given any family $\mathcal{F}$ of real-valued functions on X there is a smallest integration lattice $\mathcal{R}(\mathcal{F})$ containing $\mathcal{F}$. It is called the integration lattice generated by $\mathcal{F}$.

$\mathcal{R}(\mathcal{F})$ can be calculated as follows. Define $\mathcal{F}^0 = \mathcal{F}$ and then define $\mathcal{F}^n$ inductively:

If $n \equiv 1 \pmod 3$ put $\mathcal{F}^n = \{$set of finite linear combinations with rational coefficients, of elements of $\mathcal{F}^{n-1}\}$;

if $n \equiv 2 \pmod 3$ put $\mathcal{F}^n = \{\varphi \wedge \psi\colon \varphi, \psi \in \mathcal{F}^{n-1}\}$; and

if $n \equiv 0 \pmod 3$ put $\mathcal{F}^n = \{\varphi \wedge 1\colon \varphi \in \mathcal{F}^{n-1}\}$.

Finally, put $\mathcal{F}^\infty = \bigcup\{\mathcal{F}^n\colon n \in \mathbb{N}\}$. It is a vector space over the rationals, $\mathbb{Q}$, and is closed under taking finite suprema and infima and under $\varphi \to \varphi \wedge 1$. It is therefore called a $\mathbb{Q}$-integration lattice. If $\mathcal{F}$ is countable so is $\mathcal{F}^\infty$. $\mathcal{R}(\mathcal{F})$ lies in the uniform closure of $\mathcal{F}^\infty$. Taking real coefficients in the steps $n \equiv 1 \pmod 3$, instead, one obtains $\mathcal{F}^\infty = \mathcal{R}(\mathcal{F})$.

A similar program can be carried out for integration domains.

1.4*. The sets in an integration lattice form a clan.

LITERATURE: [10], [46]. In [40], "integration lattice" is what we call "elementary integral." Standard references for vector-valued measures are [11] and [14].

I. RIESZ SPACES AND ELEMENTARY INTEGRALS

2. Vector lattices

A. Ordered vector spaces

An ordered vector space is a real vector space $\mathcal{R}$ together with a subset $\mathcal{R}_+$, called the positive elements, satisfying

(1) $$\mathcal{R}_+ + \mathcal{R}_+ \subset \mathcal{R}_+,$$

(2) $$r\mathcal{R}_+ \subset \mathcal{R}_+ \quad \text{for} \quad 0 \leq r \in \mathbb{R}, \text{ and}$$

(3) $$\mathcal{R}_+ \cap \{-\mathcal{R}_+\} = \{0\}.$$

$\mathcal{R}_+$ is called the order cone of $\mathcal{R}$. It gives rise to the ordering

$$x \leq y \quad \text{if} \quad y - x \in \mathcal{R}_+, \qquad (x,y \in \mathcal{R}),$$

and this has the following properties:

(1') $x \leq y$ implies $x + z \leq y + z$ $(x,y,z \in \mathcal{R})$ and

(2') $x \leq y$ implies $rx \leq ry$ $(x,y \in \mathcal{R},\ r \in \mathbb{R}_+)$.

Conversely, given an order $\leq$ satisfying (1') and (2'), $\mathcal{R}_+ := \{x \in \mathcal{R}: x \geq 0\}$ is an order cone defining $\leq$.

Given $x,y \in \mathcal{R}$, it may or may not happen that there is a smallest element bigger than both x and y. If so, it is denoted by $x \vee y$ or $\sup\{x,y\}$. Similarly, $x \wedge y = \inf\{x,y\}$ is the biggest element majorized by both x and y, if such exists. Here are some useful formulae concerning infima and suprema. The equalities are to be understood to say that if one side exists, then so does the other and equality obtains. The proofs follow in parentheses.

2.1 $\quad x \wedge y + z = (x + z) \wedge (y + z)$ for all $z \in \mathcal{R}$.

(The map $x \to x + z$ is an order isomorphism.)

2.2 $\quad (x \vee y) + z = (x + z) \vee (y + z)$.

2.3 $\quad r(x \wedge y) = rx \wedge ry;\ r(x \vee y) = rx \vee ry$ if $0 \leq r \in \mathbb{R}$.

(The map $x \to rx$ is an order isomorphism.)

2.4 $\quad -(x \vee y) = (-x) \wedge (-y);\ -(x \wedge y) = (-x) \vee (-y)$.

($x \vee y \geq x, y \to -(x \vee y) \leq -x \wedge -y$; similarly, $-(x \wedge y) \geq -x \vee -y$. Replacing x, y by $-x$, $-y$ gives the other inequalities.)

2.5 $\quad x \wedge y + x \vee y = x + y$ if $x \wedge y$ or $x \vee y$ exist.

($y \leq x \vee y \to x + y - x \vee y \leq x$; similarly, $x + y - x \vee y \leq y$, hence (a) $x + y - x \vee y \leq x \wedge y$. On the other hand $y \geq x \wedge y \to x + y - x \wedge y \geq x$; as above this implies (b) $x + y - x \wedge y \geq x \vee y$. (a) and (b) result in 2.5.)

Put $x_+ = x \vee 0$, $x_- = -(x \wedge 0)$, and $|x| = x \vee -x$. $|x|$ is called the <u>absolute value</u>, <u>modulus</u>, or <u>variation</u> of x. If x_+ or x_- or $|x|$ exists then so do the others and

2.6 $\quad x = x_+ - x_-;\ |x| = x_+ + x_-;\ x_+ \wedge x_- = 0;\ |x| = |-x|$.

($x = x + 0 = x \vee 0 + x \wedge 0$ gives the first statement; $x_+ + x_- = x \vee 0 - x \wedge 0 = x \vee 0 + (-x) \vee 0 = x \vee 0 + 0 \vee x - x = 2x \vee 0 - x = x \vee (-x) = |x|$; lastly, as $x_+ \vee x_- \geq x, -x$ we have $x_+ \vee x_- \geq |x|$ and hence $0 \leq x_+ \wedge x_- = x_+ + x_- - x_+ \vee x_- \leq 0$.)

2.7 $\quad x_+ = 1/2(|x| + x);\ x_- = 1/2(|x| - x)$.

2.8 $x \vee y = 1/2(x + y + |x - y|)$; $x \wedge y = 1/2(x + y - |x - y|)$.

($x \vee y = y + (x - y) \vee 0 = y + 1/2(x - y + |x - y|) = 1/2(x+y+|x-y|)$ and similarly the other equation.)

2.9 $|x + y| \leq |x| + |y|$ (triangle inequality).

($|x| + |y| = x_+ + x_- + y_+ + y_- \geq x + y$.)

2.10 $|rx| = |r|\ |x|$.

2.11 $(x + y) \wedge z \leq x \wedge z + y \wedge z$ for $x, y, z \in \mathcal{R}_+$, if the infima exist.

($x \wedge z + y \wedge z = (y + (x \wedge z)) \wedge (z + (x \wedge z)) =$
$(y + x) \wedge (y + z) \wedge (z + x) \wedge (z + z) \geq (y + x) \wedge z$.)

It may or may not happen that a subset A of $\mathcal{R}$ has a least upper bound, i.e., a smallest element of $\mathcal{R}$ majorizing all elements of A. If it does, we denote this least upper bound by $\bigvee A$. Similarly, if a greatest lower bound exists we denote it by $\bigwedge A$. As $x \to x + b$ ($b \in \mathcal{R}$) is an order isomorphism, $\bigvee(A + b) = \bigvee A + b$. More generally,

2.12 $\bigvee(A + B) = \bigvee A + \bigvee B$, $\bigwedge(A + B) = \bigwedge A + \bigwedge B$,

provided the suprema or infima exist. The reason:
$\bigvee(A + B) = \bigvee\{\bigvee(A + b) | b \in B\} = \bigvee\{\bigvee A + b | b \in B\} = \bigvee A + \bigvee B$, and similarly for $\bigwedge$.

B. Riesz spaces

An ordered set is a <u>lattice</u> if any two elements x and y in it have a supremum $x \vee y$ and an infimum $x \wedge y$. An ordered vector space $\mathcal{R}$ is a <u>vector lattice</u>, or <u>Riesz space</u>, if it is a lattice in its order. In an ordered vector space it is not necessary to check all pairs of elements for the existence of infimum and supremum. Here are some criteria, which follow immediately from the formulae above.

2.13. Proposition. Let $\mathcal{R}$ be an ordered vector space with order cone $\mathcal{R}_+$. The following are equivalent:

(i) $\mathcal{R}$ is a vector lattice.

(ii) Any two elements of $\mathcal{R}$ have a supremum.

(iii) Any two elements of $\mathcal{R}$ have an infimum.

(iv) For each x in $\mathcal{R}$, there is a supremum $|x|$ for x and $-x$.

(v) For each x in $\mathcal{R}$, there is a supremum x_+ for x and 0.

(vi) $\mathcal{R} = \mathcal{R}_+ - \mathcal{R}_+$, and any two elements in $\mathcal{R}_+$ have a supremum in $\mathcal{R}_+$.

(vii) $\mathcal{R} = \mathcal{R}_+ - \mathcal{R}_+$, and any two elements in $\mathcal{R}_+$ have an infimum in $\mathcal{R}_+$.

The details of the proof are left to the reader. Here is an outline: Clearly, (i) implies all the other statements. If (ii) and (iii) hold, formula 2.5 defines $x \wedge y$ or $x \vee y$, whichever is lacking. If (iv) holds, formula 2.8 can be used instead. If (vi) is assumed and if $x, y \in \mathcal{R}$, then there are elements x_1, x_2, y_1, y_2 in $\mathcal{R}_+$ such that $x = x_1 - x_2$, $y = y_1 - y_2$, and $(x_1 + y_2) \vee (y_1 + x_2) - y_2 - x_2$ is then the least upper bound for x, y.

2.14. Proposition. (Riesz decomposition lemma). Let $\mathcal{R}$ be a Riesz space or an integration domain, and let $a_1, \ldots, a_m$ and $b_1, \ldots, b_n$ be positive elements of $\mathcal{R}$ for which $a_1 + \ldots + a_m = b_1 + \ldots + b_n \in \mathcal{R}$. Then there are positive elements c_{ij} in $\mathcal{R}$ $(i = 1, \ldots, m;\ j = 1, \ldots, n)$ such that

$$c_{i1} + \ldots + c_{in} = a_i \qquad (i = 1, \ldots, m)$$

and

$$c_{1j} + \ldots + c_{mj} = b_j \qquad (j = 1, \ldots, n).$$

In other words, the sums of the rows of the matrix (c_{ij}) are the a_i, and the sums of the columns are the b_j.

Proof. We deal first with the case $n = 2$. Put $u_0 = 0$, $u_i = a_1 + \dots + a_i$, and

$$c_{i1} = u_i \wedge b_1 - u_{i-1} \wedge b_1$$
$$c_{i2} = a_i - c_{i1} \qquad (i = 1,\dots,m).$$

It is clear that $c_{i1} + c_{i2} = a_i$, $c_{11} + \dots + c_{m1} = u_m \wedge b_1 = (b_1 + b_2) \wedge b_1 = b_1$, and $c_{12} + \dots + c_{m2} = b_1 + b_2 - b_1 = b_2$. As $u_i \geq u_{i-1}$ it is also clear that $c_{i1} \geq 0$. It remains to be shown that $c_{i2} \geq 0$ for $i = 1,\dots,m$. This follows from $u_{i-1} + a_i = u_i \rightarrow u_{i-1} \wedge b_1 + a_i \wedge b_1 \geq u_i \wedge b_1 \rightarrow u_i \wedge b_1 - u_{i-1} \wedge b_1 \leq a_i$. The proof for arbitrary n proceeds by induction. If $n = 1$ there is nothing to prove. Suppose the statement of the theorem is true for $n - 1$. We can find, according to the above, positive elements c_{i1} and x_i $(i = 1,\dots,m)$ such that

$$c_{i1} + x_i = a_i \qquad (i = 1,\dots,m),$$
$$c_{11} + \dots + c_{m1} = b_1, \quad \text{and}$$
$$x_1 + \dots + x_m = b_2 + \dots + b_n.$$

From the induction hypothesis there exist positive elements c_{ij} $(i = 1,\dots,m;\ j = 2,\dots,n)$ such that

$$c_{i2} + \dots + c_{in} = x_i \qquad (i = 1,\dots,m)$$
$$c_{1j} + \dots + c_{mj} = b_j \qquad (j = 2,\dots,n).$$

It is obvious that the matrix (c_{ij}) $(i = 1,\dots,m;\ j = 1,\dots,n)$ has all the desired properties. When $\mathcal{R}$ is an integration domain, we have made use of the following fact.

2.15. Lemma. Let $\mathcal{R}$ be an integration domain and let $x_1,\ldots,x_n$ be elements of $\mathcal{R}_+$ such that $x = \sum\{x_i\colon 1 \leq i \leq n\} \in \mathcal{R}$. Then $\sum\{x_i\colon 1 \leq i \leq k\} \in \mathcal{R}$ for $1 \leq k \leq n$.

Proof. The statement is obviously true for $n = 2$. Suppose it holds for $n-1$. Then $\sum\{x_i\colon 1 \leq i \leq n-1\} = x - x \wedge x_n \in \mathcal{R}$ and thus $\sum\{x_i\colon 1 \leq i \leq k\} \in \mathcal{R}$ for $1 \leq k \leq n$.#

C. Complete vector lattices

A vector lattice $\mathcal{R}$ is called complete if every subset F of $\mathcal{R}$ which is bounded from above admits a least upper bound $\vee F$.

$\mathbb{R}^S$ and the bounded functions in $\mathbb{R}^S$ are two examples of complete vector lattices. However, the space of continuous (or continuous and bounded) functions on a topological space is not, in general, a complete Riesz space. In a complete Riesz space every subset F bounded from below has a greatest lower bound (g.l.b.): $-F$ is bounded from above and $-\vee(-F)$ is a g.l.b. for F.

2.16. Proposition. For a vector lattice $\mathcal{R}$ to be complete it is sufficient that every increasingly directed and bounded subset of positive elements admit a least upper bound.

Proof. Suppose this is the case, and let F be any subset of $\mathcal{R}$ with an upper bound. Let $f \in F$. The set G_+ of positive elements of $G = -f + F$ is non-empty and bounded, and has the same supremum as G, if any. If we show that G_+ has a supremum $\vee G_+$, then F has supremum $\vee G_+ + f$, and we are finished. Now the set G' of all finite suprema of elements of G_+ is bounded and increasingly directed, and its supremum, which exists according to the assumption, is clearly also a supremum for G_+.#

2.17. Lemma. Let $\mathfrak{R}$ be a complete vector lattice, let F be bounded from above and let $a \in \mathfrak{R}$. Then $F \wedge a = \{x \wedge a | x \in F\}$ is bounded from above, and

$$(\vee F) \wedge a = \vee(F \wedge a).$$

Proof. It is clear that $(\vee F) \wedge a \geq \vee(F \wedge a)$. Also $(\vee F) \wedge a + (\vee F) \vee a = (\vee F) + a = \vee(F + a) = \vee\{x \wedge a + x \vee a | x \in F\} \leq \vee(F \wedge a) + \vee(F \vee a) \leq \vee(F \wedge a) + (\vee F) \vee a$, which, upon subtraction of $(\vee F) \vee a$, yields the reverse inequality.#

A vector subspace $\mathfrak{S}$ of the vector lattice $\mathfrak{R}$ is called a vector sublattice if it contains the infimum and supremum of each pair of its elements. Formulae 2.5 through 2.8 show that, for this to be the case, it is sufficient that $x \in \mathfrak{S}$ imply either $x_+ \in \mathfrak{S}$, or $|x| \in \mathfrak{S}$, or $x_- \in \mathfrak{S}$.

A vector sublattice $\mathfrak{S}$ of $\mathfrak{R}$ is an ideal or a solid subspace if $x \in \mathfrak{S}$ and $|y| \leq |x|$ imply $y \in \mathfrak{S}$.

A band is an ideal $\mathfrak{S}$ of $\mathfrak{R}$ that contains with every set bounded in $\mathfrak{R}$ its infimum and its supremum. From 2.6, an ideal $\mathfrak{S}$ is a band if every bounded increasingly directed subset of positive elements in $\mathfrak{S}$ has its supremum in $\mathfrak{S}$.

The intersection of an arbitrary family of sublattices (ideals, bands) is again a sublattice (ideal, band). It makes sense to talk about the sublattice (ideal, band) spanned by a subset A of $\mathfrak{R}$: it is the intersection of all sublattices (ideals, bands) containing A. The ideal and the band spanned by A will be denoted by $\{A\}$ and (A), respectively.

Examples. (1) The bounded functions on a set S form an ideal of $\mathbb{R}^S$, but not a band of $\mathbb{R}^S$, although they are a complete Riesz space in their own right.

(2) The functions in $\mathbb{R}^S$ that vanish in a fixed subset T of S form a band of $\mathbb{R}^S$.

(3) The Riesz space $C_{00}(X)$ of continuous functions with compact support on the locally compact Hausdorff space X forms an ideal of the space of all continuous functions vanishing at infinity.

Two elements x,y of a Riesz space $\mathcal{R}$ are <u>disjoint</u> or <u>orthogonal</u> if $|x| \wedge |y| = 0$. For $A \subset \mathcal{R}$, $A^{\perp}$ denotes the set of elements of $\mathcal{R}$ that are disjoint from every element of A.

<u>2.18. Theorem</u>. (F. Riesz). Let $\mathcal{R}$ be a complete vector lattice and $A \subset \mathcal{R}$.

(i) $A^{\perp}$ is a band.

(ii) $A^{\perp\perp} = (A)$.

(iii) Every element z of $\mathcal{R}$ can be written uniquely as a sum $z = x + y$ with $x \in (A)$ and $y \in A^{\perp}$. If z is positive then so are x and y.

<u>Proof</u>. Let $x,y \in A^{\perp}$, $r \in \mathbb{R}$, $a \in A$, and $|z| \leq |x|$. Then $0 \leq |z| \wedge |a| \leq |x| \wedge |a| = 0$, and hence $z \in A^{\perp}$. If $|r| \leq 1$, then $|rx| \leq |x|$ and $rx \in A^{\perp}$. Furthermore, $|x+y| \wedge |a| \leq (|x| + |y|) \wedge |a| \leq |x| \wedge |a| + |y| \wedge |a| = 0$ (2.11), and so $x + y \in A^{\perp}$. Finally, if $r > 1$, and if $|r| \leq n \in \mathbb{N}$ say, then $|rx| = |r||x| \leq n|x| \in A^{\perp}$, and $rx \in A^{\perp}$ in this case, also. This shows that $A^{\perp}$ is an ideal. Let F be a subset of $A^{\perp} \cap \mathcal{R}_+$ with least upper bound $\vee F$ in $\mathcal{R}$, and let $a \in A$. From lemma 2.17, $(\vee F) \wedge |a| = \vee(F \wedge |a|) = 0$, and so $A^{\perp}$ is, indeed, a band in $\mathcal{R}$, and (i) is proved. To prove (ii), observe that $A^{\perp\perp}$ is a band containing A, and hence $A^{\perp\perp} \supset (A)$ and $A^{\perp} \cap (A) = \{0\}$.

Let $0 \leq z \in \mathfrak{R}$, put $x = \vee\{x' \in (A): x' \leq z\}$, and let $y = z - x$. Clearly, $0 \leq x \in (A)$, and $0 \leq y$. We have to show that $y \in A^{\perp}$. To do this, let $a \in A$ and put $b = y \wedge |a| = (z - x) \wedge |a|$. Then $0 \leq b \in (A)$, and $b + x \leq z$. The maximality of x implies $b = 0$, and hence $y \in A^{\perp}$. If z is arbitrary, its positive and negative parts can each be decomposed into their components in (A) and in $A^{\perp}$. Uniqueness follows from the fact that $A^{\perp} \cap (A) = \{0\}$. This finishes the proof.

Remark. Let $\mathfrak{R}_i$ $(i = 1, 2)$ be two complete Riesz spaces, and give their vector space direct product $\mathfrak{R} = \mathfrak{R}_1 \times \mathfrak{R}_2$ the order defined by $(x,y) \geq 0$ if $x \in \mathfrak{R}_{1+}$ and $y \in \mathfrak{R}_{2+}$. Then $\mathfrak{R}$ is a complete vector lattice, and is called the ordered direct product of $\mathfrak{R}_1$ and $\mathfrak{R}_2$. From Riesz's theorem, $\mathfrak{R}$ is isomorphic in all respects with the direct sum of the bands (A), $A^{\perp}$, for each subset A of $\mathfrak{R}$.

2.19. Corollary. Let A be a subset of the complete vector lattice $\mathfrak{R}$. Then:

(i) An element x of $\mathfrak{R}$ belongs to $\{A\}$ if and only if $|x|$ is majorized by a finite sum $\sum |a_i|$, $a_i \in A$.

(ii) The positive elements of (A) are precisely the suprema of the majorized subsets of $\{A\}_+$.

Proof. The elements x of $\mathfrak{R}$ for which $|x|$ is majorized by a finite sum of absolute values of members of A obviously belong to $\{A\}$ and form an ideal. This proves (i). The suprema of bounded subsets of $\{A\}$ belong to (A). Conversely, if $x \in (A)_+$ and y is the supremum of all the elements in $\{A\}_+$ and smaller than x, then $x - y$ is disjoint from $\{A\}$, hence in $A^{\perp}$, and so zero.#

<u>2.20. Corollary</u>. If $a \in \mathfrak{R}$ and $0 \leq z = x + y$ where $x \in (a)$ and $y \perp a$, then $x = \bigvee\{n|a| \wedge z\colon n \in \mathbb{N}\}$. In particular, z is in (a) if and only if $z = \bigvee\{n|a| \wedge z\colon n \in \mathbb{N}\}$.

<u>Definition</u>. We say that x is <u>absolutely continuous</u> with respect to a, and write $x \ll a$, if $x \in (a)$.

<u>2.21. Corollary</u>. A vector subspace A of a complete Riesz space $\mathfrak{R}$ is a band if and only if

(i) $x \ll a \in A$ implies $x \in A$; and

(ii) whenever D is a subset of A consisting of mutually disjoint elements, then $\sum D$ belongs to A (provided $\sum D$ exists in $\mathfrak{R}$).

<u>Proof</u>. $\sum D$ is, of course, understood to be the supremum of the finite partial sums of D. Both conditions are evidently necessary. Suppose they are satisfied. Let $a, b \in A$. Splitting a into its part $a_1 \in (b)$ and its part a_2 orthogonal to b gives $a + b = (b + a_1) + a_2$. Both $b + a_1$ $(\ll b)$ and a_2 $(\ll a)$ belong to A, and so does their sum $a+b$. A is therefore an ideal. To show that it is a band, let $F \subset A_+$ be increasingly directed with supremum $f \in \mathfrak{R}$, and let D be a maximal collection of mutually disjoint elements of A_+ such that every finite subset of D is majorized by an element of F. The existence of D is a consequence of Zorn's lemma. It has a sum $d \in A$ smaller than f. Split f into its part f_1 absolutely continuous with respect to d and its part f_2 orthogonal to d. From formula 2.11, $f_2 = f \wedge f_2 = \bigvee\{g \wedge f_2\colon g \in F\}$. If $f_2 \neq 0$ then there is a $g \in F$ such that $g \wedge f_2 \neq 0$. The element $g \wedge f_2$ is in A, is disjoint from all elements of D, and can be adjoined to D. This contradicts the maximality of D, and so $f_2 = 0$. Finally $f = f_1$ is absolutely continuous with respect to d and so belongs to A.#

D. Homomorphisms and quotients

Let $\mathcal{R}$ and $\mathcal{S}$ be Riesz spaces. A linear map $\varphi\colon \mathcal{R} \to \mathcal{S}$ is a homomorphism of Riesz spaces if $\varphi(x \wedge y) = \varphi(x) \wedge \varphi(y)$. From formula 2.5, this implies, and is equivalent to, $\varphi(x \vee y) = \varphi(x) \vee \varphi(y)$. In fact, a linear map $\varphi\colon \mathcal{R} \to \mathcal{S}$ is a homomorphism of Riesz spaces if either $\varphi(x_+) = (\varphi(x))_+$ or $\varphi(|x|) = |\varphi(x)|$, for all x in $\mathcal{R}$ (2.8). Clearly, $x \geq 0$ implies $\varphi(x) \geq 0$.

The kernel K of a homomorphism φ is an ideal. In fact, if $\varphi(x) = 0$ then $\varphi(|x|) = 0$, and K is a sublattice. If $x \in K$ and $|y| \leq |x|$, then $0 \leq |\varphi(y)| = \varphi(|y|) \leq \varphi(|x|) = 0$ and $y \in K$. Hence K is, indeed, an ideal.

Let $\mathcal{R}$ be a Riesz space and I an ideal. We define the quotient order on the vector space $\mathcal{R}/I = \bar{\mathcal{R}}$ as follows: an element $\bar{x} = x + I$ of $\mathcal{R}/I$ is positive if there is a positive x in the class $\bar{x}$.

2.22. Proposition. With this convention, $\mathcal{R}/I$ is a Riesz space.

Proof. First, it is clear that $\bar{\mathcal{R}}_+ + \bar{\mathcal{R}}_+ \subset \bar{\mathcal{R}}_+$, and $r\bar{\mathcal{R}}_+ \subset \bar{\mathcal{R}}_+$ for all positive reals r. To show that $\bar{\mathcal{R}}_+ \cap (-\bar{\mathcal{R}}_+) = \{0\} = I$, suppose that $\bar{x} \geq 0$ and $\bar{x} \leq 0$. There are $x, y \in \bar{x}$ with $x \geq 0$ and $y \leq 0$ in $\mathcal{R}$. Then $I \ni x - y \geq x \geq 0$, and, as I is an ideal, $x \in I$. This proves that $\bar{x} = 0 = I$, and therefore $\bar{\mathcal{R}}$ is an ordered vector space with order cone $\bar{\mathcal{R}}_+$.

To show that $\mathcal{R}$ is actually a Riesz space, use criterion (v) in 2.13. If $x \in \bar{x} \in \bar{\mathcal{R}}$, then $\bar{x}_+$ $(= x_+ + I)$ certainly exceeds both $\bar{x}$ and 0. It will be proved that any other class $\bar{y}$ with this property exceeds $\bar{x}_+$, so that $\bar{x}_+ = \bar{x} \vee 0$: suppose that $\bar{y} \geq \bar{x}, 0$. If y is a positive representative in $\bar{y}$, then $y + i \geq x$ for som i in I. If necessary, i is replaced by $|i|$ and a representative $y + i$ of y is obtained which exceeds both x and 0. Hence $y + i \geq x_+$ and $\bar{y} \geq \bar{x}_+$, and so $\mathfrak{R}/I$ is, indeed, a Riesz space. The map $x \to \bar{x}$ is a Riesz space epimorphism of $\mathfrak{R}$ onto $\mathfrak{R}/I$.#

A homomorphism does not necessarily preserve suprema of arbitrary families: the injection of the continuous functions on [0,1] into all functions on [0,1] is a counterexample. If a homomorphism of Riesz spaces preserves suprema (or, equivalently, infima) of arbitrary families it is a <u>normal</u> homomorphism.

E. Supplements

2.23. The kernel of a normal homomorphism of Riesz spaces is a band. Every ideal (band) is the kernel of a (normal) homomorphism onto the quotient. An isomorphism of Riesz spaces is normal, and preserves absolute continuity and orthogonality.

2.24. If B is a band in the complete Riesz space $\mathfrak{R}$ then B is a complete Riesz space, and a band of B is a band of $\mathfrak{R}$.

3. Riesz spaces of measures

The theory of vector lattices developed in the last section is now used to study the measures on a fixed integration domain. The real-valued measures of finite variation (defined below) turn out to form a complete Riesz space in which one can distinguish bands of measures of various continuity behaviour. The S-continuous and B-continuous measures studied in (D) are the ones amenable to Lebesgue-Daniell integration, as indicated in the introduction.

A. Variation

Definition. Let $\mathcal{I}$ be an integration domain, or a Riesz space, or a part thereof, and let R be a Riesz space. A map $f: \mathcal{I} \to R$ is said to be

subadditive if $f(x+y) \leq f(x)+f(y)$ whenever x, y, and x+y belong to $\mathcal{I}$;

superadditive if $f(x+y) \geq f(x)+f(y)$ whenever x, y, and x+y belong to $\mathcal{I}$;

positively homogeneous if $f(rx) = rf(x)$ whenever $r \in \mathbb{R}_+$ and $x, rx \in \mathcal{I}$;

homogeneous if $f(rx) = rf(x)$ whenever $r \in \mathbb{R}$ and $x, rx \in \mathcal{I}$;

sublinear if it is subadditive and positively homogeneous;

additive if $f(x+y) = f(x)+f(y)$ whenever x, y and x+y belong to $\mathcal{I}$;

linear if it is additive and homogeneous;

positive if it is positive on the positive elements $\mathcal{I}_+$ of $\mathcal{I}$; and

increasing if $f(x) \geq f(y)$ whenever $x, y \in \mathcal{I}$, $x \geq y$.

Note that any map f from a clan $\mathcal{I}$ to R is homogeneous provided that $f(0) = 0$: for $x \in \mathcal{I}$, rx belongs to $\mathcal{I}$ only if $r = 0$ or $r = 1$. A similar statement can be made if $\mathcal{I}$ is an integration lattice or, more generally, a Riesz space:

3.1. Lemma. Let $\mathcal{R}$ and R be Riesz spaces and let $m: \mathcal{R}_+ \to R_+$ be an additive map. Then m is positively homogeneous.

Proof. The additivity gives $m(px) = pm(x)$ for $x \in \mathcal{R}_+$ and positive integers p. Replacing x in this equation by $p^{-1}x$, we obtain $p^{-1}m(x) = m(p^{-1}x)$ and deduce that $m(rx) = rm(x)$ for every positive rational r. Note that m is increasing on $\mathcal{R}_+$ and let s be a positive real. If r and t are positive rationals with $r \leq s \leq t$ then

$$rm(x) = m(rx) \leq m(sx) \leq m(tx) = tm(x).$$

The same inequality holds for $sm(x)$:

$$rm(x) \leq sm(x) \leq tm(x).$$

Hence $m(sx)$ and $sm(x)$ lie in the same intervals and so are equal. This shows positive homogeneity.#

3.2. Lemma. Let $\mathcal{J}$ be an integration domain (or a Riesz space) and let $m: \mathcal{J}_+ \to R_+$ be an additive map. There is a unique extension of m to a positive and additive map on all of $\mathcal{J}$. If $\mathcal{J}$ is an integration lattice, the extension is linear.

Proof. Let $x \in \mathcal{J}$. Then $0 = x - x \wedge x \in \mathcal{J}$, $-x_- = x \wedge 0 \in \mathcal{J}$, and $x_+ = x \vee 0 \in \mathcal{J}$. Every element z of $\mathcal{J}$ can therefore be written in the form $z = x - y$, with $x,y \in \mathcal{J}_+$. Define $m'(z) = m(x) - m(y)$. It does not depend on the choice of x,y; for if $z = x' - y'$ with $x',y' \in \mathcal{J}_+$, then $x + y' = x' + y$, $m(x) + m(y') = m(x') + m(y)$, and so $m(x) - m(y) = m(x') - m(y')$. Clearly, m' is additive, has restriction m to $\mathcal{J}_+$, and is uniquely defined by these properties. The last statement follows from lemma 3.1.#

Definition. Let $\mathcal{S}$ be an integration domain (or a Riesz space), let R be a complete Riesz space, and let $f: \mathcal{S}_+ \to R_+$ be a subadditive map. We say that f has finite variation if it is majorized by an additive map: i.e., if there is an additive map $n: \mathcal{S} \to R$ such that $f(x) \leq n(x)$ for all x in $\mathcal{S}_+$.

Any majorizing additive map is a positive measure (3.2). To illustrate the name "finite variation," let n be a positive measure majorizing the subadditive map $f: \mathcal{S}_+ \to R_+$. If $x, x_1, \ldots, x_k$ are any elements of $\mathcal{S}_+$ satisfying $\sum x_i \leq x$, then $\sum f(x_i) \leq \sum n(x_i) \leq n(x)$. The expression

$$|f|(x) = \sup\{\sum f(x_i): x_i \in \mathcal{S}_+, \sum x_i \leq x, \text{ the sum finite}\} \in R$$

is therefore finite for every $x \in \mathcal{S}_+$, and is smaller than $n(x)$.

3.3. Proposition. The subadditive map $f: \mathcal{S}_+ \to R_+$ has finite variation if and only if $|f|(x)$ is finite for all x in $\mathcal{S}_+$. In this instance, $|f|$ is additive on $\mathcal{S}_+$ and its unique extension to $\mathcal{S}$ is the smallest positive measure on $\mathcal{S}$ majorizing f.

Definition. The unique extension of $|f|$ to all of $\mathcal{S}$ is also denoted by $|f|$ and called the variation of f.

Proof of 3.3. The only thing left to show is that $|f|$ is additive on $\mathcal{S}_+$, if it is finite. Superadditivity is simple to prove. Let $x, x_1, \ldots, x_m$ and $y, y_1, \ldots, y_n$ be any elements of $\mathcal{S}_+$ that satisfy $\sum x_i \leq x$ and $\sum y_j \leq y$. Then $\sum x_i + \sum y_j \leq x + y$, and so

$$\sum f(x_i) + \sum f(y_j) \leq |f|(x + y).$$

On taking suprema over the families $\{x_i\}$ and $\{y_j\}$, we get $|f|(x) + |f|(y) \leq |f|(x + y)$ (2.12).

For subadditivity, let $x, y, z_1, \ldots, z_k$ be any elements of $\mathcal{I}_+$ for which $\sum z_i \leq x + y$. By adding $z_{k+1} = x + y - \sum z_i$, we may assume that $\sum z_i = x + y$ (2.15). There are elements $x_i, y_i \in \mathcal{I}_+$ such that $x_i + y_i = z_i$, $\sum x_i = x$, $\sum y_i = y$ (2.14). From the subadditivity of f,

$$\sum f(z_i) \leq \sum f(x_i) + \sum f(y_i) \leq |f|(x) + |f|(y).$$

On taking the supremum over all families $\{z_i\}$ as above, we find that $|f|(x+y) \leq |f|(x) + |f|(y)$.#

B. Variation of measures

In this section, $\mathcal{I}$ and R will be a fixed integration domain and a complete Riesz space, respectively.

Definition. An R-valued measure on $\mathcal{I}$ is an additive map from $\mathcal{I}$ to R. A measure m is positive if $m(\mathcal{I}_+) \subset R_+$; $I_+(\mathcal{I};R)$ denotes the set of such positive measures from $\mathcal{I}$ to R. A measure $m: \mathcal{I} \to R$ is of finite variation if the subadditive map $f_m: \varphi \to |m(\varphi)|$ from $\mathcal{I}_+$ to R_+ is. In this instance, the variation of m is the variation of f_m, and is denoted by $|m|$. $I_0(\mathcal{I};R)$ is the set of all measures of finite variation from $\mathcal{I}$ to R.#

From the triangle inequality 2.9, $|m(\varphi+\psi)| = |m(\varphi) + m(\psi)| \leq |m(\varphi)| + |m(\psi)|$ for $\varphi, \psi \in \mathcal{I}_+$, and so f_m is indeed subadditive. Clearly, $f_m = f_{-m}$: the variation $|m|$ is a positive measure majorizing both m and -m. Conversely, if $n: \mathcal{I} \to R$ is a positive measure majorizing both m and -m, then $n(\varphi) \geq m(\varphi)$, and $n(\varphi) \geq -m(\varphi)$; and so $n(\varphi) \geq |m(\varphi)| = f_m(\varphi)$ for all $\varphi \in \mathcal{I}_+$. $|m|$ is therefore the smallest measure majorizing both m and -m.

3.4. Theorem. A measure $m\colon \mathcal{F} \to R$ has finite variation if and only if it is the difference of two positive measures. $I_0(\mathcal{F};R)$ is a complete Riesz space in the order defined by $I_+(\mathcal{F};R)$, and the variation $|m|$ is the supremum of m and $-m$ in this order.

Proof. A positive measure on $\mathcal{F}$ coincides with its variation on $\mathcal{F}_+$, and hence on $\mathcal{F}$ (3.2). It is therefore in $I_0(\mathcal{F};R)$. The difference of two positive measures m and n is majorized by $m+n$, and so has finite variation. Conversely, suppose that $m\colon \mathcal{F} \to R$ has finite variation, $|m|$. The measure $|m| - m$ is evidently positive, and so $m = |m| - (|m| - m)$ is a representation of m as a difference of two positive measures. It is clear from the discussion above that $|m| = m \vee (-m)$ in $I_0(\mathcal{F};R)$. $I_0(\mathcal{F};R)$ is therefore a Riesz space (2.13).

To show that $I_0(\mathcal{F};R)$ is complete, let $M \subset I_+(\mathcal{F};R)$ be an increasingly directed and bounded set (2.16). For each $\varphi \in \mathcal{F}_+$ define $n(\varphi) = \sup\{m(\varphi)\colon m \in M\}$. Formula 2.12 shows that n is additive on $\mathcal{F}_+$. Its unique linear extension (3.2) is evidently the supremum of M.#

Example. $I_0(\mathcal{F};\mathbb{R}) = I_0(\mathcal{F})$ is the order dual of $\mathcal{F}$. The variation of a scalar measure m in $I_0(\mathcal{F})$ is defined by

$$|m|(\varphi) = \sup\{\textstyle\sum |m(\varphi_i)|\colon \varphi_i \in \mathcal{F}_+,\ \sum \varphi_i = \varphi, \text{ the sum finite}\},$$

for $\varphi \in \mathcal{F}_+$. Adding all the terms of the same sign, we deduce that

3.5. $$|m|(\varphi) = \sup\{m(\varphi_1) - m(\varphi_2)\colon \varphi_i \in \mathcal{F}_+,\ \varphi_1 + \varphi_2 = \varphi\}.$$

A little calculation and 2.7 give

3.6. $$m_+(\varphi) = \sup\{m(\varphi_1)\colon \varphi_1 \in \mathcal{F}_+,\ \varphi_1 \le \varphi\} \quad (\varphi \in \mathcal{F}_+).$$

In fact, it is easily seen that a scalar measure $m\colon \mathcal{F} \to \mathbb{R}$ has finite variation if and only if the supremum in 3.6 is finite for all $\varphi \in \mathcal{F}_+$. Using 2.8 instead, it can be shown equally easily that

3.7. $$(m \wedge n)(\varphi) = \inf\{m(\varphi_1) + n(\varphi_2)\colon \varphi_i \in \mathcal{F}_+,\ \varphi_1 + \varphi_2 = \varphi\}$$

for $m, n \in I_+(\mathcal{F})$, and $\varphi \in \mathcal{F}_+$.

<u>3.8. Proposition</u>. Let m, n be two positive scalar measures on $\mathcal{F}$.

(i) m and n are disjoint if and only if for all φ in $\mathcal{F}_+$ and $\varepsilon > 0$ there are φ_1, φ_2 in $\mathcal{F}_+$ such that $\varphi_1 + \varphi_2 = \varphi$ and $n(\varphi_1) + m(\varphi_2) < \varepsilon$.

(ii) The following statements are equivalent:

(a) m is absolutely continuous with respect to n (i.e., $m \in (n)$).

(b) $\vee\{m \wedge kn \mid k \in \mathbb{N}\} = m$.

(c) For every $\varphi \in \mathcal{F}_+$ and $\varepsilon > 0$ there is a $\delta > 0$ such that $0 \leq \psi \leq \varphi$ and $n(\psi) < \delta$ imply $m(\psi) < \varepsilon$.

<u>Proof</u>. (i) is an immediate consequence of 3.7; and (a) and (b) have already been proved equivalent in 2.20. Assume (b). For sufficiently large integers k, $(m - kn \wedge m)(\varphi) < \varepsilon/2$. This implies that if $0 \leq \psi \leq \varphi$ and $n(\psi) < \varepsilon/2k$, then $m(\psi) \leq \varepsilon$; and therefore (c) is proved with $\delta = \varepsilon/2k$.

Conversely, assume (c) holds. Split m into its part m_1 in (n) and m_2 orthogonal to n (2.18). Property (c) holds for m_1 and consequently for m_2, from which it will be proved that $m_2 = 0$, as desired. For $\varphi \in \mathcal{F}_+$ and arbitrary $\varepsilon > 0$, choose a $\delta > 0$ so that $0 \leq \psi \leq \varphi$ and $n(\psi) < \delta$ imply $m_2(\psi) < \varepsilon$. Next, find $\psi_1, \psi_2 \in \mathcal{F}_+$ for which $\psi_1 + \psi_2 = \varphi$ and $m_2(\psi_1) + n(\psi_2) < \delta$, using (i). Then $m_2(\psi_2) < \varepsilon$, and $m_2(\psi) = m_2(\psi_1) + m_2(\psi_2) < \varepsilon + \delta$. Thus $m_2(\varphi) = 0$, and the proposition is proved.#

Remark. For $\varphi \in \mathcal{I}$, let $\{\varphi\}_1$ denote the "unit ball" of the ideal spanned by $\{\varphi\}$ in $\mathcal{I}$:

$$\{\varphi\}_1 = \{\psi \in \mathcal{I}\colon |\psi| \leq |\varphi|\}.$$

Condition (ii c) above states that m is absolutely continuous with respect to n iff it is continuous on $\{\varphi\}_1$ with respect to the pseudo-norm $\psi \to n(|\psi|)$, for all $\varphi \in \mathcal{I}$.

C. Variation of Banach-valued measures

We fix an integration domain $\mathcal{I}$ and a real Banach space E with norm $\|\ \|$.

Definition. A measure from $\mathcal{I}$ to E is an additive map $m\colon \mathcal{I} \to E$. It is said to have finite (scalar) variation if the subadditive map $f_m\colon \varphi \to \|m(\varphi)\|$, from $\mathcal{I}_+$ to $\mathbb{R}_+$, does. In this instance, the variation of m is the variation of f_m, and is denoted by $|m|$. The set of E-valued measures of finite variation will be written as $I_0(\mathcal{I};E)$.

The variation of m is the smallest positive measure $|m|$ majorizing m, in the sense that

$$\|m(\varphi)\| \leq |m|(\varphi) \quad \text{for} \quad \varphi \in \mathcal{I}_+.$$

By inspection, a measure of finite variation is actually linear on an integration lattice (3.2), and $I_0(\mathcal{I};E)$ is a vector space.

It is now possible to extend the notions of absolute continuity and orthogonality to vector-valued measures of finite variation. Let $m\colon \mathcal{I} \to E$ and $n\colon \mathcal{I} \to F$ be such measures (F is a second Banach space). Then m is said to be absolutely continuous with respect to n (denoted by $m \ll n$), if $|m| \ll |n|$, i.e., if $|m|$ belongs to the band $(|n|)$ spanned by $|n|$. $(n)_E$ denotes the vector space of E-valued measures that are absolutely continuous with respect to n. Similarly, m and n are disjoint if $|m|$ and $|n|$ are, i.e., if $|m| \wedge |n| = 0$; and we then write $m \perp n$. Later (17.3), an analogue of Riesz's decomposition theorem 2.18 will be proved for Banach-valued measures of finite variation.

D. *-measures

Not all measures on an integration domain are accessible to the Lebesgue-Daniell integration process outlined in the introduction. Those which are will be described next. The main result is that they form a band in $I_0(\mathcal{F})$.

For the following, $\mathcal{F}$ is a fixed integration domain and E is a Banach space with norm $\| \ \|$.

Definition.

(i) A measure $m: \mathcal{F} \to E$ is called S-continuous or an S-measure, and $(\mathcal{F}, m)$ is called an elementary S-measure, if the following (Sequential, or Stone-) condition holds: whenever (φ_n) is an increasing sequence in $\mathcal{F}$ with pointwise supremum φ in $\mathcal{F}$, then $\lim m(\varphi_n) = m(\varphi)$ in E.

(ii) The measure $m: \mathcal{F} \to E$ is called B-continuous or a B-measure, and $(\mathcal{F}, m)$ is called an elementary B-measure, if the following (Bourbaki-) condition holds: whenever Φ is an increasingly directed subset of $\mathcal{F}$ with pointwise supremum φ in $\mathcal{F}$, then $\lim m(\Phi) = m(\varphi)$. (Here the limit is taken along the section filter of Φ.)#

Notation. In order to develop the theory of S- and B-measures simultaneously, let us introduce the symbol $*$, standing for S or B as the case may be. Let us make the following useful convention: The symbol

$$\mathcal{F} \supset \Phi \uparrow * \varphi \in \mathcal{F}$$

is to mean that Φ is either an increasingly directed countable set (if $* = S$) or an arbitrary increasingly directed subset (if $* = B$) of $\mathcal{F}$ converging pointwise to $\varphi \in \mathcal{F}$. A statement like $\mathcal{F} \supset \Phi \downarrow * 0$ is defined analogously.#

3.9. Lemma. The linear form $m: \mathcal{F} \to E$ is a $*$-measure $(* = S, B)$ if and only if either of the following hold:

(a) $\mathcal{F} \supset \Phi \downarrow * 0$ implies $\lim m(\Phi) = 0$

(b) $\mathcal{F}_+ \supset \Phi \uparrow * \varphi$ implies $\lim m(\Phi) = m(\varphi)$.

Proof. $*$-continuity clearly implies (b). (b) implies (a): Let $\mathcal{F} \supset \Phi * \downarrow 0$. Choose $\psi \in \Phi$, and let $\Psi = \{\varphi \in \Phi : \varphi \leq \psi\}$. Then $\mathcal{R}_+ \supset \psi - \Psi \uparrow * \psi$, hence $m(\psi) = \lim m(\psi - \Psi) = m(\psi) - \lim m(\Psi)$, and $\lim m(\Phi) = \lim m(\Psi) = 0$. (a) implies $*$-continuity: Let $\mathcal{F} \supset \Phi \uparrow * \varphi \in \mathcal{F}$. Then $\varphi - \Phi \downarrow * 0$, $\lim m(\varphi - \Phi) = m(\varphi) - \lim m(\Phi) = 0$, and $\lim m(\Phi) = m(\varphi)$.#

3.10. Theorem. A measure $m: \mathcal{F} \to E$ of finite variation $|m|$ is $*$-continuous if and only if $|m|$ is. The scalar $*$-measures of finite variation form a band, $M^*(\mathcal{F})$, in $I_0(\mathcal{F})$ $(* = S$ or $B)$.

Proof. Let $m: \mathcal{F} \to E$ be a measure of finite variation. If $|m|$ is $*$-continuous and $\mathcal{F} \supset \Phi \downarrow * 0$, then $\lim\{\|m(\varphi)\| : \varphi \in \Phi\} \leq \lim\{|m|(\varphi) : \varphi \in \Phi\} = 0$.

Conversely, suppose m is $*$-continuous, and let $\mathcal{F}_+ \supset \Phi \uparrow * \psi \in \mathcal{F}$. Let $\psi_1, \ldots, \psi_k \in \mathcal{F}_+$ satisfy $\sum \psi_i = \psi$, and define

$$\varphi_k = \varphi \wedge \sum \{\psi_i : 1 \le i \le k\} - \varphi \wedge \sum \{\psi_i : 1 \le i < k\} \quad (k = 1,\ldots,n)$$

for each $\varphi \in \Phi$. Clearly, $\sum \varphi_k = \varphi$, and $\Phi_k = \{\varphi_k : \varphi \in \Phi\}\uparrow^*\psi_k$ for $1 \le k \le n$. Since

$$|m|(\varphi) \ge \sum \|m(\varphi_k)\| \text{ tends to } \sum \|m(\psi_k)\| \le |m|(\varphi)$$

as φ increases in Φ, $\sup\{|m|(\varphi) : \varphi \in \Phi\} = |m|(\psi)$. This proves the first statement.

It is now obvious that $M^*(\mathcal{F})$ forms an ideal of $I_0(\mathcal{F})$. To show that it is a band, let F be an increasingly directed subset of $M^*_+(\mathcal{F})$ with supremum $n \in I_0(\mathcal{F})$. Let $\mathcal{F}_+ \supset \Phi\uparrow^*\psi \in \mathcal{F}$. Then

$$\sup n(\Phi) = \sup\{m(\varphi) : m \in F, \varphi \in \Phi\} = \sup\{m(\psi) : m \in F\} = n(\psi),$$

and n is $*$-continuous, and so belongs to $M^*(\mathcal{F})$.#

<u>Notation</u>. Henceforth $M^*(\mathcal{F};E)$ denotes the vector space of E-valued $*$-measures of finite (scalar) variation ($* = S$ or B). A measure of finite variation is <u>purely S-continuous</u> if its variation belongs to $M^S(\mathcal{F})$ but is disjoint from $M^B(\mathcal{F})$; it is <u>purely linear</u> if it is disjoint from $M^S(\mathcal{F})$.

The next corollary follows from F. Riesz's decomposition theorem (2.18).

<u>3.11. Corollary</u>. Any scalar measure m of finite variation has a unique decomposition $m = m_B + m_{PS} + m_{PJ}$ into a B-measure m_B, a purely S-continuous measure m_{PS}, and a purely linear measure m_{PJ}.

We shall see later (17.3) that a similar decomposition holds for Banach-valued measures of finite variation.

<u>Example</u>. Let λ denote the Riemannian content on the clan $\mathscr{C}(\mathbb{R})$ of finite unions of bounded intervals. It will now be shown that $(\mathscr{C}(\mathbb{R}),\lambda)$ is an elementary S-measure.

Let (A_n) be a countable decreasingly directed family in $\mathscr{C}$ with limit $\emptyset$. The closures $\bar{A}_n$ of the A_n are compact sets in $\mathscr{C}$ with the same content $\lambda(\bar{A}_n) = \lambda(A_n)$. They form a decreasingly directed family with an intersection $A = \cap \bar{A}_n$ consisting of countably many points at the most. This is a consequence of $\bar{A}_n - A_n$ being a finite set, for each integer n. Let $A = \{a_1, a_2, \ldots\}$. Given an $\varepsilon > 0$, choose an open interval B_n that contains a_n and has length $\varepsilon 2^{-n}$, for every integer n, and let $B = \cup B_n$. The compact sets $\bar{A}_n \setminus B$ clearly have void intersection, and so there is a natural number N such that $\bar{A}_N \setminus B = \emptyset$: i.e., $\bar{A}_N \subset B$. Since $\bar{A}_N$ is compact, it has a finite subcover $\{B_1,\ldots,B_k\}$. Finally, $\lambda(A_n) \leq \lambda(A_N) = \lambda(\bar{A}_N) \leq \sum \{\lambda(B_i): 1 \leq i \leq k\} < \varepsilon$, for $n \geq N$, and therefore $\lambda(A_n) \to 0$, as desired.#

An elementary $*$-measure $(\mathscr{S},m)$ is usually called an <u>elementary $*$-integral</u> if $\mathscr{S}$ is an integration lattice $(* = S,B)$. If $\mathscr{S}$ is a clan and $* = S$, it is called a <u>σ-additive content</u>, for in this case S-continuity is equivalent to the following: whenever (A_n) is a sequence of disjoint sets in $\mathscr{S}$ with union A in $\mathscr{S}$, then $m(A) = \sum \{m(A_n); n \in \mathbb{N}\}$.

Let $(\mathscr{C},m)$ be an elementary E-valued content. Recall that associated with it is a natural elementary integral $(\mathscr{E}(\mathscr{C}),\tilde{m})$ (1.1); it is obtained by linear extension. The following proposition shows that variation and $*$-continuity are preserved in the best possible way.

<u>3.12. Proposition</u>. The map $\sim: m \to \tilde{m}$ is linear and preserves variation. It induces a linear isomorphism of $I_0(\mathscr{C};E)$ onto $I_0(\mathscr{S}(\mathscr{C});E)$, and of $M^*(\mathscr{C};E)$ onto $M^*(\mathscr{S}(\mathscr{C});E)$ ($* = S$ or B; E a Banach space). It induces a (necessarily normal) Riesz space isomorphism of $I_0(\mathscr{C})$ onto $I_0(\mathscr{S}(\mathscr{C}))$.

<u>Proof</u>. The map $\sim$ is clearly linear. If m is a content on $\mathscr{C}$ of finite variation $|m|$, then $|m|^\sim$ majorizes $\tilde{m}$, since

$$\|\tilde{m}(\sum r_i A_i)\| \leq \sum r_i \|m(A_i)\| \leq \sum r_i |m|(A_i) = |m|^\sim(\sum r_i A_i),$$

for $\sum r_i A_i \in \mathscr{S}_+(\mathscr{C})$. Hence $\tilde{m}$ has finite variation $|\tilde{m}| \leq |m|^\sim$. Conversely, $|\tilde{m}|$ majorizes m on $\mathscr{C}$, hence $|m| \leq |\tilde{m}|$ on $\mathscr{C}$, and thus $|m|^\sim = |\tilde{m}|$ on $\mathscr{C}$. By linearity, $|\tilde{m}| = |m|^\sim$ on $\mathscr{S}(\mathscr{C})$.

Different measures m clearly have different extensions $\tilde{m}$: the map $\sim$ is one-to-one. Every measure on $\mathscr{S}(\mathscr{C})$ is the extension of its restriction to $\mathscr{C}$: the map $\sim$ is onto. It preserves order on $I_0(\mathscr{C})$ and, as we have seen, also preserves absolute value: $|m|^\sim = |\tilde{m}|$. It is thus a Riesz space isomorphism of $I_0(\mathscr{C})$ onto $I_0(\mathscr{S}(\mathscr{C}))$.

It remains to be shown that $(M^*(\mathscr{C},E))^\sim = M^*(\mathscr{S}(\mathscr{C}),E)$ for $* = S$ or $* = B$. Since the restriction of a $*$-measure on $\mathscr{S}(\mathscr{C})$ is obviously $*$-continuous, $M^*(\mathscr{S}(\mathscr{C}),E) \subset (M^*(\mathscr{C},E))^\sim$. For the reverse inclusion, let $m \in M^*(\mathscr{C},E)$. Its extension $\tilde{m}$ must be shown to be $*$-continuous on $\mathscr{S}(\mathscr{C})$: Let $\mathscr{S}(\mathscr{C}) \supset \Phi\downarrow^* 0$. Choose $\psi \in \Phi$, and define $a = \max\{\psi(x);\ x \in X\}$ and $A = [\psi > 0] \in \mathscr{C}$. Let $\varepsilon > 0$ be given. For $\varphi \in \Phi$ and $\varphi \leq \psi$,

$$\begin{aligned}\|m(\varphi)\| \leq |m|(\varphi) &= |m|(\varphi[\varphi < \varepsilon]) + |m|(\varphi[\varphi \geq \varepsilon]) \\ &\leq |m|(\varepsilon[\varphi < \varepsilon]) + |m|(a[\varphi \geq \varepsilon]) \leq \varepsilon|m|(A) + a|m|([\varphi \geq \varepsilon]).\end{aligned}$$

Since $\Phi\downarrow^* 0$, it follows that $\{[\varphi \geq \varepsilon],\ \varphi \in \Phi\}\downarrow^* 0$, and therefore $\lim |m|([\varphi \geq \varepsilon]) = 0$. This shows that $\|m(\varphi)\|$ is arbitrarily small for $\varphi \in \Phi$ sufficiently small: $\tilde{m}$ is $*$-continuous if m is.#

Corollary. The Riemann integral is S-continuous on step functions.

E. A Generalization

Definition. A Banach lattice is a Riesz space R which is also a Banach space under a norm $\| \ \|$, and which satisfies

$$\| |f| \| = \|f\| \qquad \text{for } f \in R, \text{ and}$$
$$\|f\| < \|g\| \qquad \text{for } 0 \leq f < g \text{ in } R.$$

A Banach lattice R is with order-continuous norm (or is a proper L-space) if every increasingly directed subset $F \subset R$ has a supremum g if and only if the norms $\{\|f\|: f \in F\}$ have a common upper bound, and if in this instance F converges to g in norm. (An example is $\mathbb{R}$.) A proper L-space is evidently complete as a Riesz space.

A Banach space E with norm $\| \ \|$ is said to be a Banach space over the proper L-space R if there is a subadditive map $f \to |f|$ from E to R_+, called the modulus, satisfying

$$\|f\| = \| |f| \| \qquad \text{for } f \in E. \ \#$$

Every Banach space is a Banach space over $\mathbb{R}$. The notion above generalizes this situation. Also, every proper L-space is a Banach space over itself in the obvious way. Another example will emerge in 8.13.

For the rest of this section, E is a fixed Banach space over a fixed proper L-space R.

A measure $m: \mathcal{S} \to E$ is said to be of finite R-variation if the subadditive map $f_m^R: \varphi \to |m(\varphi)|$ from $\mathcal{S}$ to R_+ is. In this instance, the variation of f_m^R is called the R-variation of m and is denoted by $|m|$.

It is the smallest measure in $I_+(\mathcal{S};R)$ majorizing m in the sense that

$$|m(\varphi)| \leq |m|(\varphi) \qquad \text{for} \quad \varphi \in \mathcal{S}_+.$$

$I_0(\mathcal{S};E,R)$ denotes the set of measures from $\mathcal{S}$ to E with finite R-variation.

The proofs of this section have been fashioned so that it is possible to read off the following results.

A measure $m: \mathcal{S} \to E$ is of finite R-variation if and only if

$$|m|(\varphi) = \sup\{\sum |m(\varphi_i)| : \varphi_i \in \mathcal{S}_+, \sum \varphi_i \leq \varphi, \text{ the sum finite}\}$$

exists in R_+ for all $\varphi \in \mathcal{S}_+$; and in this instance $|m|$ as defined by this formula coincides with the variation of m on $\mathcal{S}_+$.

A measure in $I_0(\mathcal{S};E,R)$ is $*$-continuous if and only if its R-variation--which belongs to $I_+(\mathcal{S};R)$--is ($* = S$ or B). $M^*(\mathcal{S};E,R)$ denotes the $*$-measures in $I_0(\mathcal{S};E,R)$. The $*$-measures in $I_0(\mathcal{S};R)$ form a band $M^*(\mathcal{S};R)$.

Any measure $m \in I_0(\mathcal{S};R)$ has a unique decomposition,

$$m = m_B + m_{PS} + m_{PJ},$$

into a B-measure m_B, a purely S-continuous measure m_{PS}, and a purely linear measure m_{PJ}.

Let $\mathcal{C}$ be a clan, and $\mathcal{S}(\mathcal{C})$ the step functions over $\mathcal{C}$. The natural extension by linearity of a measure m on $\mathcal{C}$ to a measure $\tilde{m}$ on $\mathcal{S}(\mathcal{C})$ preserves R-variation and $*$-continuity. It is an isomorphism of $I_0(\mathcal{C};E,R)$ with $I_0(\mathcal{S}(\mathcal{C});E,R)$, and maps $M^*(\mathcal{C};E,R)$ onto $M^*(\mathcal{S}(\mathcal{C});E,R)$ for $* = S,B$.

Remark. The results 3.6, 3.7, and 3.8 use crucially the linear ordering of the reals, and it cannot be expected that they will carry over to this setting. This much however is true: If m and n are two positive R-valued measures on $\mathcal{F}$, then $m \ll n$, i.e., $m \in (n)$, if and only if $m = \bigvee \{m \wedge kn : k \in \mathbb{N}\}$. This in turn is true if and only if for every $\varphi \in \mathcal{F}_+$ and $\varepsilon > 0$ there is a $\delta > 0$ such that $\varphi \geq \psi \in \mathcal{F}_+$ and $\|n(\psi)\| \leq \delta$ imply $\|m(\psi)\| \leq \varepsilon$ (2.20 and 3.8).

F. Supplements

3.13*. If $\mathcal{R}$ is a Riesz space or an integration domain and $m \in I_0(\mathcal{R})$, then $|m|(\varphi) = \sup\{m(\psi): \psi \in \mathcal{R}, |\psi| \leq \varphi\}$. The map $m \to |m|$ from $I_0(\mathcal{R})$ to $I_+(\mathcal{R})$ is sublinear and positively homogeneous.

3.14*. A complex-valued measure m on the integration domain $\mathcal{F}$ is also sometimes called a scalar measure. Its complex conjugate m^c is defined by $m^c(\varphi) = \overline{m(\varphi)}$, for $\varphi \in \mathcal{F}$, and is a complex measure, also. $m^r = 1/2(m+m^c)$ and $m^i = -i/2(m-m^c)$ are real measures, and $m = m^r + im^i$. m has finite variation if and only if m^r and m^i do, and then $|m| \leq |m^r| + |m^i|$.

3.15*. Suppose $\mathcal{B}$ is a uniformly dense subset of the integration lattice $\mathcal{R}$ and contains the constants. Then $m \in I_+(\mathcal{R})$ is $*$-continuous iff it is $*$-continuous on $\mathcal{B}$, i.e., $\mathcal{B} \supset \Phi \downarrow^* 0 \Rightarrow m(\Phi) \to 0$ ($* = S$ or B).

3.16*. If $\mathcal{R}$ is countably generated (1.3) and $m \in M^S(\mathcal{R};E)$ then m is B-continuous and takes values in a separable subspace of E.

3.17*. Let $T: E \to F$ be a bounded linear map of Banach spaces. If $m: \mathcal{F} \to E$ is a measure, then $T \circ m$ is also a measure: $\mathcal{F} \to F$. The map $m \to T \circ m$ maps $I_0(\mathcal{F};E)$ into $I_0(\mathcal{F};F)$, and preserves $*$-continuity ($* = S$ or B).
If $m \perp n$ in $I_0(\mathcal{F};E)$, then $Tm \perp Tn$ in $I_0(\mathcal{F};F)$.

3.18. The Riemann integral is B-continuous on $C_{00}(\mathbb{R})$; and is S-continuous, but not B-continuous, on $\mathcal{B}(\mathbb{R})$.

3.19. For measures m,n on $\mathcal{F}$ with values in a complete Riesz space R, and of finite R-variation, the following analogues of 3.6 and 3.7 hold ($\varphi \in \mathcal{F}_+$):

$$m_+(\varphi) = \sup \sum \{(m(\varphi_i))_+ : \varphi_i \in \mathcal{F}_+, \sum \varphi_i \leq \varphi\}$$

$$m \wedge n(\varphi) = \inf \sum \{m(\varphi_i) \wedge n(\varphi_i) : \varphi_i \in \mathcal{F}_+, \sum \varphi_i \leq \varphi\}.$$

LITERATURE: The first common treatment of S-measures and B-measures occurs in [46].

4. Dominated integration lattices

The dominated integration lattices $\mathfrak{R}$, to be defined here, carry a natural uniformity with respect to which all measures of finite variation are uniformly continuous. Extension by uniform continuity with respect to this uniformity is the first step in the extension theory of a measure. [17] and [26] serve as general references on uniformities.

A. Inductive limit uniformities

Let $\mathscr{K}$ be an increasingly directed set, and, for $K \in \mathscr{K}$, let S_K be a uniform space with uniformity $\mathscr{U}_K$. The system $((S_K,\mathscr{U}_K): K \in \mathscr{K})$ is called a <u>strict inductive system of uniform</u> spaces if, whenever $K < L$ in $\mathscr{K}$, then $S_K \subset S_L$, and $\mathscr{U}_K$ is the uniformity which $\mathscr{U}_L$ induces on S_K. The <u>strict inductive limit</u>

$$(S,\mathscr{U}) = \varinjlim \{(S_K,\mathscr{U}_K): K \in \mathscr{K}\}$$

is the uniform space whose underlying set is $S = \cup \{S_K: K \in \mathscr{K}\}$ and whose uniformity $\mathscr{U}$ is the finest that makes all the injections $S_K \to S$ uniformly continuous. The uniformity $\mathscr{U}$ is characterized by either of the following two lemmas.

<u>4.1. Lemma</u>. A map T from S to a uniform space E is uniformly continuous if and only if the restrictions $T|S_K$ are $\mathscr{U}_K$-uniformly continuous, for all $K \in \mathscr{K}$.

<u>Proof</u>. The condition is clearly necessary. If it is satisfied, then the uniformity $\mathscr{Y}$ on S defined by T (the coarsest one making T uniformly continuous) is coarser than $\mathscr{U}$, and so T is uniformly continuous.#

<u>4.2. Lemma</u>. A set $V \in S \times S$ belongs to $\mathscr{U}$ if and only if $V \cap \{S_K \times S_K\}$ belongs to $\mathscr{U}_K$, for every $K \in \mathscr{K}$.

<u>Proof</u>. The sets V of this description form a filter on $S \times S$, and constitute a uniformity $\mathscr{Y}$ which is evidently the finest that reduces to $\mathscr{U}_K$ on S_K.#

4.3. Lemma. For every $K \in \mathcal{K}$, let R_K be a subspace of S_K such that $R_K = R_L \cap S_K$ for $K < L$ in $\mathcal{K}$, and let $\mathcal{Y}_K$ be the trace of $\mathcal{U}_K$ on R_K. Then the trace on $R = \cup \{R_K: K \in \mathcal{K}\}$ of $\mathcal{U}$ coincides with the inductive limit uniformity, $\mathcal{Y} = \varinjlim (\mathcal{Y}_K: K \in \mathcal{K})$.

Proof. Since the maps $R_K \to S_K \to S$ are all uniformly continuous, the map $R \to S$ is also, provided that R has the $\mathcal{Y}$-uniformity. Hence $\mathcal{Y}$ is finer than the trace of $\mathcal{U}$ on R.

Conversely, given a $V \in \mathcal{Y}$, let $V_K = V \cap R_K \times R_K \in \mathcal{Y}_K$, and let

$$U_K = \cup \{U \in \mathcal{U}_K: U \cap R_K \times R_K = V_K\} \in \mathcal{U}_K,$$

for every $K \in \mathcal{K}$. Clearly, $U_L \cap S_K \times S_K = U_K$ for $K < L$, and so $U = \cup \{U_K: K \in \mathcal{K}\}$ has intersection U_K with $S_K \times S_K$, for all $K \in \mathcal{K}$. From (4.2), U is in $\mathcal{U}$ and has trace V on $R \times R$. Hence the trace of $\mathcal{U}$ on R is finer than $\mathcal{Y}$.#

B. Dominated functions and sets

Let $\mathcal{F}$ be a family of Banach-valued functions on the set X.

Definition. A subset K of X is said to be $\mathcal{F}$-dominated if there is a function $\varphi \in \mathcal{F}$ such that

$$K \leq \|\varphi\|.$$

We also say that φ majorizes K. The family of $\mathcal{F}$-dominated sets is denoted by $\mathcal{K}(\mathcal{F})$. For $K \in \mathcal{K}(\mathcal{F})$, $\mathfrak{D}_E[K]$ denotes the Banach space of bounded functions from X to the Banach space E that vanish off K, equipped with the supremum norm. Finally,

$$\mathfrak{D}_E[\mathcal{S}] = \lim_{\leftarrow} \{\mathfrak{D}_E[K] : K \in \mathcal{K}(\mathcal{S})\}$$

is the space of $\mathcal{S}$-dominated functions from X to E, and its uniformity is the uniformity of dominated uniform convergence.

Example. If $\mathcal{S}$ is a clan then a set is $\mathcal{S}$-dominated if and only if it is contained in an element of $\mathcal{S}$; and this is true if and only if it is $\mathcal{S}(\mathcal{S})$-dominated.

4.4. Example. Suppose $\mathcal{R}$ is an integration lattice or an algebra of functions on X. The inductive limit of the integration lattices or algebras

$$\mathcal{R}[K] = \mathcal{R} \cap \mathfrak{D}[K] \quad , \quad K \in \mathcal{K}(\mathcal{R}),$$

is $\mathcal{R} \cap \mathfrak{D}[\mathcal{R}]$ and its inductive limit uniformity coincides with the uniformity of $\mathcal{R}$-dominated uniform convergence (4.3). It is called the dominated integration lattice or algebra associated with $\mathcal{R}$ and is denoted by $\mathcal{R}_0$.

Its inductive limit uniformity coincides with the trace of the uniformity of $\mathfrak{D} = \mathfrak{D}_{\mathbb{R}}[\mathcal{R}]$ (4.3) and is therefore called the uniformity of dominated uniform convergence. The associated topology makes $\mathcal{R}_0$ into a topological vector space. $\mathcal{R}_0$ is large:

4.5. Proposition. If $\mathcal{R}$ is an integration lattice, then a set is $\mathcal{R}$-dominated if and only if it is $\mathcal{R}_0$-dominated; hence $(\mathcal{R}_0)_0 = \mathcal{R}_0$. Furthermore, every function $\varphi \in \mathcal{R}$ is the pointwise limit of a sequence (φ_n) in $\mathcal{R}_0$ such that $|\varphi_n| \leq |\varphi|$ for all integers n; moreover, if φ is bounded then (φ_n) can be chosen to converge uniformly; if φ is positive then (φ_n) can be chosen to be increasing.

Proof. If $K \in \mathcal{K}(\mathfrak{R})$, $K \leq \varphi \in \mathfrak{R}$ and $\varphi \leq 1$ say, then $K \leq 2\varphi - 2\varphi \wedge 1$. Since $2\varphi - 2\varphi \wedge 1$ vanishes off the $\mathfrak{R}$-dominated set $[2\varphi \geq 1]$, it belongs to $\mathfrak{R}_0$. Hence K is $\mathfrak{R}_0$-dominated, and so $\mathcal{K}(\mathfrak{R}) \subset \mathcal{K}(\mathfrak{R}_0)$. The converse inclusion is obvious.

For the second statement, define

$$\varphi_n = (-n) \vee \{(\varphi_+ - \varphi_+ \wedge n^{-1}) - (\varphi_- - \varphi_- \wedge n^{-1})\} \vee n, \quad n \in \mathbb{N}.$$

By inspection, (φ_n) converges as claimed. Since φ_n vanishes off the set $[n|\varphi| \geq 1]$, it belongs to $\mathfrak{R}_0$, for $n = 1, 2, \ldots$.

Definition. An integration lattice $\mathfrak{R}$ is said to be dominated if $\mathfrak{R} = \mathfrak{R}_0$.

Examples. If X is a locally compact space and $\mathfrak{R} = C_0(X)$, then a set $K \subset X$ is $\mathfrak{R}$-dominated if and only if it is relatively compact; and $\mathfrak{R}_0 = C_{00}(X)$. $\mathfrak{R}_0$ is closed under dominated uniform convergence.

The two standard examples of integration lattices, $\mathcal{S}(\mathcal{C})$ and $C_{00}(X)$, are dominated. #

If E is a Banach space, let $\mathfrak{R} \otimes E$ denote the vector space of functions

$$\sum \varphi_i \xi_i : x \to \sum \varphi_i(x)\xi_i,$$

where the φ_i belong to $\mathfrak{R}$, the ξ_i belong to E, and the sum is finite. From 4.3,

$$\mathfrak{R}_0 \otimes E = \lim_{\leftarrow} \{\mathfrak{R}[K] \otimes E : K \in K(\mathfrak{R})\} = \mathfrak{R} \otimes E \cap \mathfrak{D}_E[\mathfrak{R}],$$

as uniform spaces. The closure of $\mathfrak{R}_0 \otimes E$ in $\mathfrak{D}_E[\mathfrak{R}]$, with respect to dominated uniform convergence, will henceforth be denoted by $\overline{\mathfrak{R}}_E$. It will be identified as a space of uniformly continuous functions in the next section.

The following statements are verified routinely or can be found in any book on functional analysis or topological vector spaces.

4.6. Lemma. In the uniformity of dominated uniform convergence, $\mathcal{D}_E[\mathcal{R}]$, $\mathcal{R}_0 \otimes E$ and $\overline{\mathcal{R}}_E$ are topological vector spaces (i.e., the maps $(\varphi, \psi) \to \varphi + \psi$ from $\mathcal{R} \otimes E \times \mathcal{R} \otimes E$ to $\mathcal{R} \otimes E$ and $(r, \varphi) \to r\varphi$ from $\mathcal{R} \otimes E \times \mathbb{R}$ to $\mathcal{R} \otimes E$ are continuous). Moreover, a linear map U from $\mathcal{R}_0 \otimes E$ to some other Banach space is uniformly continuous if and only if it is continuous at zero.#

Let us call an increasing seminorm on $\mathcal{R}_+$ a subadditive map $M: \mathcal{R}_+ \to \mathbb{R}_+$ such that $0 \leq \varphi \leq \psi$ in $\mathcal{R}_+$ implies $M(\varphi) \leq M(\psi)$. Set $M(\varphi) = M(\|\varphi\|)$ for $\varphi \in \mathcal{R} \otimes E$.

4.7. Proposition. If M is any increasing seminorm on $\mathcal{R}$ then the M-uniformity on $\mathcal{R}_0 \otimes E$ is coarser than the uniformity of dominated uniform convergence. In particular, $\overline{\mathcal{R}}_E$ belongs to the M-closure of $\mathcal{R}_0 \otimes E$.

Proof. It has to be shown that the identity map $\mathcal{R}_0 \otimes E \to (\mathcal{R}_0 \otimes E, M)$ is uniformly continuous. From 4.1, it suffices to show that the identity map $\mathcal{R}[K] \otimes E \to (\mathcal{R}[K] \otimes E, M)$ is uniformly continuous. Let (φ_n) be a sequence in $\mathcal{R}[K] \otimes E$ converging uniformly to zero, and let $\varepsilon > 0$ be given. There is a $\varphi \in \mathcal{R}_+$ with $\varphi \geq K$. For sufficiently high integer n, $\|\varphi_n\| \leq \varphi \cdot \varepsilon / (1 + M(\varphi))$, and so $M(\varphi_n) \leq \varepsilon$.#

C. Semivariation

Definition. Let $\mathcal{R}$ be a dominated integration lattice; E,G Banach spaces; and let $U: \mathcal{R} \otimes E \to G$ be a linear map. Then U is said to have finite semivariation if it is continuous in the topology of dominated uniform convergence on $\mathcal{R} \otimes E$; and the map $\|U\|$, defined by

$$\|U\|(\varphi) = \sup\{\|U(\psi)\|: \psi \in \mathcal{R} \otimes E, \|\psi\| \leq \varphi\}, \quad \varphi \in \mathcal{R}_+ ,$$

is called __the semivariation__ of U.#

U is continuous if and only if it is uniformly continuous (4.6), and this is the case if and only if $U|\mathcal{R}[K] \otimes E$ is bounded for all $K \in \mathcal{K}(\mathcal{R})$ (4.1). Since the set

$$(\varphi)_E = \{\psi \in \mathcal{R} \otimes E: \|\psi\| \leq \varphi\} \qquad (\varphi \in \mathcal{R}_+)$$

is bounded in $\mathcal{R}[[\varphi > 0]] \otimes E$, $\|U\|(\varphi)$ is then finite.

Conversely, suppose $\|U\|(\varphi)$ is finite for all $\varphi \in \mathcal{R}_+$. If $K \in \mathcal{K}(\mathcal{R})$ and $K \leq \varphi \in \mathcal{R}$, then the unit ball of $\mathcal{R}[K] \otimes E$ lies inside $(\varphi)_E$; it follows that if $\|U\|(\varphi)$ is finite then U is continuous on $\mathcal{R}[K] \otimes E$.

__4.8. Proposition__. U has finite semivariation if and only if $\|U\|(\varphi) < \infty$ for all $\varphi \in \mathcal{R}_+$. In this instance, $\|U\|$ is an increasing seminorm on $\mathcal{R}_+$. Moreover, the semivariation of the unique extension $\tilde{U}$ of U to $\bar{\mathcal{R}}_E$ coincides with $\|U\|$ on $\mathcal{R}_+$.

__Proof__. The first statement has been proved above. The proof of the last two is deferred to the next section (5.9).#

The following stronger condition on U will be considered:

__Definition__. Let $\mathcal{R}$ be a dominated integration lattice and let $U: \mathcal{R} \otimes E \to G$ be a linear map. Then U is said to have __finite variation__ if there is a positive measure n on $\mathcal{R}$ majorizing U; i.e.,

$$\|U(\varphi)\| \leq n(\|\varphi\|) \quad \text{for all} \quad \varphi \in \mathcal{R} \otimes E.$$

If U is majorized by $n \in I_+(\mathcal{R})$ then clearly so is the sublinear map $\|U\|: \mathcal{R}_+ \to \mathbb{R}_+$. Hence $\|U\|$ has finite variation smaller than n.

Definition. If $U: \mathcal{R} \otimes E \to G$ has finite variation, then the variation of $\|U\|$ is called the variation of U and is denoted by $|U|$.#

From the definitions, the following formula is evident.

(4.9)
$$|U|(\varphi) = \sup\{\sum \|U\|(\varphi_i): 0 \leq \varphi_i \in \mathcal{R}, \sum \varphi_i = \varphi\}$$
$$= \sup\{\sum \|U(\varphi_i)\|: \varphi_i \in \mathcal{R} \otimes E, \sum \|\varphi_i\| \leq \varphi\},$$

where $\varphi \in \mathcal{R}_+$ and where the sums are finite.

4.10. Proposition. If $U: \mathcal{R} \otimes E \to G$ has finite variation then

$$\|U\| \leq |U| \quad \text{on} \quad \mathcal{R}_+,$$

and both U and $|U|$ are continuous in the topology of dominated uniform convergence.

A real-valued measure m on $\mathcal{R}$ has finite variation if and only if it is continuous in the topology of dominated uniform convergence, and in this instance $\|m\| = |m|$.

Proof. Both $\|U\|$ and $|U|$ are increasing norms with respect to which U is continuous (4.8). The result follows from 4.7. Formulae 4.9 and 3.13 show that m has finite variation if and only if it has finite semi-variation, and that $\|m\| = |m|$.#

4.11. Example. Let X be a locally compact space and let F be a Banach space. An F-valued Radon measure on X is a linear map $m: C_{oo}(X) \to F$ of finite semivariation.

A Radon measure is B-continuous. A measure of finite variation is a Radon measure.

The second statement follows from the proposition above. The first is an immediate consequence of the next result.

4.12. Dini's theorem. Let X be a topological space and Φ a decreasingly directed family of continuous real-valued functions, with infimum zero.

Then Φ converges to zero uniformly on every compact subset K of X.

Proof. Given an $\varepsilon > 0$, consider the sets $[\varphi < \varepsilon]$, $\varphi \in \Phi$. They form an increasingly directed open cover of K. There is therefore a $\varphi_0 \in \Phi$ such that $K \subset [\varphi_0 < \varepsilon]$. Clearly, $\varphi < \varepsilon$ on K for all $\varphi \in \Phi$ which are smaller than φ_0.#

D. *-continuous linear maps

The notion of *-continuity, introduced for measures in section 3D, can be generalized to linear maps $U: \mathfrak{R} \otimes E \to G$. As for measures, such a condition is needed to insure a "good," Lebesgue integration theory for U. In order to introduce the weakest such condition that will do, we lay down the following definition.

Definition. A subset V of the strong dual G' of G is said to be norming, if

$$\|\xi\| = \sup\{ |\langle \xi ; v \rangle| : v \in V, \|v\| \leq 1\}, \text{ for all } \xi \in G.$$

The topology on G determined by the functions $v \in V$ is called the V-weak topology, $\sigma(G,V)$; if $V = G'$ then the prefix V- is omitted.

Examples. (1) E' is norming; E is a proper norming subspace of E'',

(2) If E and F are Banach spaces then $L(E,F)$ denotes the Banach space of bounded linear operators from E to F, equipped with the operator norm. For $\xi \in E$ and $\eta' \in F'$,

$$T \to \langle T\xi;\eta'\rangle = \langle v_{\xi\eta'};T\rangle \quad , \quad T \in L(E,F),$$

defines an element $v_{\xi\eta'}$ of $(L(E,F))'$. The set V of linear combinations of the $v_{\xi\eta'}$ can be identified with the algebraic tensor product $E \otimes F'$, and, is a norming subspace of $(L(E,F))'$. The norm of $v_{\xi\eta'}$ in $(L(E,F))'$ is $\|\xi\| \|\eta'\|$.

Definition. Let V be a norming subspace of G' and let $* = S$ or $* = B$. The map $U: \mathcal{R} \otimes E \to G$ is said to be V-weakly $*$-continuous if, whenever $\mathcal{R} \supset \Phi\downarrow{*}0$ and $v \in V$ then

$$\lim(\langle U(\varphi\xi);v\rangle: \varphi \in \Phi) = 0$$

uniformly for all ξ in the unit ball E_1 of E. (The suffix $_1$ shall always denote the unit ball of the normed space to which it is affixed.)

4.13. Example. If X is a locally compact space, denote by $C_{ooE}(X)$ the set of continuous E-valued functions of compact support. Let $U: C_{ooE}(X) \to G$ be a linear map, continuous in the topology of dominated uniform convergence. Then U is B-continuous.

Indeed, if $C_{oo}(X) \supset \Phi\downarrow^{B}0$, then Φ converges uniformly to zero, due to Dini's theorem (4.12), and so $\lim\|U(\Phi E_1)\| = 0$.

Suppose that $U: \mathcal{R} \otimes E \to G$ has finite semivariation and is V-weakly *-continuous, where V is a fixed norming subspace of G'. For every element $\eta \in G'$ define a measure $U\eta$ on $\mathcal{R}$, with values in E', by

$$\langle \xi; U\eta(\varphi) \rangle = \langle U(\varphi\xi); \eta \rangle \quad ; \quad \varphi \in \mathcal{R},\ \xi \in E.$$

4.14. <u>Proposition</u>. $U\eta$ is an E'-valued measure of finite variation and

$$|U\eta|(\varphi) \leq \|\eta\| \cdot \|U\|(\varphi) \quad , \quad \text{for} \quad \varphi \in \mathcal{R}_+.$$

Moreover, if U is V-weakly *-continuous then Uv is *-continuous for all $v \in V$ (* = S or B).

<u>Proof</u>. From $|\langle U(\varphi\xi);\eta\rangle| \leq \|\xi\| \cdot \|\eta\| \cdot \|U\|(\varphi)$ we see that the map $\xi \to \langle U(\varphi\xi);\eta\rangle$ defines a continuous linear functional $U\eta(\varphi)$ on E, for each $\varphi \in \mathcal{R}$ and $\eta \in G'$.

To see that $U\eta$ has finite variation smaller than $\|\eta\| \cdot \|U\|$, let $\varphi, \varphi_1, \varphi_2, \ldots, \varphi_n$ be positive elements of $\mathcal{R}$ such that $\sum \varphi_i = \varphi$. Given an $\varepsilon > 0$, there are unit vectors $\xi_1, \ldots, \xi_n$ in E such that

$$\sum \|U\eta(\varphi_i)\| \leq \varepsilon + \sum \langle \xi_i; U\eta(\varphi_i)\rangle = \varepsilon + \sum \langle U(\varphi_i\xi_i);\eta\rangle = \varepsilon + \langle U(\sum \varphi_i\xi_i);\eta\rangle.$$

Since $\|\sum \varphi_i\xi_i\| \leq \varphi$, the number to the right is smaller than $\varepsilon + \|\eta\| \cdot \|U\|(\varphi)$. Taking the supremum over all finite collections $\varphi_1, \ldots, \varphi_n$, and taking into account the arbitrariness of ε, we obtain $|U\eta|(\varphi) \leq \|\eta\| \cdot \|U\|(\varphi)$.

Suppose, then, that U is V-weakly *-continuous, and let $\mathcal{R} \supset \Phi \downarrow {*0}$ and $v \in V$. Then

$$\sup\{\langle \xi; Uv(\varphi)\rangle : \xi \in E_1\} = \|Uv(\varphi)\|$$

is arbitrarily small if φ is, and hence $\lim Uv(\Phi) = 0$.#

E. Supplements

4.15. If $X \in \mathcal{S}$ then the uniformity of dominated uniform convergence coincides with the uniformity of uniform convergence. If $X \notin \mathcal{S}$ and if $\mathcal{K}(\mathcal{S})$ has a cofinal countable set then these uniformities are different. They may coincide on $\overline{\mathcal{R}}_E$ if $X \notin \mathcal{S}$ and if $\mathcal{K}$ has no countable cofinal set.

4.16. If $\mathcal{K}(\mathcal{R})$ has a finite or countable cofinal subset, then a subset of $\mathcal{D}_E[\mathcal{R}]$ or of $\overline{\mathcal{R}}_E$ is bounded if and only if it is contained and bounded in one of the $\mathcal{D}_E[K]$ or $\overline{\mathcal{R}}_E[K]$, respectively.

4.17. Let $U \to \tilde{U}$ denote the map which associates with every linear map $U: \mathcal{R} \otimes E \to G$ of finite semivariation its unique continuous extension $\tilde{U}: \overline{\mathcal{R}}_E \to G$. The map $U \to \tilde{U}$ preserves semi-variation, variation, and V-weak *-continuity (V is some norming subset of G'). It is an isomorphism of $I_0(\mathcal{R};F)$ onto $I_0(\overline{\mathcal{R}};F)$ preserving variation and *-continuity (F is some Banach space).

4.18. A scalar Radon measure has finite variation.

4.19. If X is a locally compact space, then any measure in $I_0(C_{00}(X);E,R)$ is B-continuous.

LITERATURE: $\mathcal{K}(\mathcal{R})$ is a domination system in the sense of [45]. $\mathcal{R}_0$ was introduced in [2]. For the standard definition of semivariation see [11], [14].

5. The $\mathfrak{R}$-uniformity

With an integration lattice $\mathfrak{R}$ on a set X there is associated a natural uniformity on X, the coarsest one that makes all the functions of $\mathfrak{R}$ uniformly continuous, as functions from X to the extended reals. It is the natural ingredient in the definition of measurability (Chapter III). Its completion is a compact space, which can be used to great advantage in the investigation of the measures on $\mathfrak{R}$: the strong results on (locally) compact spaces, e.g., Urysohn's lemma, Tietze's extension theorem, the decomposition of the identity, etc., turn into powerful tools via the Bauer-Gelfand transform defined below. As general references to uniformities serve [26] and [17].

A. Review of the extended reals

The extended reals consist of the set $\overline{\mathbb{R}} = \mathbb{R} \cup \{-\infty, \infty\}$. The order of $\mathbb{R}$ is extended to $\overline{\mathbb{R}}$ in such a way that $-\infty < r < \infty$ for all r in $\mathbb{R}$. $\overline{\mathbb{R}}$ is a complete lattice in this order. The algebraic operations are defined on $\mathbb{R}$ as usual and additionally by $r\infty = (-r)(-\infty) = \infty$ for $0 < r \in \overline{\mathbb{R}}$; $r\infty = (-r)(-\infty) = -\infty$ for $0 > r \in \overline{\mathbb{R}}$; $r + \infty = \infty$, $r - \infty = -\infty$, $r + (-\infty) = r - \infty$, $r/\infty = r/-\infty = 0$ for $r \in \mathbb{R}$. The symbols $\infty - \infty$, ∞/∞ etc. are not defined. We do, however, define $\infty/(1+\infty) = 1$, $-\infty/(1+\infty) = -1$.

On $\overline{\mathbb{R}}$ we introduce the metric

$$\Delta(x,y) = \left| \frac{x}{1+|x|} - \frac{y}{1+|y|} \right| \qquad (x,y \in \overline{\mathbb{R}}).$$

Under Δ, $\overline{\mathbb{R}}$ is a compact metric space: the map $x \to x/(1+|x|)$ is an isometry of $\overline{\mathbb{R}}$ with the compact interval $[-1,1]$.

Let A be a bounded subset of $\mathbb{R}$, $A \subset [-a,a]$ say. The metric induced on A by Δ is then equivalent to the normal metric $d(x,y) = |x-y|$ on A: using the mean value theorem one finds easily that

$$d(x,y) \geq \Delta(x,y) \geq 2(1+a)^{-2} d(x,y) \quad \text{for } x,y \in A.$$

A function f from a set X with values in $\overline{\mathbb{R}}$ is called <u>numerical</u>. A numerical function $f: X \to \overline{\mathbb{R}}$ is the uniform limit in $\overline{\mathbb{R}}^X$ of the bounded functions $f_n = (-n \vee f) \wedge n$, since

$$\Delta(f_n(x), f(x)) \leq (1+n)^{-1} \quad \text{for all } x \text{ in } X.$$

B. The uniformity generated by a family of functions

Let X be a set, and let $\mathcal{F}$ be any family of numerical functions on it. The <u>uniformity</u> $\mathcal{U}(\mathcal{F})$ <u>generated by</u> $\mathcal{F}$ is the coarsest uniformity on X for which all the functions $\varphi: X \to \overline{\mathbb{R}}$ in $\mathcal{F}$ are uniformly continuous. Its gauge has a basis given by the following family of pseudo-metrics Δ_φ:

$$\Delta_\varphi(x,y) := \Delta(\varphi(x), \varphi(y)) = \left| \frac{\varphi(x)}{1+|\varphi(x)|} - \frac{\varphi(y)}{1+|\varphi(y)|} \right| ; \ \varphi \in \mathcal{F}; \ x, y \in X.$$

A family of functions from X to the reals can be trivially identified with a family of numerical functions, and so generates a uniformity.

<u>Warning</u>: This uniformity is coarser than the one that makes the functions in $\mathcal{F}$ uniformly continuous <u>as functions to</u> $\mathbb{R}$.

A map ψ from X to some uniform space Y is $\mathcal{F}$-uniformly continuous if and only if for every set V in the uniformity of Y there are finitely many functions $\varphi_1, \ldots, \varphi_n$ in $\mathcal{F}$ and a $\delta > 0$ such that

$$\Delta_{\varphi_i}(x,y) < \delta \quad \text{for } 1 \leq i \leq n \quad \text{implies} \quad (\psi(x), \psi(y)) \in V.$$

Conversely, a map ψ from Y to X is uniformly continuous if and only if $\varphi \circ \psi: Y \to \overline{\mathbb{R}}$ is, for all $\varphi \in \mathcal{F}$.

Let $\varphi: X \to \overline{\mathbb{R}}$ be an $\mathcal{F}$-uniformly continuous function. Since $r \to -n \vee r \wedge n$ is a uniformly continuous map of $\overline{\mathbb{R}}$ into $[-n,n]$ for each positive real n, $\varphi_n := -n \vee \varphi \wedge n$ is uniformly continuous from X to $[-n,n]$. Note that the uniformities on $[-n,n]$ inherited from $(\overline{\mathbb{R}},\Delta)$ and from $(\mathbb{R},d)$ coincide since $[-n,n]$ is compact. Now, φ is the uniform limit in $\overline{\mathbb{R}}^X$ of the sequence (φ_n), and so the $\mathcal{F}$-uniformity is the coarsest uniformity on X that makes all the functions $\{\varphi_n: \varphi \in \mathcal{F},\ n \in \mathbb{N}\}$ uniformly continuous from X to $\mathbb{R}$ (with the normal metric d on $\mathbb{R}$!). In other words, the $\mathcal{F}$-uniformity is also generated by the bounded functions $\{\varphi_n: \varphi \in \mathcal{F},\ n \in \mathbb{N}\}$, and it does not matter whether they are viewed as functions from X to $\mathbb{R}$ or to $\overline{\mathbb{R}}$:

The bounded uniformly continuous functions from X to $\mathbb{R}$ form a uniformly closed algebra and integration lattice $\mathcal{U}_b(\mathcal{F}, \mathbb{R})$, which in turn generates the $\mathcal{F}$-uniformity.

Example 1. Let $\mathcal{C}$ be a clan on X. The $\mathcal{C}$-uniformity coincides with the uniformity generated by the step functions $\mathcal{S}(\mathcal{C})$ over $\mathcal{C}$. Conversely, if $\mathcal{R}$ is a full integration lattice then the $\mathcal{R}$-uniformity is generated by the sets $\mathcal{C}(\mathcal{R})$ in $\mathcal{R}$ (6.3).

5.1. Example 2. Let $\mathcal{R}$ be an integration lattice on X. The $\mathcal{R}$-uniformity is generated by $\mathcal{R}_b$, and also by the dominated integration lattice $\mathcal{R}_0$. Indeed, we know that every function in $\mathcal{R}$ is the uniform limit (in $\overline{\mathbb{R}}^X$!) of functions in $\mathcal{R}_0$ (4.5). The functions in $\mathcal{R} \otimes E$, $\mathcal{R}_b \otimes E$, and $\mathcal{R}_0 \otimes E$ are all uniformly continuous (E is a Banach space).

Let us look at an arbitrary $\mathcal{R}$-uniformly continuous map f from X to a metric space (E,d). For every natural number n, there are functions

$\varphi_n^1,\ldots,\varphi_n^{i(n)}$ in $\mathcal{R}_0$ such that $|\varphi_n^i(x) - \varphi_n^i(y)| < 1$ (for $x,y \in X$ and $1 \leq i \leq i(n)$) implies $d(f(x),f(y)) < n^{-1}$. It follows that f is constant outside the countable union $\cup\{[\varphi_n^i \neq 0]: n \in \mathbb{N}, 1 \leq i \leq i(n)\}$ of dominated sets. If (E,d) is complete, more can be said: the limit

$$\lim\{f(X - K): K \in \mathcal{K}(\mathcal{R})\} = f_0$$

exists along the increasingly directed set $\mathcal{K}(\mathcal{R})$, and so the map f is said to <u>converge at infinity</u>. If E is a Banach space, we arrive at the following: A uniformly continuous function f is the sum of a constant f_0 and a function vanishing at infinity.#

C. Completion and Spectrum

We consider again an arbitrary family $\mathcal{F}$ of numerical functions on X. A <u>character</u> of $\mathcal{F}$ is a map $t: \mathcal{F} \to \overline{\mathbb{R}}$ that respects all algebraic and order relations in $\mathcal{F}$. To wit,

$$(1)\qquad \left.\begin{aligned} t(\varphi+\psi) &= t(\varphi) + t(\psi)\\ t(\varphi-\psi) &= t(\varphi) - t(\psi)\\ t(\varphi\wedge\psi) &= t(\varphi) \wedge t(\psi)\\ t(\varphi\vee\psi) &= t(\varphi) \vee t(\psi)\\ t(\varphi\cdot\psi) &= t(\varphi) \cdot t(\psi)\\ t(\varphi\wedge r) &= t(\varphi) \wedge r\\ t(\varphi\vee r) &= r(\varphi) \vee r \end{aligned}\right\} \quad \begin{array}{l}\text{For all } \varphi,\psi \in \mathcal{F} \text{ and } r \in \mathbb{R},\\ \text{provided that both sides of}\\ \text{the equations are defined.}\end{array}$$

For example, if $\varphi\psi$ is not in $\mathcal{F}$, or if $t(\varphi)t(\psi)$ is not defined in $\overline{\mathbb{R}}$, then the fifth equation places no restraint on t. We denote by $\hat{X} = \hat{X}(\mathcal{F})$ the set of all characters of $\mathcal{F}$ and give it the topology of pointwise convergence:

A net $(t_i)_{i\in I}$ converges to t in $\hat{X}$ if $(t_i(\varphi))$ converges to $t(\varphi)$ for all φ in $\mathcal{F}$. The topological space $\hat{X}(\mathcal{F})$ is called the <u>character space</u> of $\mathcal{R}$. $\hat{X}$ is a topological subspace of the space $\overline{\mathbb{R}}^{\mathcal{F}}$ of all maps from $\mathcal{F}$ to $\overline{\mathbb{R}}$ with the product topology. By Tychonoff's theorem the latter space is compact.

We want to show that $\hat{X}$ is closed in $\overline{\mathbb{R}}^{\mathcal{F}}$, and so is also compact. We have to prove that if (t_i) is a net of characters converging pointwise to a map $p: \mathcal{F} \to \overline{\mathbb{R}}$, then p is a character also; i.e., that p satisfies the seven equations (1) above. This is verified for the first such equation, only. Let $\varphi, \psi \in \mathcal{F}$. If $p(\varphi) = \infty$ and $p(\psi) = -\infty$, the first condition is satisfied since then $p(\varphi) + p(\psi)$ is undefined. The same is true if $p(\varphi) = -\infty$ and $p(\psi) = \infty$, or if $\varphi+\psi$ is not in $\mathcal{F}$. In any other case the sums in the following equation are defined for sufficiently high indices i:

$$p(\varphi) + p(\psi) = \lim_i t_i(\varphi) + \lim_i t_i(\psi) = \lim_i t_i(\varphi+\psi) = p(\varphi+\psi).$$

To summarize:

<u>The character space of $\mathcal{F}$ is a compact Hausdorff space</u>.

Besides the zero character, ∞, mapping all of $\mathcal{F}$ to zero, there are the following special ones. Every element $x \in X$ defines a character $\pi(x)$ by

$$\pi(x)(\varphi) = \varphi(x) \quad , \quad \text{for } \varphi \in \mathcal{F}.$$

The map π throws X into $\hat{X}$. It is one-to-one if $\mathcal{F}$ separates the points of X; and $\infty \notin \pi(X)$ if the functions of $\mathcal{F}$ do not all vanish simultaneously at any point of X.

To every function φ in $\mathscr{F}$ there corresponds the continuous function

$$\check{\varphi}: t \to t(\varphi) \qquad t \in \hat{X}$$

on $\hat{X}$, called the (generalized) <u>Gelfand transform</u> of φ. The set $\check{\mathscr{F}}$ of all Gelfand transforms has exactly the same algebraic and order structures as $\mathscr{F}$. If, for instance, $\mathscr{F}$ is an algebra of bounded functions then so is $\check{\mathscr{F}}$. The map $\varphi \to \check{\varphi}$ preserves the sup-norms on bounded functions and has the inverse

$$\check{\varphi} \to \check{\varphi} \circ \pi \ .$$

All this is immediately obvious from the equations (1).

Both $\mathscr{F}$ and $\check{\mathscr{F}}$ give rise to uniformities on their respective domains, X and $\hat{X}$. The uniformity of $\hat{X}$ is evidently its compact uniformity. The map $\pi: X \to \hat{X}$ is uniformly continuous, and the $\mathscr{F}$-uniformity on X is the coarsest one such that this is the case.

Recall that every uniform space E has a unique <u>Hausdorff completion</u>. This is a complete Hausdorff space $\tilde{E}$ together with a uniformly continuous map $i: E \to \tilde{E}$ such that the following universal property holds: whenever $f: E \to D$ is a uniformly continuous map with values in a complete Hausdorff space D, there is a unique uniform map $\tilde{f}: \tilde{E} \to D$ such that $f = \tilde{f} \circ i$.

<u>5.2. Proposition</u>. Let $\tilde{S} = \tilde{S}(\mathscr{F})$ denote the closure of $\pi(X)$ in $\hat{X}$. Then $\pi: X \to \tilde{S}(\mathscr{F})$ is the completion of X in the $\mathscr{F}$-uniformity.

<u>Proof</u>. Let f be a uniform map from X to the complete Hausdorff space D. If $\pi(x) = \pi(y)$ for $x, y \in X$, then (x,y) lies in every member of the $\mathscr{F}$-uniformity, and so $f(x) = f(y)$; f is constant on the fibres

$\pi^{-1}(t)$, $t \in \hat{X}$, of π and can be viewed as an $\check{\mathcal{F}}$-uniformly continuous map from $\pi(X)$ to D. There is therefore a unique uniformly continuous extension $\tilde{f}$ to the closure $\tilde{S}$ of $\pi(X)$ in $\hat{X}$.#

Now, put $X_\infty = \pi^{-1}(\infty)$ and $X_0 = X - X_\infty$. X_0 consists of all points in X at which not all the functions of $\mathcal{F}$ vanish, and is called the <u>essential part</u> of X. Furthermore, $S(\mathcal{F}) = S$ denotes the set of all non-zero characters in the completion $\tilde{S}$ of X, and is called <u>the spectrum</u> of $\mathcal{F}$. It is a locally compact space in its topology, and equals $\tilde{S}$ if it is actually compact. If S is not compact then $\tilde{S}$ is its one-point-compactification. The restriction of π to X_0 maps X_0 onto a dense subset of $S(\mathcal{F})$. It is again denoted by π.

D. <u>The Gelfand transform</u>

For any uniformly continuous function f on X we again call the restriction of $\tilde{f}$ to $S = S(\mathcal{F})$ <u>the Gelfand transform</u> of f and denote it by $\hat{f}$. The Gelfand transforms are continuous on $S(\mathcal{F})$ and vanish at infinity. The set of such transforms will be denoted by $\hat{\mathcal{F}}$.

Suppose that $\mathcal{R}$ is an integration lattice. From the fact that $\mathcal{R}$, $\mathcal{R}_b$, and $\mathcal{R}_0$ all give rise to the same uniformity on X we deduce that the corresponding completions are the same, and thus

$$S(\mathcal{R}) = S(\mathcal{R}_b) = S(\mathcal{R}_0).$$

We specialize henceforth to the case that $\mathcal{F}$ is an integration lattice $\mathcal{R}$ or an algebra of bounded functions $\mathcal{R}$. Then so is $\hat{\mathcal{R}}$, on the spectrum S of $\mathcal{R}$. With the notation $C_0(S)$ for the continuous functions on S vanishing at infinity, we obtain the following result.

5.3. Proposition. $\hat{\mathcal{R}}$ is uniformly dense in $C_0(S)$, and the uniform closure $\overline{\mathcal{R}}^u$ of $\mathcal{R}$ is an algebra and integration lattice isometrically isomorphic with $C_0(S)$. The isomorphism is the Gelfand transform $\varphi \to \hat{\varphi}$, and has the inverse $\hat{\varphi} \to \hat{\varphi}\circ\pi$.

Proof. The Gelfand transform preserves algebraic structure and sup-norm, and so does $\varphi \to \varphi\circ\pi$. $\hat{\mathcal{R}}$ is an algebra or integration lattice on S separating the points of S, so it is uniformly dense in $C_0(S)$ due to the theorem of Stone-Weierstrass.#

5.4. Corollary. A subset K of X is $\mathcal{R}$-dominated if and only if $\pi(K)$ is relatively compact in the spectrum of $\mathcal{R}$.

Proof. If $K \leq \varphi \in \mathcal{R}$ then $\pi(K) \leq \hat{\varphi} \in C_0(S)$ and $\pi(K)$ is relatively compact.

Conversely, if $\pi(K)$ has compact closure $\overline{\pi(K)}$ in S then there is a function $\psi \in C_0(S)$ such that $\psi \geq \overline{\pi(K)}$ (Ursysohn's lemma). There is a $\hat{\varphi} \in \hat{\mathcal{R}}$ such that $|\psi - \hat{\varphi}| < 1/2$; evidently K is majorized by $2\varphi \in \mathcal{R}$. #

5.5. Lemma. Suppose $\mathcal{R}$ is a dominated integration lattice and $\varphi: X \to \mathbb{R}$ a uniformly continuous function with dominated carrier K. Then there exists a sequence (φ_n) in $\mathcal{R}[K]$ which approximates φ uniformly and satisfies $|\varphi_n| \leq |\varphi|$ for all integers n.

Proof. Splitting φ into its positive and negative parts and approximating them separately will prove the result; we may therefore assume that φ is positive. Once we have found a sequence (φ_n) in $\mathcal{R}[K]$ with $|\varphi_n - \varphi| \leq 1/n$ we are finished; for then the sequence $\varphi_n' = \varphi_n - \varphi_n \wedge 1/n$ in $\mathcal{R}[K]$ will approximate φ uniformly and satisfy $\varphi_n' \leq \varphi$ for all integers n.

Put $a = \sup\{\varphi(x): x \in X\}$ and let an $\varepsilon > 0$ be given. Put $\hat{K} = [\hat{\varphi} \geq \varepsilon]$ and $\hat{U} = [\hat{\varphi} > 0]$. Then K is compact and contained in the open set $\hat{U} \subset S(\mathcal{R})$, and due to Urysohn's lemma there exists a function $\psi \in C_{oo}(S)$ such that $0 \leq \psi \leq 1$, $\psi = 1$ on $\hat{K}$, and $\psi = 0$ outside $\hat{U}$. There exists a $\psi' \in \mathcal{R}$ such that $|\hat{\psi}' - \psi| < 1/4$ (5.3). Put

$$\psi'' = 1 \wedge 2(\psi' - \psi' \wedge 1/4) \in \mathcal{R}.$$

Then $0 \leq \psi'' \leq 1$, $\hat{\psi}'' = 1$ on $\hat{K}$ and $\hat{\psi}'' = 0$ outside $\hat{U}$. Choose now a $\varphi' \in \mathcal{R}_+$ so that $|\varphi - \varphi'| < \varepsilon$ and put

$$\varphi'' = \varphi' \wedge a\psi'' \in \mathcal{R}_+ .$$

Clearly $|\varphi'' - \varphi| < \varepsilon$. Hence $\varphi''' = \varphi'' - \varphi'' \wedge \varepsilon$ belongs to $\mathcal{R}[K]$ and approximates φ uniformly to within 2ε.#

5.6. Theorem. Suppose $\mathcal{R}$ is a dominated integration lattice or algebra. Then $\hat{\mathcal{R}}$ is dense in $C_{oo}(S)$ in the topology of dominated uniform convergence, and the closure $\bar{\mathcal{R}}$ of $\mathcal{R}$ in $\mathcal{D}[\mathcal{R}]$ is an algebra and integration lattice uniformly isomorphic with $C_{oo}(S)$ via the Gelfand transform. $\bar{\mathcal{R}}$ consists exactly of all $\mathcal{R}$-uniformly continuous functions that vanish outside some dominated set.

Proof. If $\mathcal{R}$ is an integration lattice, this follows from the lemma. Suppose, then, that $\mathcal{R}$ is a dominated algebra, and let φ be a uniformly continuous function with dominated carrier K. There is a $\psi \in C_{oo}(S)$ with $2\psi \geq \pi(K)$ (5.4 and Urysohn's lemma). By 5.3, ψ may be assumed to belong to $\hat{\mathcal{R}}$. The function

$$\hat{\varphi}\psi'^{-1} = \hat{\varphi}(\psi' \vee 1/2)^{-1}$$

belongs to $C_{oo}(S)$ and so is the uniform limit of a sequence $(\hat{\varphi}_n)$ in $\hat{\mathcal{R}}$. Hence φ is the uniform limit of the sequence $(\varphi_n \cdot \psi' \circ \pi)$ in $\mathcal{R}[[\psi' \circ \pi > 0]]$, which converges in the uniformity of dominated uniform convergence.#

A subset of a uniform space E is said to be <u>precompact</u> if its image in the completion of E is relatively compact. The notions of precompactness and relative compactness coincide if E is complete.

<u>5.7. Corollary</u>. Suppose $\mathcal{R}$ is a dominated integration lattice and K any subset of X. If $f: K \to E$ is a uniformly continuous map having values in a uniform space E, then $f(K)$ is precompact in E.

If E is, in particular, a Banach space and if K is dominated, then there is a dominated set L containing K and a sequence (φ_n) in $\mathcal{R}[L] \otimes E$ which converges uniformly on K to f.

<u>Proof</u>. f can be viewed as a uniformly continuous function from $\pi(K) \subset S(\mathcal{R})$ to the completion $\tilde{E}$ of E, and so has a continuous extension $\tilde{f}$ to the compact closure $\overline{\pi(K)}$ of $\pi(K)$ in $\tilde{S}(\mathcal{R})$. Hence $f(K)$ lies in the compact subset $\tilde{f}(\overline{\pi(K)}) \subset \tilde{E}$. This proves the first statement. Note that $\overline{\pi(K)}$ lies in the spectrum $S(\mathcal{R})$ of $\mathcal{R}$ if K is dominated (5.4).

Suppose that E is a Banach space. Find a compact neighborhood L' of $\overline{\pi(K)}$ and a function $\psi \in C_{oo+}(S)$ which equals one on $\overline{\pi(K)}$ and vanishes outside L' (Urysohn's lemma). Let $\varepsilon > 0$ be given. There is a finite open cover $\{U_i : 1 \leq i \leq n\}$ of $\overline{\pi(K)}$ such that $\tilde{f}$ oscillates by less than ε on every one of the sets U_i. Put $U_0 = S - \overline{\pi(K)}$ and find a decomposition of the identity subordinate to the cover $\{U_i : 0 \leq i \leq n\}$;

i.e., a collection of positive continuous functions φ_i on S, of sum 1, such that $\varphi_i \leq U_i$ for $0 \leq i \leq n$. For $1 \leq i \leq n$, select a vector $\xi_i \in \tilde{f}(U_i \cap \overline{\pi(K)})$ and put

$$\varphi = \sum \psi \varphi_i \xi_i \in C_{oo}(S) \otimes E \ .$$

If $x \in K$, $x \in U_1$ say, then

$$\|\tilde{f}(x) - \varphi(x)\| = \|\sum \varphi_i(x)(f(x) - \xi_i)\| \leq \sum \varphi_i(x)\|\tilde{f}(x) - \xi_i\| \leq \varepsilon,$$

for if $\varphi_i(x) \neq 0$ then $\|\tilde{f}(x) - \xi_i\| \leq \varepsilon$. Therefore $\|\tilde{f} - \varphi\| \leq \varepsilon$ uniformly on K. With $A = \sup\{\|\tilde{f}(x)\|: x \in K\}$, we have clearly $\|\varphi\| \leq A + \varepsilon$. We may now approximate each of the functions $\psi \varphi_i$ uniformly to within ε/A by a function $\hat{\varphi}'_i \in \hat{\mathcal{R}}[L']$ (5.5). Clearly

$$\varphi' = \sum \varphi'_i \xi_i \in \mathcal{R}[L] \qquad (L = \pi^{-1}(L'))$$

approximates f uniformly on K to within 2ε and satisfies $\|\varphi'\| \leq A + 2\varepsilon$. The proof is finished.#

<u>5.8. Corollary</u>. If $\mathcal{R}$ is a dominated integration lattice and E a Banach space, then $\overline{\mathcal{R}}_E$ consists exactly of all E-valued $\mathcal{R}$-uniformly continuous functions of dominated support and is isomorphic with $C_{ooE}(S)$ via the Gelfand transform. In particular, $\overline{\mathcal{R}} \cdot \overline{\mathcal{R}}_E \subset \overline{\mathcal{R}}_E$.

More precisely, if $\varphi: X \to E$ is a uniformly continuous map with dominated carrier L, then there exists a sequence (φ_n) in $\mathcal{R}[L] \otimes E$ which approximates φ uniformly and satisfies $\|\varphi_n\| \leq \|\varphi\|$ for all integers n.

Proof. Everything will follow from the last statement. Given an $\varepsilon > 0$, we put $K = [\|\varphi\| > \varepsilon]$ and $L' = [\|\hat{\varphi}\| > 0]$, and procure a function

$$\varphi' = \sum \varphi_i' \xi_i \in \mathcal{R}[L]$$

as in the proof of the theorem, which approximates φ on K to within ε/A. Replacing φ_i' by $\varphi_i' - \varphi_i' \wedge \varepsilon$, we arrive at a function $\varphi'' \in \mathcal{R}[L]$ which approximates φ to within 2ε and satisfies $\|\varphi''\| \leq \|\varphi\|$.#

We are now in the position to finish the proof of proposition 4.8 of the last section. Let $U: \mathcal{R} \otimes E \to G$ be a linear map of finite semivariation. The formula

$$\hat{U}(\hat{\varphi}) = U(\varphi) \quad , \quad \varphi \in \mathcal{R} \otimes E,$$

defines a map $\hat{U}: \hat{\mathcal{R}} \otimes E \to G$, which we call the Gelfand-Bauer transform of U.

As the variation and the semivariation are defined purely algebraically, without reference to the set underlying $\mathcal{R}$, and as $\mathcal{R} \longleftrightarrow \hat{\mathcal{R}}$ is an isomorphism algebraically, U has finite variation or semivariation if and only if $\hat{U}$ has; and, indeed

$$|\hat{U}| = |U|^{\wedge} \quad \text{and} \quad \|\hat{U}\| = \|U\|^{\wedge}$$

in the obvious sense. Moreover, if $\tilde{U}$ denotes the unique extension of U to $\bar{\mathcal{R}}_E$ by continuity in dominated uniform convergence, then $\hat{\tilde{U}}$ is the Gelfand-Bauer transform of $\tilde{U}$, and

$$\|\tilde{U}\|^{\wedge} = \|\hat{\tilde{U}}\| \quad \text{and} \quad |\tilde{U}|^{\wedge} = |\hat{\tilde{U}}| \ .$$

<u>5.9. Corollary</u>. If $U: \mathcal{R} \otimes E \to G$ has finite semivariation then $\|U\|$ is an increasing seminorm and $\|\tilde{U}\| = \|U\|$ on $\mathcal{R}_+$.

<u>Proof</u>. From the discussion above, we may assume that $X = S(\mathcal{R})$ is locally compact. It is clear that $\|\tilde{U}\| \geq \|U\|$ on $\mathcal{R}_+$. To see the reverse inequality, let a $\varphi \in \mathcal{R}_+$ and an $\varepsilon > 0$ be given, and let $a = \|\tilde{U}\|(\varphi)$. There is a $\psi \in \bar{\mathcal{R}}_E (= C_{ooE}(S))$ such that $\|\psi\| \leq \varphi$ and $\|\tilde{U}(\psi)\| \geq a - \varepsilon$. From the last part of the corollary 5.8, there is a sequence (φ_n) in $\mathcal{R} \otimes E$ such that $\|\varphi_n\| \leq \varphi$ and $\varphi_n \to \psi$ uniformly. As $\tilde{U}$ is continuous (4.7), $\lim\|\tilde{U}(\varphi_n)\| = \|\tilde{U}(\psi)\| \geq a - \varepsilon$ and so $\|\tilde{U}\|(\varphi) \geq a - \varepsilon$.

To see that $\|\tilde{U}\|$ is subadditive, let $\varphi_1, \varphi_2 \in \mathcal{R}_+$ and an $\varepsilon > 0$ be given. If $\|\tilde{U}\|(\varphi_1 + \varphi_2) > a \in \mathbb{R}$, then there is a $\psi \in \bar{\mathcal{R}}_E$ such that $\|\tilde{U}(\psi)\| > a$ and $\|\psi\| \leq \varphi_1 + \varphi_2$. The functions

$$\psi_i = \psi \varphi_i / (\varphi_1 + \varphi_2) \quad , \quad i = 1, 2,$$

are clearly continuous on the compact set $[\varphi_1 + \varphi_2 > 0]^-$, hence are uniformly continuous. They satisfy $\|\psi_i\| \leq \varphi_i$, and so

$$\|\tilde{U}\|(\varphi_1) + \|\tilde{U}\|(\varphi_2) \geq \|\tilde{U}(\psi_1)\| + \|\tilde{U}(\psi_2)\| \geq \|\tilde{U}(\psi_1 + \psi_2)\| = \|\tilde{U}(\psi)\| > a. \#$$

E. Supplements

The Gelfand transform seems to have been introduced (not under this name) by Bauer, and the theory as exposed here was systematically studied in [2].

5.10*. (i) The $\mathcal{R}$-uniformity is metrizable if and only if $\mathcal{R}$ is countably generated. (ii) If $\mathcal{R}$ is an arbitrary integration lattice and f an $\mathcal{R}$-uniformly continuous map with values in a metric space, then there exists a countable subset $\mathcal{R}^c$ of $\mathcal{R}$ such that f is $\mathcal{R}^c$-uniformly continuous.

5.11*. Let $\mathcal{R}_\uparrow^B$ denote the family of numerical functions that are the pointwise suprema of subfamilies of $\mathcal{R}$; and let $\mathcal{U}^B(\mathcal{R})$ denote the sets, i.e., the idempotents in $\mathcal{R}_\uparrow^B$. (i) A function $h > 0$ belongs to $\mathcal{R}_\uparrow^B$ if and only if $[h > 0]$ belongs to $\mathcal{U}^B(\mathcal{R})$ for all $r > 0$, i.e., if h is lower semicontinuous in the $\mathcal{R}$-topology. (ii) $\mathcal{U}^B(\mathcal{R}) \cup \{X\}$ is the $\mathcal{R}$-topology, i.e., the topology associated with the $\mathcal{R}$-uniformity; and it consists exactly of the sets $\pi^{-1}(U)$, where U is open in $S(\mathcal{R})$. (iii) The sets $[\varphi > 0]$, $\varphi \in \mathcal{R}$, form a basis for the $\mathcal{R}$-topology.

5.12. Suppose X is a normal topological space. Then the spaces: $C_b(X)$ of real-valued continuous and bounded functions; $C_o(X)$ of continuous functions vanishing at infinity; and $C_{oo}(X)$ of continuous functions of compact support, are integration lattices. The $C_b(X)$-topology on X coincides with the given topology on X; and the spectrum of $C_b(X)$ is the Stone-Čech-compactification of X.

5.13*. Let $\mathcal{R}_\uparrow^S$ denote the family of those numerical functions that are the pointwise suprema of countable subfamilies of $\mathcal{R}$; and let $\mathcal{U}^S(\mathcal{R})$ denote the family of sets in $\mathcal{R}_\uparrow^S$. (i) $\mathcal{R}_\uparrow^S = \mathcal{R}_\uparrow^B$ if $\mathcal{R}$ is countably generated. (ii) $\mathcal{U}^S(\mathcal{R})$ is closed under the formation of countable unions. (iii) a function $h > 0$ belongs to $\mathcal{R}_\uparrow^S$ if and only if $[h > r]$ belongs to $\mathcal{U}^S(\mathcal{R})$ for all $r > 0$. (iv) The sets in $\mathcal{U}^S(\mathcal{R})$ are $\mathcal{R}$-open and are exactly the pre-images $\pi^{-1}(U)$ of the open K_σ's U in $S(\mathcal{R})$.

5.14*. For $* = S$ or B, put $\mathcal{R}_\downarrow^* = -\mathcal{R}_\uparrow^*$ and let $\mathcal{K}^*(\mathcal{R})$ denote the sets in $\mathcal{R}_\downarrow^*$. (i) $\mathcal{K}^B(\mathcal{R})$ consists exactly of the closed and precompact subsets of X; and $\mathcal{K}^S(\mathcal{R})$ consists of the closed and precompact G_δ's. (ii) A dominated function $k > 0$ belongs to $\mathcal{R}_\downarrow^*$ if and only if $[k \geq r]$ belongs to $\mathcal{K}^*(\mathcal{R})$ for all $r > 0$.

5.15*. If $\mathcal{R}_{\uparrow o}^*$ denotes the bounded and dominated functions in $\mathcal{R}_\uparrow^*$, then $\mathcal{R}_{\uparrow o}^* - \mathcal{R}_{\uparrow o}^*$ is an integration lattice ($* = S$ or B).

5.16. If $\mathcal{C}$ is an algebra of sets (6.11) on X then the spectrum of $\mathcal{C}$ is the Stone space of the Boolean ring $\mathcal{C}$.

LITERATURE: [2].

6. Scalar measures on full integration domains

A. S-closure

Definition. Let X be a set and E a Banach space. A family $\mathcal{F}$ of E-valued functions on X is said to be full or S-closed if it is closed under pointwise limits of majorized sequences; that is to say, if φ is the pointwise limit of a sequence (φ_n) in $\mathcal{F}$ such that $\|\varphi_n\| \leq \|\psi\|$ for some $\psi \in \mathcal{F}$ and all integers n, then φ belongs to $\mathcal{F}$.#

The intersection of any collection of full families $\mathcal{F}$, of functions from X to E, is evidently also full. Since E^X is full, every family $\mathcal{F}$ is contained in a smallest full family, which is called the full span or the S-closure of $\mathcal{F}$ and is denoted by $\mathcal{F}^S$.

Given a family $\mathcal{F} \subset E^X$, we can calculate its S-closure by induction: Put $\mathcal{F}^o = \mathcal{F}$. If $\mathcal{F}^a$ has been defined for all ordinals $a < b$, let $\mathcal{F}^b$ be the set of all pointwise limits of those sequences that are in

$$\tilde{\mathcal{F}}^b = \cup\{\mathcal{F}^a : a < b\}$$

and that are majorized by a function in $\mathcal{F}$. Then $\mathcal{F}^S = \mathcal{F}^c$ for all ordinals c that are bigger or equal to the first uncountable ordinal $\aleph_1$.

Indeed, it is clear that $\mathcal{F}^c \subset \mathcal{F}^S$ for all ordinals c. Conversely, if (φ_n) is a majorized sequence in $\tilde{\mathcal{F}}^{\aleph_1}$ with pointwise limit φ, then $\varphi_n \in \mathcal{F}^{a_n}$ for some $a_n < \aleph_1$ and, since $b = \sup\{a_n : n \in \mathbb{N}\} < \aleph_1$, $\varphi \in \mathcal{F}^b \subset \tilde{\mathcal{F}}^{\aleph_1}$. Therefore $\tilde{\mathcal{F}}^{\aleph_1}$ is full and so equals $\mathcal{F}^S$.

Note that every element of $\mathcal{F}^S$ is majorized by an element of $\mathcal{F}$.

6.1. Proposition. If $\mathcal{I}$ is an integration lattice or a clan, so is its full span.#

If $\mathcal{R}$ is an integration lattice, then $(\mathcal{R} \otimes E)^S$ is a vector space containing $\mathcal{R}^S \otimes E$. If $\mathcal{R}$ is dominated then $(\mathcal{R} \otimes E)^S$ is $\mathcal{R}$-dominated.

Proof. It is easy to see, by induction, that the $\mathcal{I}^a$ are all integration lattices or clans.

The set of functions in $(\mathcal{R} \otimes E)^S$ that are majorized by an element of $\mathcal{R}$ is S-closed and forms a vector space.#

Definition. Let $\mathcal{R}$ be a dominated integration lattice. The functions in $(\mathcal{R} \otimes E)^S$ are called the dominated E-valued $\mathcal{R}$-Baire functions. The sets in $\mathcal{R}^S$ are the dominated $\mathcal{R}$-Baire sets, and their collection is written $\mathfrak{B}(\mathcal{R}^S)$.#

Evidently $\mathfrak{B}(\mathcal{R}^S)$ is a full clan, or δ-ring.

B. Full integration domains

6.2. Lemma. An integration domain $\mathcal{I}$ is full if and only if it is closed under pointwise limits of majorized decreasing (or increasing) sequences.

Proof. Both conditions are evidently necessary. Conversely, assume that $\mathcal{I}$ is closed under pointwise limits of majorized decreasing sequences.

It will first be shown that $\mathcal{I}$ is also closed under majorized increasing sequences. Indeed, if (φ_n) is such a sequence in $\mathcal{I}$, majorized by $\psi \in \mathcal{I}_+$, then $(\psi - \varphi_n)$ is a decreasing sequence in $\mathcal{I}_+$ majorized by ψ, so that $\lim(\psi - \varphi_n) = \psi - \lim(\varphi_n) \in \mathcal{I}$. Hence $\lim(\varphi_n) = \psi - \psi \wedge (\psi - \lim(\varphi_n)) \in \mathcal{I}$.

Now, let (φ_n) be any majorized sequence in $\mathcal{I}$, converging to φ. The functions $\psi_n = \varphi_n \vee \varphi_{n+1} \vee \dots$ are in $\mathcal{I}$, being limits of increasing majorized sequences, and so is $\varphi = \psi_1 \wedge \psi_2 \dots$, being the limit of a decreasing majorized sequence. Hence $\mathcal{S}$ is full. The proof that $\mathcal{S}$ is full when it is closed under pointwise limits of majorized increasing sequences is analogous.#

Example. A clan $\mathcal{C}$ is full if and only if it is closed under countable intersections. Full clans are usually called δ-rings.

Example. Consider a full integration lattice $\mathcal{R}$. The following statements show that there are many sets in $\mathcal{R}$:

(i) For $\varphi \in \mathcal{R}$, $0 < r \in \mathbb{R}$, the set $[\varphi > r]$ belongs to $\mathcal{R}$.

Indeed, $[\varphi > r]$ is the pointwise limit of the sequence $\varphi_n = (n(\varphi - \varphi \wedge r)) \wedge 1$, in $\mathcal{R}$ which is majorized by $r^{-1}|\varphi| \in \mathcal{R}$.

(ii) For $\varphi \in \mathcal{R}$, $0 < r \in \mathbb{R}$, the set $[\varphi \geq r]$ belongs to $\mathcal{R}$.

Indeed, it is the limit of the decreasing sequence $([\varphi > r - n^{-1}])_{n \in \mathbb{N}}$ in $\mathcal{R}_+$.

(iii) For $\varphi \in \mathcal{R}$ and $0 > r \in \mathbb{R}$, the sets $[\varphi < r]$ and $[\varphi \leq r]$ belong to $\mathcal{R}$.

For example, $[\varphi < r] = [-\varphi > -r]$.

(iv) For $\varphi \in \mathcal{R}$ and $r, s \in \mathbb{R}$ with $0 \notin [r,s]$, the sets $[r < \varphi < s]$, $[r < \varphi \leq s]$, $[r \leq \varphi < s]$, and $[r \leq \varphi \leq s]$ belong to $\mathcal{R}$.

For example, $[r < \varphi < s] = [r < \varphi] - [s \leq \varphi] = [\varphi < s] - [\varphi \leq r]$.

(v) The sets in $\mathcal{R}$ clearly form a full clan, i.e., a δ-ring $\mathcal{C}(\mathcal{R})$.

If φ is a positive function in $\mathcal{R}$, then the following sequence of step functions over $\mathfrak{F}(\mathcal{R})$,

$$s_n = \sum_{m=0}^{4^n} m2^{-n}[m2^{-n} \leq \varphi < (m+1)2^{-n}] \quad , \quad (n \in \mathbb{N}),$$

clearly increases and converges to φ, uniformly so if φ is bounded. To summarize:

6.3. Proposition. If $\mathcal{R}$ is a full integration lattice then every function φ of $\mathcal{R}$ is the pointwise limit of a sequence (s_n) of step functions over the δ-ring of sets in $\mathcal{R}$. If φ is positive, (s_n) can be chosen to be increasing; and if φ is also bounded, (s_n) can be chosen to approximate φ uniformly.#

6.4. Corollary. If φ and ψ are elements of the full integration lattice $\mathcal{R}$ and if either is bounded, then their product $\varphi\psi$ is in $\mathcal{R}$. A full and dominated integration lattice is therefore an algebra.

Proof. If both φ and ψ are idempotents, the statement follows from $\varphi\psi = \varphi \wedge \psi$. By linearity, it therefore holds if both are step functions. Suppose next that both φ and ψ are positive, with $\psi \leq A$. Let (s_n) and (t_n) be sequences of step functions converging increasingly to φ and ψ, respectively. The sequence $(s_n \cdot t_n)$ converges to $\varphi\psi$ and is majorized by $A\varphi \in \mathcal{R}$, and so $\varphi\psi$ belongs to $\mathcal{R}$. The general case is reduced to this one by decomposing φ and ψ into their positive and negative parts.#

C. Scalar S-measures

6.5. Proposition. Every scalar S-measure m on a full integration lattice $\mathcal{R}$ has finite variation.

Proof. Decomposing m into its real and imaginary parts and proving the contention for them separately will yield it for m (3.10), and so m can be assumed to be real-valued. From 3.6, it only need be shown that the expression

$$m_+(\varphi) = \sup\{m(\psi): \varphi \geq \psi \in \mathcal{R}_+\}$$

if finite for all $\varphi \in \mathcal{R}_+$. Suppose this is not so and let (ψ_n) be a sequence in $\mathcal{R}_+$ majorized by φ such that $m(\psi_n) \geq 2^n$ for every integer n. The function $\psi = \sum 2^{-n}\psi_n \leq \varphi$ is in $\mathcal{R}_+$, and satisfies $m(\psi) = \lim_{N\to\infty} m(\sum_{n=1}^{N} 2^{-n}\psi_n) = \lim N = \infty$, a contradiction.#

Henceforth $\mathcal{I}$ denotes a full clan or integration lattice, $\mathcal{C}$ the sets in $\mathcal{I}$, and $\mathcal{S}$ the step functions over $\mathcal{C}$.

6.6. Theorem of Hahn. Two positive scalar S-measures m and n on $\mathcal{I}$ are disjoint if and only if every set A in $\mathcal{C}$ is the disjoint union of two sets B and C in $\mathcal{C}$ such that $m(B) = n(C) = 0$.

Proof. Assume first that m and n are disjoint, and let A be a set in $\mathcal{C}$. Let $\mathcal{B}$ denote the family of subsets D of A in $\mathcal{C}$ such that $m(D) = 0$ and put $b = \sup\{n(D); D \in \mathcal{B}\}$. There is an increasing sequence (D_k) in $\mathcal{B}$ such that $b = \sup\{n(D_k); k \in \mathbb{N}\}$. We put $B = \bigcup\{D_k; k \in \mathbb{N}\}$. As both m and n are S-continuous, we find $m(B) = 0$: B is an m-null subset of A of maximal n-measure. We set $C = A - B$ and shall show that $n(C) = 0$. By way of contradiction, assume that $n(C) > 0$. Using the characterization of disjointness 3.8, we construct by induction two sequences (φ_k) and (ψ_k) in $\mathcal{I}_+$ as follows: Put $\varphi_0 = C$ and $\psi_0 = 0$. If the sequences are defined up to the index k such that $n(\varphi_i) > 0$ for $i = 1,\ldots,k$, then select φ_{k+1} and ψ_{k+1} in $\mathcal{I}_+$ so that

$$\varphi_{k+1} + \psi_{k+1} = \varphi_k, \; m(\varphi_{k+1}) < 3^{-k} n(\varphi_k), \quad \text{and} \quad n(\psi_{k+1}) < 3^{-k} n(\varphi_k).$$

Clearly, (φ_k) is decreasing and satisfies $n(\varphi_{k+1}) > (1 - 3^{-k}) n(\varphi_k)$ for all natural numbers k. If $\varphi = \lim \varphi_k = \inf \varphi_k$ then $m(\varphi) = 0$ and $n(\varphi) \geq \Pi \{(1 - 3^k) : k \in \mathbb{N}\} \cdot n(C) > 0$. Note that φ_k and φ are sets if $\mathcal{J}$ is a clan. The set $[\varphi > 0] \subset A$ belongs to $\mathcal{C}$ and satisfies $m([\varphi > 0]) = 0$, as it is the pointwise limit of the increasing sequence of functions $(k\varphi \wedge 1) \leq k\varphi$ which are majorized by C. On the other hand, $n([\varphi > 0]) > 0$, and so $B \cup [\varphi > 0]$ is an m-null set of n-measure bigger than b, a contradiction: we have $n(C) = 0$, as desired.

To prove the converse implication, assume that m and n are such that every set A in $\mathcal{J}$ decomposes into two disjoint sets B and C of $\mathcal{C}$ with $m(B) = n(C) = 0$. Then $(m \wedge n)(A) = (m \wedge n)(B) + (m \wedge n)(C) \leq m(B) + n(C) = 0$ for all sets A in $\mathcal{R}$. This shows that m and n are disjoint if $\mathcal{J}$ is a clan. In the case that $\mathcal{J}$ is an integration lattice, it implies that $(m \wedge n)(\varphi) = 0$ for all positive step functions over $\mathcal{C}$ and, since every positive function in $\mathcal{J}$ is the limit of an increasing sequence of such step functions (6.3), $(m \wedge n)(\varphi) = 0$ for all positive functions in $\mathcal{J}$. Hence $m \wedge n = 0$, and m and n are disjoint.#

Our next aim is to characterize absolute continuity. In preparation, observe that if n is a positive scalar measure on $\mathcal{J}$ and $g \in \mathcal{B}$, then

$$m(\varphi) = \tilde{n}(\varphi g)$$

defines another positive measure m on $\mathcal{J}$. Here $\tilde{n}$ denotes the unique measure on $\mathcal{B}$ that coincides with n on $\mathcal{C}$ (1.1). If $\mathcal{J}$ is an integration lattice then $\tilde{n}$ is the restriction of n to $\mathcal{B} \subset \mathcal{J}$. Since

$m(\varphi) = \sup\{\tilde{n}(\varphi(g \wedge k)) : k \in \mathbb{N}\}$ and since $\tilde{n}(\varphi(g \wedge k)) \leq kn(\varphi)$, m is absolutely continuous with respect to n (3.8). m is called the <u>measure with base</u> n <u>and density</u> g and is denoted by gn.

<u>6.7. Theorem</u>. Let m and n be two positive scalar measures on the full integration domain $\mathcal{I}$. Then m is absolutely continuous with respect to n if and only if one of the following conditions holds:

(a) $n(A) = 0$ implies $m(A) = 0$ for all $A \in \mathcal{C} = \mathcal{C}(\mathcal{I})$.

(b) If $A \in \mathcal{C}$ has non-zero measure $m(A) > 0$, then there is a $g \in \mathcal{S}$ such that $m \geq gn > 0$.

(c) $m = \bigvee\{gn: g \in \mathcal{S}_+, gn \leq m\}$.

<u>Proof</u>. Assume first that $m \ll n$, i.e., $m = \bigvee\{kn \wedge m: k \in \mathbb{N}\}$ (3.8). If $A \in \mathcal{C}$ with $n(A) = 0$ then $m(A) \leq \sup\{kn(A): k \in \mathbb{N}\} = 0$, and (a) follows.

Next, assume (a) holds. For any natural number k split $m - k^{-1}n$ into its positive and negative parts: $m - k^{-1}n = r_k - s_k$ with $r_k, s_k \in M_+^S(\mathcal{I})$ and $r_k \perp s_k$. Fix a set $A \in \mathcal{C}$ so that $m(A) > 0$. There are subsets $B_k \in \mathcal{C}$ of A such that $r_k(B_k) = s_k(A - B_k) = 0$ (6.6). If $B = \bigcap\{B_k: k \in \mathbb{N}\}$ then $(m - k^{-1}n)(B) \leq 0$ for all $k \in \mathbb{N}$, and so $m(B) = 0$. Since thus $m(A - B) > 0$, (a) implies that $n(A - B) > 0$. Due to the S-continuity of n there is a natural number k such that $n(A - B_k) > 0$. Put $g = k^{-1}(A - B_k)$. Evidently $g \in \mathcal{S}$, $gn > 0$, and $m - gn = m - k^{-1}(A - B_k)n \geq (A - B_k)(m - k^{-1}n) = (A - B_k)r_k - (A - B_k)s_k = r_k \geq 0$. Hence $0 < gn \leq m$, and (b) follows.

Next assume (b) and set $m' = \bigvee\{gn: g \in \mathcal{S}_+, gn \leq m\}$. If $m - m' > 0$ then there is a set $A \in \mathcal{C}$ with $(m - m')(A) > 0$. Due to (b) there is a function $g \in \mathcal{S}_+$ such that $m - m' \geq gm > 0$. This implies $m \geq m' + gm > m'$, in contradiction to the definition of m'. Hence $m = m'$, and (c) follows.

Lastly, if (c) holds then $m \ll n$ from (3.8).#

D. Supplements

6.8. Every element in the full span $\mathcal{I}^S$ of an integration domain $\mathcal{I}$ lies in the full span of a countably generated subdomain.

6.9. Let $\mathfrak{C}$ be a clan. Every function f in $(\mathcal{S}(\mathfrak{C}))^S$ is the majorized uniform limit of a sequence in $\mathcal{S}(\mathfrak{C}^S)$. A positive function f belongs to $(\mathcal{S}(\mathfrak{C}))^S$ if it is majorized by a function in $\mathcal{S}(\mathfrak{C})$ and $[f > r] \in \mathfrak{C}^S$ for all $r > 0$.

6.10. A σ-ring on the set X is a clan $\mathfrak{C}$ which is closed under countable intersections. A σ-ring is a δ-ring. Any set in the σ-ring spanned by a collection F of sets lies in the σ-ring spanned by a countable subcollection of F, and is covered by countably many sets in F.

6.11. An algebra on X is a clan containing X. The algebra spanned by a clan $\mathfrak{C}$ equals $\mathfrak{C} \cup \{X - K : K \in \mathfrak{C}\}$.

6.12. A σ-algebra, or tribe, on X is a σ-ring which is an algebra. A δ-ring containing X is a σ-algebra. The σ-algebra spanned by a ring $\mathfrak{C}$ is the algebra generated by the σ-ring spanned by $\mathfrak{C}$.

6.13. The full closure of a dominated integration lattice $\mathcal{R}$ is dominated and coincides with

$$\cup \{ (\mathcal{R}[K])^S : K \in \mathcal{K}(\mathcal{R}) \}.$$

6.14*. If two S-measures on $\mathcal{R}^S$ coincide on $\mathcal{R}$ then they are equal.

6.15. A Banach valued S-measure on a full integration domain has finite semivariation.

6.16*. A family $\mathcal{F}$ of Banach-valued functions is Σ-closed if it is closed under pointwise limits of sequences.

There exists a smallest Σ-closed family $\mathcal{F}^\Sigma$ containing a given family $\mathcal{F}$. It is called the Σ-closure and can be calculated in first-uncountably many steps as in 6A above. Let $\mathcal{R}$ be a dominated integration lattice. The Σ-closure of $\mathcal{R} \otimes E$ coincides with the Σ-closure of $\mathcal{R}_E$ and consists of the E-valued $\mathcal{R}$-Baire functions. Everyone of them lies in the Σ-closure of a countable subfamily of $\mathcal{R} \otimes E$ and vanishes outside the union of countably many $\mathcal{R}$-dominated sets.

The dominated Baire functions are exactly the Baire functions that are majorized by some element in $\mathcal{R}$.

6.17*. Suppose $\mathcal{R}$ is a full integration lattice and $m_1, m_2, \ldots$ a countable family of mutually disjoint positive measures on $\mathcal{R}$ such that $\sum m_i$ exists. Then every set $K \in \mathcal{R}$ is the disjoint union of countably many sets $K_1, K_2, \ldots$ in $\mathcal{R}$ such that $m_i(K) = m_i(K_i)$ for $i = 1, 2, \ldots$.

6.18. $\mathcal{R}^S_{\uparrow +}$ consists exactly of the positive lower semicontinuous (with respect to the $\mathcal{R}$-topology) $\mathcal{R}$-Baire functions (use 6.8).

II. UPPER GAUGES AND EXTENSION THEORY

This chapter describes the extension of an elementary *-integral under an upper gauge, as sketched in the introduction. In order to emphasize the point that virtually all results in Lebesgue-Daniell's integration theory are results on upper gauges, and since upper gauges are of great interest in themselves, the chapter starts laying out their theory, and only later turns to their application in the extension theory of an elementary measure.

§1. Upper S-norms and upper gauges

7. Upper S-norms

A. Definition. An upper norm on the set X is a map from the set $\overline{\mathbb{R}}_+^X$ of all positive numerical functions on X to $\overline{\mathbb{R}}_+$ which is increasing, subadditive, and positively homogeneous (3A).

An upper norm M on X is S-continuous or an upper S-norm, if the following (Sequential) condition holds:

$$\sup M(f_n) = M(\sup f_n)$$

for all increasing sequences (f_n) in $\overline{\mathbb{R}}_+^X$. #

Unless stated otherwise, M denotes henceforth an upper S-norm on the set X. Here are a few preliminary consequences of the definition.

A subset K of X is called M-negligible if $M(K) = 0$. It follows from the defining properties of M that any subset of a negligible set and any union of countably many negligible sets are negligible. To see the latter, let $K_n, n \in \mathbb{N}$, be negligible sets. Then

$$\cup\{K_n : n \in \mathbb{N}\} \leq \Sigma\{K_n : n \in \mathbb{N}\} = \sup\{\sum_{n=1}^{N} K_n ; N \in \mathbb{N}\}, \quad \text{and so}$$

$$M(\cup K_n) \leq M(\sup_N \sum_{n=1}^{N} K_n) = \sup_N M(\sum_{n=1}^{N} K_n) \leq \sup_N \sum_{n=1}^{N} M(K_n) = 0.$$

A property $P(x)$ of the points x of X is said to hold M-almost everywhere (M-a.e. for short) if the set of points x where $P(x)$ does not hold is M-negligible. Accordingly, a function defined on a subset A of X is defined almost everywhere if $X - A$ is negligible. Let f and g be two functions defined almost everywhere. We say that f equals g almost everywhere, and write

$$f \doteq g(M) \quad \text{or} \quad f \doteq g,$$

if the set of points where both f and g are defined and coincide has negligible complement. It is easily seen that $f \doteq g$ is an equivalence relation for functions defined a.e. and with values in the same set. The class of f modulo $\doteq$ will henceforth be denoted by $f^{\cdot}$.

For functions f, g defined a.e. with values in $\mathbb{R}$ or in $\overline{\mathbb{R}}$, we write

$$f \dot{\leq} g(M) \quad \text{or} \quad f \dot{\leq} g$$

if the inequality $f \leq g$ holds a.e. Clearly, if $f' \doteq f \dot{\leq} g \doteq g'$ then $f' \dot{\leq} g'$. This defines an order, $f^{\cdot} \leq g^{\cdot}$, for equivalence classes of numerical functions.

7.1. Lemma. A function $f: X \to \overline{\mathbb{R}}_+$ vanishes almost everywhere if and only if $M(f) = 0$.

Proof. Assume first that $M(f) = 0$. From $[f > 0] = \sup\{nf \wedge 1: n \in \mathbb{N}\}$ we get $M([f > 0]) = \sup M(nf \wedge 1) \leq \sup\{nM(f): n \in \mathbb{N}\} = 0$. Conversely, if $M([f > 0]) = 0$ then $M(f) = M(f[f > 0]) = \sup\{M(n \wedge f[f > 0]): n \in \mathbb{N}\}$ $\leq \sup\{nM([f > 0]): n \in \mathbb{N}\} = 0$. #

7.2. Corollary. Let E be a normed vector space with norm $\| \ \|$, and let f,g be two functions defined a.e. on X and with values in E. Then $f \doteq g(M)$ if and only if $M(\|f-g\|) = 0$.#

For f any function defined a.e. and with values in E, the function $\|f\|$ is defined a.e., and we may put

$$M(f) := M_1(f) := M(\|f\|).$$

If $f \doteq g(M)$ then $M(f) = M(g)$ (Cf. 7.4 below), and so M is also defined unambiguously on the classes $\dot{f}$ of almost everywhere defined functions f with values in E. The set of functions with values in E defined M-a.e. on X will be denoted by $E^X[M]$.

This extension of the upper S-norm M to functions and classes of functions on X satisfies the following additional properties:

7.3 $$M(f+g) \le M(f) + M(g).$$

7.4 $$|M(f) - M(g)| \le M(f-g).$$

7.5 $$M(rf) = |r| M(f) \quad , \quad (r \in \mathbb{R}).$$

7.6 $$M(\sum_{n=1}^{\infty} f_n) \le \sum_{n=1}^{\infty} M(f_n) \quad \text{(countable convexity)}.$$

7.7 If $M(f) < \infty$, then f is finite a.e. $(f \in \overline{\mathbb{R}}^X[M])$.

Here f, g, f_n are functions in either $E^X[M]$ or $\overline{\mathbb{R}}^X[M]$, or are classes of such functions. The proofs of these properties are straightforward, and so only three of them are given.

<u>Proof</u> of 7.4, for functions f,g in $E^X[M]$: f and g may be assumed to be defined everywhere. From $\|f\| \le \|f-g\| + \|g\|$ and the subadditivity

of $\cdot M$ on $\overline{\mathbb{R}}_+^X$, $M(f) \leq M(f-g) + M(g)$, and $M(f) - M(g) \leq M(f-g)$. Reversing the roles of f and g yields $M(g) - M(f) \leq M(f-g)$, whence 7.4.#

Proof of 7.6, for classes $f_n^{\cdot}$: Let $(f_n^{\cdot})$ be a sequence of classes of functions f_n in $\overline{\mathbb{R}}^X[M]$. The sum $\sum f_n$ is defined a.e. and its class does not depend on the choice of f_n within $f_n^{\cdot}$, $n = 1, \ldots,$ and so $(\sum f_n)^{\cdot} = \sum f_n^{\cdot}$ is well defined. We may assume that the f_n are everywhere defined, and then

$$M(\sum f_n^{\cdot}) = M(\sum f_n) \leq M(\sum |f_n|) = M(\sup_N \sum_{n=1}^{N} |f_n|)$$

$$= \sup_N M(\sum_{n=1}^{N} |f_n|) \leq \sup_N \sum_{n=1}^{N} M(|f_n|) = \sum M(f_n^{\cdot}) \; . \; \#$$

Proof of 7.7. For all integers $n \geq 1$, $n[|f| = \infty] \leq |f|$ and hence $nM([|f| = \infty]) \leq M(f) < \infty$. This shows that $M([|f| = \infty]) = 0$.#

B. The spaces $\mathcal{F}_E(M)$

For M an S-norm on X and E a Banach space, denote by $\mathcal{F}_E(M)$ the set of all functions f, defined almost everywhere from X to E, and such that $M(f)$ is finite. Similarly, $\mathcal{F}(M) = \{f \in \overline{\mathbb{R}}^X[M]: M(f) < \infty\}$. The difference of two functions f, g in $\mathcal{F}(M)$ is well defined, since the set of points $x \in X$ for which both $f(x)$ and $g(x)$ are infinite and of the same sign is negligible. Formulae 7.3 and 7.5 show that both $\mathcal{F}_E(M)$ and $\mathcal{F}(M)$ are vector spaces, and that M is a semi-norm on them. Convergence with respect to the seminorm M is often called convergence in M-mean.

<u>7.8. Proposition</u>. Let M be an upper S-norm on the set X and E a Banach space.

(i) If (f_n) is a sequence in $\mathcal{F}_E(M)$, or in $\mathcal{F}(M)$, such that $\sum M(f_n) < \infty$ then $\sum f_n(x)$ converges almost everywhere absolutely and in M-mean to a function f in $\mathcal{F}_E(M)$, or in $\mathcal{F}(M)$, respectively.

(ii) If (g_n) is an M-Cauchy sequence in $\mathcal{F}_E(M)$, or in $\mathcal{F}(M)$, then (g_n) converges in M-mean to an element g of $\mathcal{F}_E(M)$, or $\mathcal{F}(M)$, respectively. Moreover, there exists a subsequence $(g_{n(k)}: k \in \mathbb{N})$ which converges pointwise almost everywhere to g.

<u>Proof</u>. (i): Let A'_n denote the set where f_n is defined $(n \in \mathbb{N})$. $A' = \bigcap \{A'_n : n \in \mathbb{N}\}$ has M-negligible complement (7.6). Consider the numerical function $h = \sum \|f_n\| \in \overline{\mathbb{R}}_+^X$: It is defined M-almost everywhere, namely on A', and satisfies $M(h) \leq \sum M(f_n) < \infty$ (7.6), and hence is finite almost everywhere (7.7). Denote by A the subset of A' of points x where $h(x)$ is finite. A has negligible complement, and $\sum \{f_n(x); n \in \mathbb{N}\}$ converges absolutely for $x \in A$. Let $f = \sum f_n$, defined on A. We have $M(f) \leq \sum M(f_n) < \infty$, and so $f \in \mathcal{F}_E(M)$ (or $f \in \mathcal{F}(M)$, as the case may be). Furthermore, $\|f - \sum_{k=1}^{n} f_k(x)\| \leq \sum_{k=n}^{\infty} \|f_k\|$, and hence

$$M(f - \sum_{k=1}^{n} f_k(x)) \leq \sum_{k=n}^{\infty} M(f_k) \to 0 \quad \text{as} \quad n \to \infty ,$$

so that $\sum f_n$ converges to f in M-mean.

(ii): Select indices $n(k)$, $k \in \mathbb{N}$, such that $M(g_{n(k+1)} - g_{n(k)}) \leq 2^{-k}$. The sequence $f_k = g_{n(k+1)} - g_{n(k)}$ has the properties dealt with under (i) above. Let f be its sum, and put $g = f + g_{n(1)}$. The function g is almost everywhere defined and in $\mathcal{F}_E(M)$ (or in $\mathcal{F}(M)$), and is the limit M-a.e. and in M-mean of the subsequence $(g_{n(k)})$ of (g_n). As (g_n) is a Cauchy sequence with convergent subsequence, it converges, and $\lim g_n = g$ in $\mathcal{F}_E(M)$, in M-mean.#

7.9. Corollary. $\mathcal{F}_E(M)$ and $\mathcal{F}(M)$ are complete semi-normed spaces.#

Let us denote by $\mathcal{N}_E(M)$ the functions in $\mathcal{F}_E(M)$ of vanishing norm, i.e., vanishing almost everywhere. Similarly $\mathcal{N}(M) = \{f \in \overline{\mathbb{R}}^X : f \doteq 0 \ (M)\}$. $F_E(M)$ (or $F(M)$) denotes the quotient space $\mathcal{F}_E(M)/\mathcal{N}_E(M)$ (or $\mathcal{F}(M)/\mathcal{N}(M)$). It is the space of classes modulo $\doteq$ of the elements of $\mathcal{F}_E(M)$ (or $\mathcal{F}(M)$). The following is obvious.

7.10. Corollary. $F_E(M)$ is a Banach space, and $F(M)$ is a Banach lattice.#

C. Integrable functions

Let $\mathcal{R}$ be an integration lattice on X and let E be a Banach space. If the upper norm M is finite on $\mathcal{R}$ then it is finite on $\mathcal{R} \otimes E$, the set of all functions of the form

$$\varphi = \sum \varphi_i \xi_i \qquad (\varphi_i \in \mathcal{R},\ \xi_i \in E, \text{ the sum finite}).$$

Indeed, $\|\sum \varphi_i \xi_i\| \le \sum \|\xi_i\| |\varphi_i| \in \mathcal{R}_+$, and so $M(\|\sum \varphi_i \xi_i\|) < \infty$.

Definition. Suppose M is an upper norm on X, $\mathcal{R}$ is an integration lattice on X, and M is finite on $\mathcal{R}_+$.

The closure of $\mathcal{R} \otimes E$ in $\mathcal{F}_E(M)$ is denoted by $\mathcal{L}^1_E(\mathcal{R},M)$, or by $\mathcal{L}^1_E$ for short, and its elements are the E-valued $(\mathcal{R},M)$-integrable functions. Similarly, $\mathcal{L}^1 = \mathcal{L}^1(\mathcal{R},M)$ is the closure of $\mathcal{R}$ in $\mathcal{F}(M)$, and consists of the numerical integrable functions. The term "Bochner-integrable" functions is also in use.#

7.11. Lemma. Suppose $\mathcal{R}$ is a dominated integration lattice. If $f \in \mathcal{L}^1_E$ and $g \in \mathcal{L}^1$ and if either f or g is bounded, then $fg \in \mathcal{L}^1_E$.

Proof. Suppose f is bounded, $\|f\| \leq A$ say. If an $\varepsilon > 0$ is given, find a $\varphi \in \mathcal{R}$ such that $M(g-\varphi) \leq \varepsilon/2A$, and put $B = \sup\{|\varphi(x)|: x \in X\}$. Find a $\psi \in \mathcal{R} \otimes E$ such that $M(f-\psi) \leq \varepsilon/2B$. From 5.8, $\varphi\psi$ is in $\overline{\mathcal{R}}_E$ and from 4.7, $\varphi\psi$ is integrable (7.17). Since

$$M(fg - \psi\varphi) \leq M(f(g-\varphi)) + M(\varphi(\psi - f)) \leq \varepsilon/2 + \varepsilon/2 = \varepsilon,$$

fg is integrable as well. The proof for bounded g is similar.#

7.12. Theorem. Let $\mathcal{R}$ be an integration lattice on X, and M an upper S-norm that is finite on $\mathcal{R}_+$:

(i) $\mathcal{L}^1_E(\mathcal{R},M)$ and $\mathcal{L}^1(\mathcal{R},M)$ are complete semi-normed vector spaces, and contain all negligible functions. Every function in $\mathcal{L}^1_E$ (or in $\mathcal{L}^1$) is the sum M-almost everywhere and in M-mean of functions φ_k $(k = 0, 1, \ldots)$ in $\mathcal{R} \otimes E$ (or in $\mathcal{R}$, respectively) such that $\sum_{k=0}^{\infty} M(\varphi_k) < \infty$.

(ii) Every Cauchy sequence (f_n) in $\mathcal{L}^1_E$ (in $\mathcal{L}^1$) contains a subsequence that converges almost everywhere to $\lim f_n$.

Proof. (i), for E: As a closed subspace of the complete space $\mathcal{F}_E(M)$, $\mathcal{L}^1_E(\mathcal{R},M)$ is complete, and as the closure of a vector subspace it is itself a vector space. Let $f \in \mathcal{L}^1_E(\mathcal{R},M)$ and let (ψ_n) be a sequence in $\mathcal{R} \otimes E$ converging to f in $\mathcal{F}_E(M)$. By assumption and by proposition 7.8, the Cauchy sequence (ψ_n) contains a subsequence $(\psi_{n(k)})$ which converges pointwise almost everywhere to f, and satisfies $\sum\{M(\psi_{n(k+1)} - \psi_{n(k)}): k \in \mathbb{N}\} < \infty$. Put $\varphi_0 = \psi_{n(0)}$ and $\varphi_k = \psi_{n(k+1)} - \psi_{n(k)}$, for $k = 1, 2, \ldots$. Clearly, $\varphi_k \in \mathcal{R} \otimes E$, and $f = \sum \varphi_k$ in M-mean and pointwise M-almost everywhere.#

7.13. Corollary. If $f \in \mathcal{L}^1_E(\mathcal{R},M)$, then $\|f\| \in \mathcal{L}^1(\mathcal{R},M)$. A numerical function f is integrable if and only if f_+ and f_- are. The real-valued, everywhere defined, integrable functions form an integration lattice.

Proof. Let $\varphi \in \mathcal{R} \otimes E$ be such that $M(\varphi - f) < \varepsilon$. From $|\,\|\varphi\| - \|f\|\,| \leq \|\varphi - f\|$, $M(\|\varphi\| - \|f\|) < \varepsilon$. Hence if $f \in \mathcal{L}^1_E$ then $\|f\| \in \mathcal{L}^1$. The remaining statements follow similarly.#

$\mathcal{L}^1 = \mathcal{L}^1(\mathcal{R},M)$ contains functions that are infinite at some points of X and not defined at others, and this is strictly speaking not an integration lattice. It has all the formal properties of an integration lattice, though, and so we will extend the notion of an integration lattice so as to include spaces like $\mathcal{L}^1$ and $\mathcal{F}(M)$.

7.14. Corollary. Let $L^1_E(\mathcal{R},M)$ denote the classes modulo $\doteq$ of the functions in $\mathcal{L}^1_E(\mathcal{R},M)$. Similarly, $L^1(\mathcal{R},M) = \mathcal{L}^1(\mathcal{R},M)/\mathcal{N}(M)$. Then $L^1_E(\mathcal{R},M)$ is a Banach space, and $L^1(\mathcal{R},M)$ is a Banach lattice.#

Let M and N be two upper S-norms on X finite on $\mathcal{R}_+$ and such that $M \leq N$, i.e., $M(f) \leq N(f)$ for $f \in \overline{\mathbb{R}}^X_+$. Then $\mathcal{F}_E(N) \subset \mathcal{F}_E(M)$, and $\mathcal{L}^1_E(\mathcal{R},N) \subset \mathcal{L}^1_E(\mathcal{R},M)$. As $f \doteq g$ (N) implies $f \doteq g$ (M), there is a natural injection of $L^1_E(\mathcal{R},N)$ into $L^1_E(\mathcal{R},M)$.

7.15. Corollary. Let $\mathcal{R}$ be an integration lattice on X, and let M,N be two upper S-norms on X that are finite on $\mathcal{R}$. If $M \leq N$ then $L^1_E(\mathcal{R},N) \subset L^1_E(\mathcal{R},M)$, and if $M = N$ on $\mathcal{R}_+$ then $L^1_E(\mathcal{R},M)$ is isomorphic with $L^1_E(\mathcal{R},M)$. In this instance, every class in $L^1_E(\mathcal{R},M)$ contains a function which is defined and finite everywhere and both $(\mathcal{R},M)$- and $(\mathcal{R},N)$-integrable.

Proof. If $R(f) = M(f) + N(f)$ for $f: X \to \overline{\mathbb{R}}_+^X$, then R is an upper S-norm on X, and is finite on $\mathcal{R}_+$. If $f^{\cdot} \in L^1_E(\mathcal{R},M)$, and (φ_n) is a Cauchy sequence in $\mathcal{R} \otimes E$ converging in M-mean to $f' \in \dot{f}$, then (φ_n) is an R-Cauchy sequence. Some subsequence converges in R-mean and R-almost everywhere to a function $f \in \mathcal{L}^1_E(\mathcal{R},R)$. Clearly, $f \in f^{\cdot}$, and $f \in \mathcal{L}^1_E(\mathcal{R},M) \cap \mathcal{L}^1_E(\mathcal{R},N)$, and so $L^1_E(\mathcal{R},M) = L^1_E(\mathcal{R},R) = L^1_E(\mathcal{R},N)$.#

Let us call two upper S-norms M and N essentially equal if they are finite and equal on $\mathcal{R}_+$. The second part of the corollary states that two essentially equal upper S-norms M and N have isomorphic spaces $L^1_E(\mathcal{R},M)$ and $L^1_E(\mathcal{R},N)$.

D. Supplements

7.16. Suppose $\mathcal{R}$ is an integration lattice and m is a positive scalar measure on $\mathcal{R}$. By

$$m^J(f) = \inf\{m(\varphi): f \le \varphi \in \mathcal{R}\}, \qquad f: X \to \overline{\mathbb{R}}_+$$

(and $m^J(f) = \infty$ if f fails to be majorized by an element of $\mathcal{R}$), the Jordan upper norm m^J is defined. It is an upper norm finite on $\mathcal{R} \otimes E$ (E is a Banach space). $\mathcal{L}^1_E(\mathcal{R},m^J)$ is a vector space which contains $\mathcal{R} \otimes E$ densely. For $\varphi = \sum \varphi_i \xi_i \in \mathcal{R} \otimes E$, define $\int \varphi\, dm = \sum m(\varphi_i)\xi_i$. Then $\|\int \varphi\, dm\| \le m^J(\|\varphi\|)$, and there is a unique linear extension

$$f \to \int f\, dm, \qquad f \in \mathcal{L}^1_E(\mathcal{R},m^J)$$

which satisfies $\|\int f\, dm\| \le m^J(\|f\|)$. It is the Jordan integral, defined on the Jordan integrable functions.

If $m: \mathcal{R} \to F$ is Banach-valued and of finite variation, then there is a unique extension $f \to \int f\, dm$ satisfying $\|\int f\, dm\| \le \int |f|\, d|m| \le |m|^J(|f|)$, to all of $\mathcal{L}^1(\mathcal{R}, |m|^J)$.

7.17*. If $\mathcal{R}$ is dominated and M an upper norm finite on $\mathcal{R}$, then $\overline{\mathcal{R}}_E \subset \mathcal{L}^1_E(\mathcal{R},M)$ for all Banach spaces E (4.7).

7.18. If M is an upper S-norm finite on $\mathcal{R}$ and E is a Banach space, then every element of $\mathcal{L}^1_E(\mathcal{R},M)$ is equivalent to a Baire function.

7.19. A bounded linear map $T: E \to F$ of Banach spaces induces a bounded linear map $f \to T \circ f$ from $\mathcal{L}^1_E(\mathcal{R},M)$ to $\mathcal{L}^1_F(\mathcal{R},M)$.

7.20*. Definition. Let $\mathcal{R}$ be a dominated integration lattice and let M,N be two upper S-norms finite on $\mathcal{R}$. Then M is said to be absolutely continuous with respect to N on $\mathcal{R}$, denoted $M \ll N$, if, for every $K \in \mathcal{K}(\mathcal{R})$ and every bounded set $B \subset \mathcal{R}[K]$, M is continuous at zero with respect to N (i.e., for every $\varepsilon > 0$ there is a $\delta > 0$ such that $N(\varphi) < \delta$ implies $M(\varphi) < \varepsilon$ for all $\varphi \in B$).

If $h \in \mathcal{L}^1(\mathcal{R},M)$ then $hM: f \to M(hf)$ is an upper S-norm, finite and absolutely continuous with respect to M on $\mathcal{R}$.

7.21*. The pointwise supremum of a family of upper S-norms is itself an upper S-norm.

7.22. Let M be an upper S-norm on X, and let E_x be a Banach space for each $x \in X$. A cross section is a map f from X to $\cup \{E_x: x \in X\}$ such that $f(x) \in E_x$. The cross sections f with $M(f) < \infty$ form a complete pseudometric space $\mathcal{F}(E,M)$. Let $\mathcal{R}$ be a vector space of cross sections on which M is finite. The closure of $\mathcal{R}$ under M is called the direct integral of the Banach spaces E_x, and is denoted by $\int_{\mathcal{R}}^{\oplus} E_x dM$. It is a complete pseudonormed vector space, and each of its elements is almost everywhere, and in mean, the sum of cross sections in $\mathcal{R}$.

7.23. An outer S-norm on the set X is an increasing, subadditive, and positively homogeneous map M from the power set $\mathcal{P}(X)$ of X to $\overline{\mathbb{R}}_+$, satisfying: $A_n \uparrow A$ implies $M(A_n) \uparrow M(A)$.

M is extended by $M(A - B) := M_1(A - B) := M(|A - B|) = M(A \Delta B)$, and satisfies $M(\emptyset) = 0$, $M(A \cup B) \leq M(A) + M(B)$, $M(A - C) \leq M(A - B) + M(B - C)$, and $M(\cup A_n) \leq \sum M(A_n)$.

If $\mathcal{F} = \{A \subset X: M(A) < \infty\}$, then $\mathcal{F}$ is a clan and $(\mathcal{F},M)$ is a complete pseudometric space.

If $\mathcal{C}$ is a clan on X on which M is finite, then its closure $S(\mathcal{C},M)$ in this norm is a complete subclan of $\mathcal{F}$, and every Cauchy sequence (A_n) in $S(\mathcal{C},M)$ contains a subsequence which converges almost everywhere to $\lim(A_n)$.

An outer S-norm is often called an outer measure in the sense of Caratheodory.

LITERATURE: The proofs above are from Bourbaki [5], Ch. IV. For 7.22, Cf. [12].

8. Upper gauges

A. Weak upper gauges

Definition. A weak upper gauge $(\mathcal{R},M)$ on X is a pair consisting of an integration lattice $\mathcal{R}$ on X and an upper S-norm M finite on $\mathcal{R}$ such that the following holds:

(WUG) Whenever (φ_n) is an increasing sequence in $\mathcal{R}_+$ such that $f = \sup(\varphi_n)$ is majorized by an integrable function, then f is integrable and $f = \lim(\varphi_n)$ in M-mean.#

8.1. Lemma. (Weak monotone convergence theorem). Suppose $(\mathcal{R},M)$ is a weak upper gauge and (f_n) an increasing sequence of positive integrable functions, majorized by the integrable function h. Then $f = \sup(f_n)$ is integrable and (f_n) converges to f in mean.

Proof. Let $\varepsilon > 0$ be given. We construct inductively an increasing sequence (φ_n) in $\mathcal{R}_+$ such that $M(f_n - \varphi_n) < \varepsilon(1 - 2^{-n})$ for all integers $n \geq 1$. Suppose $\varphi_1, \ldots, \varphi_{n-1}$ are so chosen. Find a $\psi \in \mathcal{R}_+$ such that $M(f_n - \psi) < \varepsilon 2^{-n}$ and put $\varphi_n = \psi \vee \varphi_{n-1}$. The elementary inequality

$$|f_n - \psi \vee \varphi_{n-1}| \leq |f_n - \psi| + |f_{n-1} - \varphi_{n-1}|$$

yields $M(f_n - \varphi_n) < \varepsilon(1 - 2^{-n+1}) + \varepsilon 2^{-n} \leq \varepsilon(1 - 2^{-n})$ as desired. Clearly (φ_n) is increasing and satisfies $M(f_n - \varphi_n) < \varepsilon$ for all integers $n > 1$. Choose now a $\varphi \in \mathcal{R}_+$ such that $M(h - \varphi) < \varepsilon$. From

$$M(f_n - f_m) \leq M(f_n - f_n \wedge \varphi) + M(f_n \wedge \varphi - \varphi_n \wedge \varphi) + M(\varphi_n \wedge \varphi - \varphi_m \wedge \varphi)$$
$$+ M(\varphi_m \wedge \varphi - f_m \wedge \varphi) + M(f_m \wedge \varphi - f_m)$$

we can see that (f_n) is a Cauchy sequence. Indeed, since $f_n - f_n \wedge \varphi \leq h - h \wedge \varphi$ and $M(h - h \wedge \varphi) < \varepsilon$, the first and the last term on the right are smaller than ε; since $|f_n \wedge \varphi - \varphi_n \wedge \varphi| \leq |f_n - \varphi_n|$, the two adjacent terms are smaller than ε; and since $(\varphi_n \wedge \varphi)$ is convergent in mean, and therefore Cauchy, we may arrive at $M(f_n - f_m) < 5\varepsilon$ by choosing m and n sufficiently big. Hence (f_n) is Cauchy, and its limit in mean equals f, as a subsequence is known to converge to the limit pointwise a.e.#

<u>8.2. Corollary</u>. If $(\mathcal{R},M)$ is a weak upper gauge then so is $(\mathcal{L}^1,M)$.#

<u>8.3. Theorem</u>. (Lebesgue's weak dominated convergence theorem). Suppose $(\mathcal{R},M)$ is a <u>weak</u> upper gauge and E is a Banach space. Suppose (f_n) is a sequence of $(\mathcal{R},M)$-integrable functions with values in E or in $\overline{\mathbb{R}}$, that converges M-a.e. to a function f.

If there is an <u>integrable</u> function $h \in \mathcal{L}^1(\mathcal{R},M)$ such that $\|f_n\| \leq h$ for all $n = 1, 2, \ldots$, then (f_n) converges in M-mean to f. In particular, f is then integrable itself.

<u>Proof</u>. First alter f,h, and the f_n on a common negligible set so that they are defined and finite everywhere, and that $\lim f_n = f$ and $\|f_n\| \leq h$ everywhere. The functions

$$g_k = \sup\{\|f_n - f_m\| : k \leq n,m\}, \qquad k = 1, 2, \ldots ,$$

are the suprema of the increasing and majorized (by 2h) sequences

$$g_k^N = \sup\{\|f_n - f_m\| : k \leq n,m \leq N\}$$

of integrable functions, and so are integrable. They are all majorized by 2h and converge pointwise and decreasingly to zero. Hence $g_1 - g_k \uparrow g_1$ and thus

$$M(g_k) = M(g_1 - (g_1 - g_k)) \to 0.$$

This shows that (f_n) is an M-Cauchy sequence and so has a limit in M-mean. Since a subsequence converges pointwise a.e., this limit equals f a.e. (7.12). From 6.3 and 4.5 we get the following consequence.

<u>8.4. Corollary</u>. Let $(\mathcal{R},M)$ be a weak upper gauge. Then so is $(\mathcal{L}^1,M)$, and $\mathcal{L}^1$ is a full integration lattice. $\mathcal{L}^1_E$ contains $(\mathcal{R} \otimes E)^S$ and thus $\mathcal{R}^S \otimes E$. The $(\mathcal{R},M)$-integrable sets form a δ-ring $\mathcal{C}(\mathcal{R},M)$, and every Banach-valued or numerical integrable function f is the limit a.e. and in mean of a sequence (s_n) of step functions over $\mathcal{C}(\mathcal{R},M)$. Moreover, if f is positive then (s_n) can be chosen to be increasing, and if f is bounded then (s_n) can be chosen to converge uniformly.

Furthermore, $\mathcal{R}_0$ is dense in $\mathcal{L}^1(\mathcal{R},M)$ and $\mathcal{R}_0 \otimes E$ is dense in $\mathcal{L}^1_E(\mathcal{R},M)$.

Let (f_n) be a sequence of numerical integrable functions. A necessary and sufficient condition for $\sup(f_n)$ (or $\inf(f_n)$) to be integrable is that there exist an integrable function h such that $f_n \leq h$ (or $f_n \geq h$, respectively) for all integers $n \geq 1$.#

<u>Remark</u>. The richness of Lebesgue's integration theory, as compared with Riemann's, is almost entirely a consequence of the dominated convergence theorem above, and of the resulting abundance of integrable sets (8.4). Note that this result holds in the presence of a <u>weak</u> upper gauge.

<u>8.5. Corollary</u>. If $f \in \mathcal{L}^1_E$ and $g \in \mathcal{L}^1(\mathcal{R},M)$, and if either f or g is bounded, then fg belongs to $\mathcal{L}^1_E$.

<u>Proof</u>. The dominated functions are dense in $\mathcal{R}$ (cf. 7.11).#

8.6. Corollary. If $f, g \in \mathcal{L}^1_E$ with $M(f-g) > 0$, then there is a number $a > 0$ and an integrable set K with $M(K) > 0$ such that $\|f-g\| > a$ on K.

Proof. The sets $[\|f-g\| > n^{-1}]$, $n = 1, 2, \ldots$, are integrable. If they were all negligible, we would have $f \doteq g$ (M) and therefore $M(f-g) = 0$ (7.2).#

Definition. An $(\mathcal{R}, M)$-adequate cover of integrable sets is a family $C \subset \mathcal{C}(\mathcal{R}, M)$ such that every integrable set A is contained almost everywhere in the union of a countable subfamily C' of C. C' is also said to cover A adequately if $A \setminus \cup C'$ is negligible.#

8.7. Corollary. The dominated $\mathcal{R}$-Baire sets form an $(\mathcal{R}, M)$-adequate cover, for any weak upper gauge $(\mathcal{R}, M)$.

An $(\mathcal{R}, M)$-integrable function vanishes a.e. outside the union of some countable subfamily, of any $(\mathcal{R}, M)$-adequate cover.

Proof. $\mathcal{C}(\mathcal{R}, M)$ is the closure of $\mathcal{C}(\mathcal{R}^S)$ under M (8.4). Given any $A \in \mathcal{C}(\mathcal{R}, M)$, there is therefore a sequence (K_n) of dominated Baire sets such that $\sum K_n = A$ M-a.e. and in M-mean (7.12). $A - \cup K_n$ is clearly M-negligible, and so the first statement is proved. If $f: X \to E$ is integrable, then so are the sets $A_r = [\|f\| \geq r^{-1}]$, for all integers $r \geq 1$. Given an adequate cover C, there is a countable family $C_r \subset C$ which covers A_r adequately; and $[f \neq 0] = \cup A_r$ is adequately covered by the countable family $\cup C_r \subset C$.#

Definition. Let $(\mathcal{R}, M)$ be a weak upper gauge and A an $(\mathcal{R}, M)$-integrable set. A family F of $(\mathcal{R}, M)$-integrable sets is said to be $(\mathcal{R}, M)$-dense in A if, for every $\varepsilon > 0$, there is a set $K \in F$ such that $K \subset A$ and $M(A - K) < \varepsilon$. F is $(\mathcal{R}, M)$-dense if it is dense in every $(\mathcal{R}, M)$-integrable set.#

By theorem 8.3, F is dense in $A \in \mathscr{C}(\mathscr{R},M)$ if and only if there is a sequence (K_n) in F of subsets of A such that $\lim K_n = A$ in M-mean and M-almost everywhere. The last corollary has the following consequence.

8.8. Corollary. If $(\mathscr{R},M)$ is a weak upper gauge, then the family $\mathscr{C}_o(\mathscr{R},M)$ of dominated $(\mathscr{R},M)$-integrable sets is dense.#

B. Strong upper gauges and upper integrals

Definition. Let $\mathscr{R}$ be an integration lattice on X and M an upper S-norm finite on $\mathscr{R}_+$. Then $(\mathscr{R},M)$ is said to be an upper gauge if it has the following property.

(UG) Whenever (φ_n) is an increasing sequence in $\mathscr{R}_+$ such that $\sup M(\varphi_n) < \infty$ then (φ_n) is an M-Cauchy sequence.

An upper gauge $(\mathscr{R},M)$ is said to be strong provided that

(SUG) if $0 \leq f \leq g$ are two integrable functions with $M(f) = M(g)$ then $M(g - f) = 0$.#

Example. If M is an upper S-norm that is finite and additive on $\mathscr{R}_+$, then $(\mathscr{R},M)$ is called an upper integral.

An upper integral is a strong upper gauge. Indeed, if (φ_n) is an increasing sequence in $\mathscr{R}_+$ such that $M(\varphi_n) < \infty$, then $(M(\varphi_n))$ is a Cauchy sequence in $\mathbb{R}$ and so

$$\lim\{M(\varphi_n - \varphi_m): n \geq m \to \infty\} = \lim\{M(\varphi_n) - M(\varphi_m): \ldots\} = 0.$$

Hence (φ_n) is M-Cauchy, and $(\mathscr{R},M)$ satisfies (UG). Next, since M is continuous with respect to M on $\mathscr{L}^1$ and additive on the dense subset $\mathscr{R}_+$ of $\mathscr{L}^1_+$, it is additive on $\mathscr{L}^1_+$. Hence if $0 \leq f \leq g$ in $\mathscr{L}^1_+$ and $M(f) = M(g)$ then $M(g - f) = M(g) - M(f) = 0$, and $(\mathscr{R},M)$ satisfies (SUG).#

8.9. Theorem. (Monotone convergence theorem). Let $(\mathcal{R},M)$ be an upper gauge, and let (f_n) be an increasing or decreasing sequence in $\mathcal{L}^1$. In order that $f = \lim(f_n)$ belong to $\mathcal{L}^1$ it is necessary and sufficient that

$$\sup\{M(f_n): n \in \mathbb{N}\} < \infty.$$

In this instance, (f_n) converges to f in M-mean.

Proof. The necessity of the condition is obvious, and so let us assume it is satisfied. We consider at first the case that all the f_n are positive and that (f_n) is increasing.

Let $\varepsilon > 0$ be given. We construct as in the proof of 8.1 an increasing sequence (φ_n) in $\mathcal{R}_+$ such that $M(f_n - \varphi_n) < \varepsilon$ for all n. Since $\sup M(\varphi_n) < \infty$, clearly, (φ_n) is Cauchy; and thus so is (f_n). Clearly $f = \lim(f_n)$ a.e. and in mean.

The case of an arbitrary increasing sequence (f_n) is reduced to the one dealt with by considering the sequence $(|f_1| + f_n)$; the case of a decreasing sequence (f_n) is reduced to that case by considering $(-f_n)$.#

8.10. Corollary. If $(\mathcal{R},M)$ is a (strong) upper gauge then so is $(\mathcal{L}^1,M)$. An upper gauge is a weak upper gauge.#

8.11. Corollary. (Fatou's lemma). Let $(\mathcal{R},M)$ be an upper gauge. If (f_n) is a sequence of positive functions in $\mathcal{L}^1$ such that $\liminf M(f_n) < \infty$, then $\liminf(f_n)$ is integrable and

$$M(\liminf(f_n)) \leq \liminf(M(f_n)).$$

Proof. If $g_n = \inf\{f_{n+k}; k \in \mathbb{N}\}$, $n = 1, 2, \ldots$, then g_n is in $\mathcal{L}^1$ with $M(g_n) \leq \inf\{M(f_{n+k}); k \in \mathbb{N}\} \leq \sup_n \inf_k (M(f_{n+k})) = \liminf M(f_n) < \infty$, and so $\liminf(f_n) = \sup(g_n)$ is in $\mathcal{L}^1(\mathcal{R},M)$ and satisfies the inequality in question.#

Using the monotone convergence theorem instead of its weak version in the proof of the weak Lebesgue dominated convergence theorem, we obtain the following result.

8.12. Theorem. (Lebesgue's dominated convergence theorem). Suppose $(\mathcal{R},M)$ is an upper gauge and E is a Banach space. Suppose (f_n) is a sequence of $(\mathcal{R},M)$-integrable functions with values in E or in $\overline{\mathbb{R}}$ that converges M-a.e. to a function f.

If there is a function $h: X \to \overline{\mathbb{R}_+}$ with $M(h) < \infty$ such that $\|f_n\| \leq h$ for all $n = 1, 2, \ldots$, then (f_n) converges in M-mean to f. In particular, f is then integrable itself.#

8.13. Proposition. If $(\mathcal{R},M)$ is a strong upper gauge then the space $L^1 = L^1(\mathcal{R},M)$ is a Banach lattice with order-continuous norm; and $L^1_E(\mathcal{R},M)$ is a Banach space over $L^1(\mathcal{R},M)$ with modulus $f \to \|f\|$. (Here E is a Banach space with norm $\| \ \|$.)

Proof. L^1 is a Banach lattice (7.14). To see that its norm is order-continuous, consider an increasingly directed subset F of L^1_+. If F has a supremum g, then clearly $\sup M(F) \leq M(g) < \infty$. Conversely, let us assume that $\sup M(F) = s < \infty$. There is an increasing sequence $(f^\bullet_n)$ in F such that $\sup M(f^\bullet_n) = s$. We can pick $f_n \in f^\bullet_n$ for each $n = 1, 2, \ldots$ so that $f_{n+1} \geq f_n$. The monotone convergence theorem asserts that $f = \sup(f_n)$ is integrable. It still has to be shown that the class $f^\bullet$ is the supremum of F, and that F converges to $f^\bullet$ in mean. If $g \in g^\bullet \in F$ then $g^\bullet \leq f^\bullet$. Indeed, $M(g \vee f) = \sup M(g \vee f_n) = s = M(f)$, and by (SUG) $g \vee f = f$ almost everywhere. Therefore $g^\bullet \leq f^\bullet$, and $f^\bullet$ is an upper bound for F.

Let $\varepsilon > 0$ be given. As $(f_n^\bullet)$ converges in mean to $f^\bullet$ (8.9), there is a natural number N such that $M(f^\bullet - f_N^\bullet) < \varepsilon$. For $g^\bullet \in F$ and $g^\bullet \geq f_N^\bullet$, it is clear that $M(f^\bullet - g^\bullet) < \varepsilon$, and so F converges to $f^\bullet$. From this it follows easily that $f^\bullet$ is the least upper bound of F: For if $g^\bullet < f^\bullet$ is another upper bound then $M(f^\bullet - g^\bullet) > 0$ due to (SUG), and so $M(f - f_n^\bullet) \geq M(f^\bullet - g^\bullet) > 0$ for all n. This is a contradiction.

The last statement is obvious.#

C. B-continuous upper gauges

For this important class of upper gauges the monotone convergence theorem can be somewhat strengthened, with the result that there are more integrable functions.

Definition. The weak upper gauge $(\mathcal{R}, M)$ is said to be B-continuous if, for every increasingly directed subset Φ of $\mathcal{R}_+$,

$$M(\sup \Phi) = \sup M(\Phi). \#$$

The family of all functions that are suprema of arbitrary families in $\mathcal{R}$ will be denoted by $\mathcal{R}_\uparrow^B$, and $\mathcal{R}_{\uparrow+}^B$ will denote those that are positive (5.11). For later use we define $\mathcal{R}_\uparrow^S$ to be the family of functions that are suprema of countable subsets of $\mathcal{R}$, and $\mathcal{R}_{\uparrow+}^S$ the positive functions in $\mathcal{R}_\uparrow^S$. Extending the convention of 3D we write $\mathcal{R}_\uparrow^*$ and $\mathcal{R}_{\uparrow+}^*$, where $*$ can take the value S or B.

8.14. Lemma. If $(\mathcal{R}, M)$ is a weak B-continuous upper gauge, then

$$M(\sup H) = \sup M(H)$$

for all increasingly directed subsets H of $\mathcal{R}_{\uparrow+}^B$.

Proof. Suppose that $H \subset \mathcal{R}^B_{\uparrow +}$ is increasingly directed with supremum h and denote by Φ the family of $\varphi \in \mathcal{R}_+$ that are majorized by some element of H. Clearly Φ is increasingly directed and has supremum h, and so $M(h) \geq \sup M(H) \geq \sup M(\Phi) = M(h)$.#

8.15. Proposition. Suppose $(\mathcal{R},M)$ is a B-continuous (weak) upper gauge and let Φ be an increasingly or decreasingly directed subset of $\mathcal{R}$ such that $\sup M(\Phi) < \infty$ (or $\sup |\Phi| \leq h$ for some $h \in \mathcal{L}^1$, if $(\mathcal{R},M)$ is weak).

Then $\sup \Phi$ is integrable and Φ converges--along its section filter--in M-mean to $\sup \Phi$.

If $H \subset \mathcal{R}^B_{\uparrow}$ is increasingly directed with supremum h, then h is integrable and H converges in mean to h, provided that $M(h) < \infty$ (or that h is majorized by some integrable function, if $(\mathcal{R},M)$ is weak).

Proof. As in the proof of 8.9, everything is reduced to the case that $\Phi \subset \mathcal{R}_+$ is increasingly directed. Let $h = \sup \Phi$ with $M(h) < \infty$ (or $h \leq g \in \mathcal{L}^1$, if $(\mathcal{R},M)$ is weak). It has to be shown that $\inf\{M(h - \varphi) : \varphi \in \Phi\} = 0$.

Suppose this is not true, so that $M(h - \varphi) > a$ for all φ in Φ and some $a > 0$ in $\mathbb{R}$. Since $M(h) > a$, there is a $\varphi_1 \in \Phi$ such that $M(\varphi_1) > a$. Since $a < M(h - \varphi_1) = \sup\{M(\varphi - \varphi_1) : \varphi_1 \leq \varphi \in \Phi\}$, there is a $\varphi_2 \geq \varphi_1$ in Φ such that $M(\varphi_2 - \varphi_1) > a$. Proceeding by induction, we establish the existence of an increasing sequence (φ_n) in Φ such that $M(\varphi_{n+1} - \varphi_n) > a$, for $n = 1, 2, \ldots$. This is a contradiction to property (WUG) of $(\mathcal{R},M)$, because (φ_n) converges in mean to $\sup(\varphi_n)$ and thus is a Cauchy sequence.

For the second statement, we choose Φ as in the proof of 8.14, and obtain $\inf\{M(h - f) : f \in H\} \leq \inf\{M(h - \varphi) : \varphi \in \Phi\} = 0$.#

Denote by $\mathcal{R}^*_\downarrow = -\mathcal{R}^*_\uparrow$ the set of all functions that are infima of arbitrary $(* = B)$ or countable $(* = S)$ subsets of $\mathcal{R}$. The following can be proved similarly to the last proposition.

<u>8.16. Corollary</u>. If K is a decreasingly directed set in $\mathcal{R}^B_\downarrow$ with infimum k, and if $M(k) < \infty$, then k is integrable and the limit in mean of K, provided $(\mathcal{R},M)$ is a B-continuous upper gauge.

If $(\mathcal{R},M)$ is a weak B-continuous upper gauge, the same conclusion holds, provided there is an integrable function $h \leq k$.

Denote by $\mathcal{R}^B_{\uparrow o}$ the elements of $\mathcal{R}^B_\uparrow$ that are majorized by a function in $\mathcal{R}$, by $\mathcal{R}^B_{\uparrow\downarrow}$ the integration lattice $\mathcal{R}^B_{\uparrow o} - \mathcal{R}^B_{\uparrow o}$ (5.15), and by $\mathcal{R}^B$ the full span of $\mathcal{R}^B_{\uparrow\downarrow}$. $\mathcal{R}^B$ consists of the <u>dominated $\mathcal{R}$-Borel functions</u>, and the sets $\mathfrak{C}(\mathcal{R}^B)$ in $\mathcal{R}^B$ form the δ-ring of <u>dominated $\mathcal{R}$-Borel sets</u> (6A). The functions in $\mathcal{R}^B_{\uparrow\downarrow}$ are $(\mathcal{R},M)$-integrable for every weak B-continuous upper gauge $(\mathcal{R},M)$ on $\mathcal{R}$, and Lebesgue's theorem (8.3) yields the following.

<u>8.17. Corollary</u>. If $(\mathcal{R},M)$ is a B-continuous weak upper gauge, then the functions in $\mathcal{R}^B$ and in $\mathcal{R}^B \otimes E$ are $(\mathcal{R},M)$-integrable (E is some Banach space). Every dominated $\mathcal{R}$-Borel set is $(\mathcal{R},M)$-integrable.

D. Supplements

8.18*. Definition. Two weak upper gauges $(\mathcal{R},M)$ and $(\mathcal{R},N)$ are said to be equivalent, written $(\mathcal{R},M) \approx (\mathcal{R},N)$ or $M \approx N$, if they have the same dominated negligible sets.

Equivalent weak upper gauges have the same dominated integrable sets. If $\mathcal{R}$ is full, then $(\mathcal{R},M)$ and $(\mathcal{R},N)$ are equivalent if either is absolutely continuous with respect to the other (7.20).

8.19. The dominated integration lattice associated with $\mathcal{L}^1 = \mathcal{L}^1(\mathcal{R},M)$, $\mathcal{L}^1_0$, consists exactly of all bounded functions $f \in \mathcal{L}^1$ such that $[f \neq 0]$ is integrable. $\mathcal{K}(\mathcal{L}^1)$ consists of all subsets of integrable sets.

8.20. If F is a dense family and G a family dense in every member of F, then G is dense.

8.21*. If F is a dense family for the weak upper gauge $(\mathcal{R},M)$, then every integrable set A can be written as the disjoint union of countably many sets in F and a negligible set.

8.22. A family $F \subset \mathcal{C}(\mathcal{R},M)$ closed under finite unions is dense if for every $A \subset \mathcal{C}(\mathcal{R},M)$ with $M(A) > 0$ there is a $K \in F$ with $K \subset A$ and $M(K) > 0$.

8.23. The dominated $\mathcal{R}$-Baire sets coincide with the δ-ring spanned by the family $\mathcal{K}^S(\mathcal{R})$ of sets in $\mathcal{R}^S_\downarrow$. The dominated $\mathcal{R}$-Borel sets coincide with the δ-ring spanned by the family $\mathcal{K}^B(\mathcal{R})$ of sets in $\mathcal{R}^B_\downarrow$.

8.24. Lebesgue's theorem 8.3 and 8.12 remain valid if (f_n) is replaced by a pointwise convergent filter $\mathcal{F}$ in $\mathcal{L}^1_E$ having a countable basis and such that $\|f\| \leq g$ a.e. for all $f \in F \in \mathcal{F}$ and some g with $g \in \mathcal{L}^1$ or $M(g) < \infty$, respectively.

LITERATURE: The proofs above are adapted from [5], Ch. IV.

9. The standard upper integrals

One major step is still missing in the integration process outlined in the introduction: given a measure m, we have to find an upper gauge $(\mathcal{R},M)$ majorizing m: $\|m(\varphi)\| \leq M(\varphi)$ for $\varphi \in \mathcal{R}_+$. Once this is done, we can extend m to all of $\mathcal{L}^1(\mathcal{R},M)$ by the simple expedient of extension by continuity.

A necessary condition for the existence of a majorizing upper gauge $(\mathcal{R},M)$ is this: whenever (φ_n) is an increasing sequence in $\mathcal{R}_+$ with supremum $\varphi \in \mathcal{R}_+$, then $m(\varphi_n)$ must converge to $m(\varphi)$; for (φ_n) then converges to φ in M-mean and by continuity, $m(\varphi_n)$ converges to $m(\varphi)$. That is, m must be S-continuous. If we want $(\mathcal{R},M)$ to be B-continuous m must be so also, for a similar reason (8.16).

A. The case of positive measures

It will now be shown that a positive S-measure m on an integration lattice $\mathcal{R}$ can be majorized by an upper gauge $(\mathcal{R},M)$, and that $(\mathcal{R},M)$ can be chosen B-continuous if $(\mathcal{R},m)$ is.

In order to be as general as possible we consider henceforth a positive *-measure m on $\mathcal{R}$ having values in a fixed proper L-space L (* = S or B, Cf. 3D). The norm on L will be denoted by $\| \ \|$, and the modulus $x \vee -x$ of $x \in L$ by $|x|$. (If $L = \mathbb{R}$ we have $|x| = \|x\|$.)

As with the reals, we adjoin to L a biggest element ∞ and a smallest element $-\infty$, and put $\overline{L} = \{-\infty\} \cup L \cup \{\infty\}$. We set $\|-\infty\| = \|\infty\| = \infty$, and $x + \infty = \infty$, $x - \infty = -\infty$ for $x \in L$.

The following construction of m*, the standard upper gauge associated with a positive *-measure m, is very similar for * = S and * = B;

and so it will be done simultaneously for both, reinvigorating the convention concerning $*$ (3D, 8C). In the first step m^* is defined on $\mathcal{R}^*_\uparrow$ as follows:

$$m^*(h) = \sup\{m(\varphi): h \geq \varphi \in \mathcal{R}\} \quad \text{for} \quad h \in R^*_\uparrow .$$

A moment's inspection shows that the map $m^*: \mathcal{R}^*_\uparrow \to \overline{L}$ is increasing, positively homogeneous, and coincides with m on $\mathcal{R}$. Let us show further:

(1) m^* <u>is *-continuous on</u> $\mathcal{R}^*_\uparrow$. This will be proved only in the slightly more difficult case that $* = S$. Let (h_n) be a countable increasingly directed family in $\mathcal{R}^S_\uparrow$, having supremum h. Clearly $\sup m^S(h_n) \leq m^S(h)$.

To show the reverse inequality, consider a $\varphi \in \mathcal{R}$ smaller than h. For every natural number n let $\{\varphi_n^m: m \in \mathbb{N}\}$ be a countable increasing family in $\mathcal{R}$ with supremum h_n. With $\psi_n = \varphi_1^1 \vee \dots \vee \varphi_n^n$, (ψ_n) is an increasing sequence in $\mathcal{R}$ with supremum h, and $\sup m^S(h_n) \geq \sup m(\psi_n) \geq \sup m(\psi_n \wedge \varphi) = m(\varphi)$. Hence $\sup m^S(h_n) \geq \sup\{m(\varphi): \varphi \leq h\} = m^S(h)$. #

(2) m^* <u>is additive</u>. Let h_1 and h_2 be elements of $\mathcal{R}^*_\uparrow$ and select increasing families Φ_1 and Φ_2 in $\mathcal{R}$ such that $h_1 = \sup \Phi_1$, $h_2 = \sup \Phi_2$. If $* = S$ choose both families countable. Then $m^*(h_1) + m^*(h_2) = \sup\{m(\varphi_1): \varphi_1 \in \Phi\} + \sup\{m(\varphi_2): \varphi_2 \in \Phi_2\} = \sup\{m(\varphi): \varphi \in \Phi_1 + \Phi_2\} = m^*(h_1 + h_2)$, since $\Phi_1 + \Phi_2$ is increasingly directed with supremum $h_1 + h_2$ (and countable if $* = S$). #

In the second step, m^* is now defined on <u>all</u> numerical functions f on X, as follows,

$$m^*(f) = \inf\{m^*(h): f \leq h \in \mathcal{R}^*_\uparrow\} \qquad (f: X \to \overline{\mathbb{R}}).$$

If f fails to be majorized by an element of $\mathcal{R}^*_\uparrow$, then $m^*(f) = \infty$.

A moment's inspection shows that $m^*: \overline{\mathbb{R}}^X \to \overline{L}$ is increasing, positively homogeneous and equal to m on $\mathcal{R}$. Furthermore,

(3) m^* <u>is subadditive</u>. Let $f_1, f_2: X \to \overline{\mathbb{R}}$. If $m^*(f_1) = \infty$ or $m^*(f_2) = \infty$ there is nothing to prove. If both elements of $\overline{L}$ are smaller than ∞, then $m^*(f_1) + m^*(f_2) = \inf\{m^*(h_1): f_1 \leq h_1 \in \mathcal{R}^*_\uparrow\} + \inf\{m^*(h_2): f_2 \leq h \in \mathcal{R}^*_\uparrow\} \geq \inf\{m^*(h): f_1 + f_2 \leq h \in \mathcal{R}^*_\uparrow\} = m^*(f_1 + f_2)$. #

(4) m^* <u>is</u> S-<u>continuous on</u> $\overline{\mathbb{R}}^X_+$. Consider an increasing sequence (f_n) of positive numerical functions on X with supremum f. Clearly $\sup m^*(f_n) \leq m^*(f)$. To prove the reverse inequality, we may assume that $\sup m^*(f_n) < \infty$, since otherwise there is nothing to prove. Given a fixed $\varepsilon > 0$, select a sequence h_n in $\mathcal{R}^*_\uparrow$ such that $f_n \leq h_n$ and

$$\|m^*(h_n) - m^*(f_n)\| \leq \varepsilon 2^{-n} \quad \text{for each} \quad n = 1, 2, \ldots ,$$

and put $g_n = h_1 \vee \ldots \vee h_n$. Then $f_n \leq g_n \in \mathcal{R}^*_\uparrow$ for each n and $f \leq \sup g_n \in \mathcal{R}^*_\uparrow$. We shall show by induction that

$$\|m^*(g_n) - m^*(f_n)\| \leq \varepsilon(1 - 2^{-n}), \quad \text{for} \quad n = 1, 2, \ldots .$$

This is obviously true for $n = 1$. Suppose it is proved for n. Taking into account the equality

$$g_n + h_{n+1} = g_n \vee h_{n+1} + g_n \wedge h_{n+1} = g_{n+1} + g_n \wedge h_{n+1}$$

and the additivity of m^* on $\mathcal{R}^*_\uparrow$,

$$\begin{aligned} \|m^*(h_{n+1}) - m^*(g_{n+1})\| &= \|m^*(g_n \wedge h_{n+1}) - m^*(g_n)\| \\ &= \|m^*(g_n) - m^*(f_n) - \{m^*(g_n \wedge h_{n+1}) - m^*(f_n)\}\| \\ &\leq \|m^*(g_n) - m^*(f_n)\| \leq \varepsilon(1 - 2^{-n}) , \end{aligned}$$

using the fact that the norm is strictly increasing on L_+. Hence

$\|m^*(f_{n+1}) - m^*(g_{n+1})\| \leq \|m^*(f_{n+1}) - m^*(h_{n+1})\| + \|m^*(h_{n+1}) - m^*(g_{n+1})\|$
$\leq \varepsilon 2^{-n-1} + \varepsilon(1 - 2^{-n}) = \varepsilon(1 - 2^{-n-1})$, as claimed.

It follows that $\|m^*(g_n) - m^*(f_n)\| < \varepsilon$ for all n and so $\|\sup m^*(g_n) - \sup m^*(f_n)\| \leq \varepsilon$. Now $\sup m^*(g_n) = m^*(\sup g_n) \geq m^*(f)$ and thus $\|m^*(f) - \sup m^*(f_n)\| \leq \varepsilon$. As ε was arbitrary, we must have $m^*(f) = \sup m^*(f_n)$, as claimed.#

Finally, the <u>standard upper gauge associated with the positive *-measure</u> m is defined by

$$\|m^*\|(f) = \|m^*(f)\| \quad \text{for} \quad f: X \to \overline{\mathbb{R}}_+ .$$

<u>9.1. Theorem</u>. $\|m^*\|$ is a *-continuous strong upper gauge on $\mathcal{R}$ majorizing m. It is the biggest *-continuous upper gauge M on $\mathcal{R}$ satisfying $M(\varphi) = \|m(\varphi)\|$ for $\varphi \in \mathcal{R}_+$. If the norm is additive on L_+ (in particular if $L = \mathbb{R}$), then $\|m^*\|$ is an upper integral.

<u>Definition</u>. If $m: \mathcal{R} \to L$ is a positive measure and if $(\mathcal{R},M)$ is an upper gauge, then $(\mathcal{R},M)$ is said to be an <u>upper gauge for</u> m if $M(\varphi) = \|m(\varphi)\|$ for all $\varphi \in \mathcal{R}_+$.

<u>Proof of 9.1</u>. We start by showing that $\|m^*\|$ is maximal. Let $(\mathcal{R},M)$ be another *-continuous upper gauge for m, and let $0 \leq h \in \mathcal{R}^*_\uparrow$. If $\mathcal{R}^*_{\uparrow +} \supset \Phi \uparrow^* h$, then $m^*(h) = \sup\{m(\varphi);\ \varphi \in \Phi\}$, and so $\|m^*\|(h) = \sup\{\|m(\varphi)\|;\ \varphi \in \Phi\} = M(h)$. If $f \in \overline{\mathbb{R}}^X_+$, then $m^*(f) = \inf\{m^*(h);\ f \leq h \in \mathcal{R}^*_\uparrow\}$ so that $\|m^*\|(f) = \inf\{\|m^*\|(h);\ f \leq h \in \mathcal{R}^*_\uparrow\}$ $= \inf\{M(h);\ f \leq h \in \mathcal{R}^*_\uparrow\} \geq M(f)$.

The last statement of the theorem is trivial.

Let us check (UG). If (φ_n) is an increasing sequence in $\mathcal{R}_+$ with $\sup \|m^*\|(\varphi_n) < \infty$, then $m^*(\varphi_n)$ converges, and hence $\|m^*\|(\varphi_n - \varphi_m) = \|m^*(\varphi_n) - m^*(\varphi_m)\|$ tends to zero as $n, m \to \infty$.

As for (SUG), m^* is additive on $\mathcal{R}_+$ and continuous with respect to $\|m^*\|$, and so is additive on $\mathcal{L}^1_+(\mathcal{R}, \|m^*\|)$. Hence if $0 \leq f \leq g$ in $\mathcal{L}^1$, then $m^*(g) - m^*(f) = m^*(g - f) \geq 0$. If $\|m^*(g)\| = \|m^*(f)\|$ then $m^*(g) = m^*(f)$, since the norm is strictly increasing on L_+ (3E).

Let us show finally that $\|m^*\|$ majorizes m on $\mathcal{R}$. For $\varphi \in \mathcal{R}$, $m(\varphi) = m(\varphi_+) - m(\varphi_-) \leq m(|\varphi|)$. Replacing φ by $-\varphi$ yields $|m(\varphi)| \leq m(|\varphi|)$, and so $\|m(\varphi)\| = \| \, |m(\varphi)| \, \| \leq \|m(|\varphi|)\| = \|m^*\|(|\varphi|)$ for all φ in $\mathcal{R}$.#

<u>9.2. Corollary</u>. For $f: X \to \overline{\mathbb{R}}_+$,

$$\|m^*\|(f) = \inf\{\sup\{\|m(\varphi)\|: 0 \leq \varphi \leq h, \ \varphi \in \mathcal{R}_+\}: f \leq h \in \mathcal{R}^*_\uparrow\}. \quad \#$$

A positive $*$-measure m on $\mathcal{R}$ may be extended to $\mathcal{L}^1(\mathcal{R}, \|m^*\|)$ in the manner described in the introduction. If m is given on a clan $\mathcal{C}$ it can be extended by linearity to a $*$-measure $\tilde{m}$ on $\mathcal{S}(\mathcal{C})$ (3.12), and from there to $\mathcal{L}^1(\mathcal{S}(\mathcal{C}), \|\tilde{m}^*\|)$. The restriction of $\int \cdot \, d\tilde{m}$ to $\mathcal{C}(\mathcal{S}(\mathcal{C}), \|\tilde{m}^*\|)$ is an extension of the original measure m to a δ-ring containing $\mathcal{C}$. This procedure is called the <u>standard extension</u>, or <u>Lebesgue-Daniell extension</u>, of an elementary $*$-integral or a $*$-content, respectively.

<u>Example</u>. Suppose $L = L^1(\mathcal{S}, N)$ where $(\mathcal{S}, N)$ is an upper gauge on a set Y, and $m: \mathcal{R} \to L$ is a positive $*$-measure. It may happen that m does not have finite scalar variation. Yet $\|m^*\|$ is an upper gauge on $\mathcal{R}$ under which m can be extended. If $(\mathcal{S}, N)$ is an upper integral then $(\mathcal{R}, \|m^*\|)$ is one also, and vector-valued functions can be integrated with respect to it (10.11)

B. Characterization of the $(\mathcal{R}, \|m^*\|)$-integrable functions

For $k \in \mathcal{R}^*_\downarrow = -\mathcal{R}^*_\uparrow$ define

$$m_*(k) = -m^*(-k) = \inf\{m(\varphi): k \leq \varphi \in \mathcal{R}\},$$

and for an arbitrary numerical function f on X,

$$m_*(f) = -m^*(-f) = \sup\{m_*(k): f \geq k \in \mathcal{R}^*_\downarrow\}.$$

9.3. Proposition. Let $(\mathcal{R},m)$ be a positive *-integral on X having values in a proper L-space as above.

(i) An element of $\mathcal{R}^*_\uparrow$ is $(\mathcal{R}, \|m^*\|)$-integrable if and only if $m^*(h) < \infty$, and in this instance

$$m^*(h) = \int hdm.$$

Furthermore, if $* = B$, then m^B and $\|m^B\|$ are B-continuous on $\mathcal{R}^B_\uparrow$

(ii) An element k of $\mathcal{R}^*_\downarrow$ is $(\mathcal{R}, \|m^*\|)$-integrable if and only if $m_*(k) > -\infty$, and in this instance

$$m_*(k) = \int kdm = m^*(k).$$

Furthermore, m_* is *-continuous on $\mathcal{R}^*_\downarrow$:

$$\mathcal{R}^*_\downarrow \supset \Phi\downarrow^* k \in \mathcal{R}^*_\downarrow \quad \text{implies} \quad m_*(\Phi)\downarrow^* m_*(k).$$

(iii) For any numerical function f,

$$m_*(f) \leq m^*(f),$$

and f is integrable if and only if $m_*(f) = m^*(f)$; i.e., if and only if for any $\varepsilon > 0$ there are a $k \in \mathcal{R}^*_\downarrow$ and an $h \in \mathcal{R}^*_\uparrow$ such that

$$k \leq f \leq h \quad \text{and} \quad \|m^*(h) - m_*(k)\| < \varepsilon.$$

In this instance,

$$m_*(f) = m^*(f) = \int f dm.$$

Proof. (i) Let $h \in \mathcal{R}^*_\uparrow$ with $m^*(h) < \infty$. It may be assumed that h is positive. Let $R \supset \Phi \uparrow^* h$. Since $(\mathcal{R}, m^*)$ is $*$-continuous, h is integrable and Φ converges in m^*-mean to h (8.9, 8.15); hence

$$m^*(h) = \lim\{m^*(\varphi);\ \varphi \in \Phi\} = \lim\{m(\varphi);\ \varphi \in \Phi\} = \int h dm.$$

(ii) The proof is the same as for (i), with h replaced by $-k$.

(iii) Let $k \in \mathcal{R}^*_\downarrow$, and $h \in \mathcal{R}^*_\uparrow$ with $k \leq f \leq h$. We claim that $m_*(k) \leq m^*(h)$. If $m_*(k) = -\infty$ or $m^*(h) = +\infty$ there is nothing to prove. If both numbers are finite then $m^*(h) - m_*(k) = \int hdm - \int kdm = \int (h - k)dm \geq 0$, and so $m_*(f) \leq m^*(f)$. If f is integrable then $m^*(f) = \int fdm = -\int -fdm = m_*(f)$. Assume then that $m_*(f) = m^*(f) = a$ is finite. Given $\varepsilon > 0$, there is a $k \in \mathcal{R}^*_\downarrow$ and an $h \in \mathcal{R}^*_\uparrow$ such that $k \leq f \leq h$, $\|m_*(k)\| \geq \|a\| - \varepsilon$, and $\|m^*(h)\| \leq \|a\| + \varepsilon$. Hence, $k \in \mathcal{L}^1(\mathcal{R}, m^*)$, $|f - k| \leq h - k$, and $\|m^*(h - k)\| = \|\int (h - k)dm\| = \|m^*(h)\| - \|m^*(k)\| < 2\varepsilon$, and so f is integrable.#

Let $\mathcal{U}^* = \mathcal{U}^*(\mathcal{R})$ denote the sets in $\mathcal{R}^*_\uparrow$, and $\mathcal{K}^* = \mathcal{K}^*(\mathcal{R})$ the sets in $\mathcal{R}^*_\downarrow$ (5.11 - 5.14).

9.4. Corollary. Let $(\mathcal{R};m)$ be a positive $*$-integral on X. A subset A of X is $(\mathcal{R}, m^*)$-integrable if and only if for all $\varepsilon > 0$ there is a $K \in \mathcal{K}^*(\mathcal{R})$ and a $U \in \mathcal{U}^*(\mathcal{R})$ such that

$$K \subset A \subset U \quad \text{and} \quad \int Udm - \int Kdm < \varepsilon.$$

In particular, $\int Adm = \sup\{\int Kdm;\ A \supset K \in \mathcal{K}^*\}$. Furthermore $\int dm$ is *-continuous on $\mathcal{K}^*$: If

$$\mathcal{K}^* \supset \Gamma{\downarrow}{*}0, \quad \text{then} \quad \lim\{\int Kdm\colon K \in \Gamma\} = 0.$$

Proof. The condition is clearly sufficient. To show that it is necessary, let $k \in \mathcal{R}^*_{\downarrow}$ and $h \in \mathcal{R}^*_{\uparrow}$ be such that $k \leq A \leq h$ and $m^*(h - k) < \varepsilon$. It may be assumed that $k \geq 0$ and $h \leq 1$. The sets $K_n = [k \geq 1/n]$, $n = 1, 2, \ldots$, are in $\mathcal{K}^*$ (5.14) and satisfy $k \leq \sup K_n \leq A$. For sufficiently high n, $\int kdm \leq \varepsilon + \int K_n dm$ (8.9). Similarly, the sets $U_n = [h > 1 - 1/n]$ are in $\mathcal{U}^*$ and for sufficiently high integer j, $\int hdm \geq \varepsilon + \int U_j dm$. Clearly, $K_n \subset A \subset U_j$, and $\int U_j dm - \int K_n dm < 3\varepsilon$. The last statement follows from 8.16.#

C. Banach-valued measures

We now have a satisfactory extension theory of positive *-measures (* = S or B) with values in $\mathbb{R}$ or in a proper L-space, whether defined on an integration lattice or on a clan. It remains to be seen what can be done in the vector-valued case.

Let $(\mathcal{R}, m)$ be an elementary *-integral having values in a Banach space F and having finite scalar variation, $|m|$ (* = S or B). $(\mathcal{R}, |m|)$ is *-continuous (3.10) and is majorized by the upper integral $(\mathcal{R}, |m|^*)$. Since m is evidently majorized by $|m|^*$, too, it can be extended to $\mathcal{L}^1(\mathcal{R}, |m|^*)$. Functions in $\mathcal{L}^1_E(\mathcal{R}, |m|^*)$ can be integrated, as in 10.11. If m is given on a clan $\mathcal{C}$, it can be extended to a measure $\tilde{m}$ of finite variation $|\tilde{m}| = |m|^{\sim}$ on $\mathcal{S}(\mathcal{C})$ (3.12), and then, under the upper integral $(\mathcal{S}(\mathcal{C}), |\tilde{m}|^*)$, to $\mathcal{L}^1(\mathcal{S}(\mathcal{C}), |\tilde{m}|^*)$.

The situation generalizes as in 3E to the case that F is a Banach space over the proper L-space R and $m\colon \mathcal{R} \to F$ has finite R-variation, $|m|$. Then m can be extended under the upper gauge $(\mathcal{R}, \| |m|^* \|)$ to all of $\mathcal{L}^1(\mathcal{R}, \| |m|^* \|)$.

Notation. Where no confusion can occur, we shall write simply m^* for $\|m^*\|$ or for $\| |m|^* \|$, in the sequel.

D. Supplements

9.5*. The product of two positive functions in $\mathcal{R}^*_\uparrow$ belongs to $\mathcal{R}^*_\uparrow$.

9.6*. The families $\mathcal{K}^*(\mathcal{R})$ and $\mathcal{U}^*_0$ (of dominated sets in $\mathcal{U}^*(\mathcal{R})$) are $(\mathcal{R},M)$-adequate covers for every *-continuous upper gauge $(\mathcal{R},M)$. $\mathcal{K}^*(\mathcal{R})$ is dense for the upper gauges $(\mathcal{R},m^*)$, $m \in M^*(\mathcal{R};E,R)$.

9.7. The map $m \to m^*$ is positively homogeneous and additive from $M^*_+(\mathcal{R})$ to the set of *-continuous upper integrals.

9.8. The functions in $\mathcal{L}^{1*}_E = \cap\{\mathcal{L}^1_E(\mathcal{R},m^*): m \in M^*_+(\mathcal{R})\}$ are called <u>universally</u> *-<u>integrable</u> (* = S or B). Clearly $\mathcal{R}^*_E \subset \mathcal{L}^{1*}_E$, and $\mathcal{L}^{1*}_E$ is S-closed. The map $m \to \int \cdot\, dm$ is a Riesz space isomorphism of $M^S(\mathcal{R})$ onto $M^S(\mathcal{L}^{1S})$. It is a Riesz space isomorphism of $M^B(\mathcal{R})$ onto the band of <u>regular</u> measures in $M^S(\mathcal{L}^{1B})$, i.e., of those measures m that satisfy

$$m(f) = \sup\{m(k): f \geq k \in \mathcal{R}^B_\downarrow\} = \inf\{m(h): f \leq h \in \mathcal{R}^B_\uparrow\}.$$

9.9. Let m,n be Banach-valued measures on $\mathcal{R}$ of finite variation. Then $m \ll n$ if and only if $m^* \ll n^*$ (* = J, S, B according as $|m|, |n| \in M^*(\mathcal{R})$). (Cf. (7.20)).

9.10. If $(\mathcal{R},M)$ is a weak upper gauge satisfying (SUG) but not (UG), then $L^1(\mathcal{R},M)$ is a Banach lattice with weakly order continuous norm (WWOCN) in the sense that if Φ is an increasingly directed set which is majorized by an element of L^1, then sup Φ exists and Φ converges to sup Φ in norm.

Suppose L is a Banach lattice WWOCN and $m: \mathcal{R} \to L$ is a positive *-measure. If m^* and $\|m^*\|$ are defined as above, then $(\mathcal{R}, \|m^*\|)$ is a weak upper gauge satisfying (SUG).

9.11. Suppose m is a positive S-measure on a clan $\mathcal{C}$, $\tilde{m}$ its extension by linearity to $\mathcal{S}(\mathcal{C})$. Then

$$\tilde{m}^S(A) = \inf\{\sup m(K_n): \cup K_n \supset A;\ K_n \in \mathcal{C}\}$$

for all sets $A \subset X$. The set function $A \to m^S(A)$ defined by the right hand side of this equation is an outer S-norm in the sense of 7.23. It is an outer gauge in the sense that $\mathcal{C} \supset (K_n) \uparrow A$ and $m^S(A) < \infty$ imply $m^S(A - K_n) \to 0$.

9.12. Let $(\mathcal{R},M)$ be a *-continuous strong upper gauge. Then, for every $A \in \mathcal{C}(\mathcal{R}^*)$,

$$M(A) = \sup\{M(K): A \supset K \in \mathcal{K}^*(\mathcal{R})\},$$

i.e., M is <u>inner regular</u> on the δ-ring $\mathcal{C}(\mathcal{R}^*)$ (* = S,B; Cf. 6A and p. 92).

LITERATURE: [46], [5] Ch. IV, [49], [40], [27].

§2. Integration of vector-valued functions

10. Extension under an upper norm

A. The integral

Definition. Let E and G be Banach spaces and let $U: \mathfrak{R} \otimes E \to G$ be a linear map. The upper norm M is said to majorize U if

$$\|U(\varphi)\| \leq M(\varphi) < \infty \quad \text{for all} \quad \varphi \in \mathfrak{R} \otimes E.$$

In this instance, there exists a unique continuous extension of U to all of $\mathcal{L}_E^1(\mathfrak{R},M)$, which is called the integral[1)] of U and is denoted by

$$f \to \int f dU = \int f(x) dU(x) \ , \quad \text{for} \quad f \in \mathcal{L}_E^1(\mathfrak{R},M).$$

In order that U be majorized by an upper norm it must clearly have finite semivariation (4.7).

Example. If $m: \mathfrak{R} \to F$ has finite variation and is $*$-continuous ($*$ = S or B), then it is majorized by the upper integral m^* and so has an extension by continuity to all of $\mathcal{L}^1(\mathfrak{R},m^*)$. This is the standard, or Daniell-extension of m. The fact that $(\mathfrak{R},m^*)$ is an upper gauge entrains the all-important Lebesgue-continuity property: if (f_n) is a sequence with limit f m^*-a.e., of integrable functions that are all majorized by the same function h of finite upper integral $m^*(h)$, then (f_n) converges to f in m^*-mean, and therefore f is integrable and

$$\int f dm = \lim \int f_n dm. \#$$

1) The term Bochner integral is also in use. For the (little) functional analysis used in the sequel see [9], [14], [15], [21], [35], [38], or [48].

We return to the more general data of the definition. The map $f \to \int f dU$ is evidently linear from $\mathcal{L}^1_E$ to G and satisfies $\|\int f dU\| \leq M(f)$. Let $T: G \to H$ be a linear map of norm $\|T\|$ into some Banach space H. Then

$$T \circ U: \mathcal{R} \otimes E \to H$$

is clearly majorized by the upper norm $\|T\| M$.

<u>10.1. Proposition</u>. $T(\int f dU) = \int f d(T \circ U)$, for $f \in \mathcal{L}^1_E$.

<u>Proof</u>. The equation holds for $f \in \mathcal{R} \otimes E$ and both sides depend continuously on f, with respect to the norm $\|T\| M$. #

Let D be another Banach space and let $S: D \to E$ be a linear map of norm $\|S\|$. By

$$S(\sum \varphi_i \xi_i) = \sum \varphi_i S(\xi_i) = S \circ \sum \varphi_i \xi_i \ , \quad \sum \varphi_i \xi_i \in \mathcal{R} \otimes D,$$

it induces a linear map S from $\mathcal{R} \otimes D$ to $\mathcal{R} \otimes E$ such that $\|S(\varphi)\| \leq \|S\| \, \|\varphi\|$. The linear map

$$\varphi \to U \circ S(\varphi) \ , \quad \varphi \in \mathcal{R} \otimes D$$

is evidently majorized by $\|S\| M$, and with the same argument as above we obtain the following.

<u>10.2. Proposition</u>. $\int f d(U \circ S) = \int S \circ f dU$, for $f \in \mathcal{L}^1_D(\mathcal{R}, M)$. #

<u>Example</u>. Let m be a positive scalar measure on the dominated integration lattice $\mathcal{R}$, majorized by the upper norm M. Then m can be extended to $\mathcal{L}^1(\mathcal{R}, M)$ by continuity.

On the other hand, by

$$m(\sum \varphi_i \xi_i) = \sum m(\varphi_i)\xi_i , \quad (\sum \varphi_i \xi_i \in \mathcal{R} \otimes E),$$

m can be viewed as a linear map from $\mathcal{R} \otimes E$ to E. To see that m is well defined in this way, let $\psi \in \mathcal{R}_+$ be such that $\|\sum \varphi_i \xi_i\| \leq \psi$. For any $\eta \in E_1'$, we have $\sum \varphi_i \langle \xi_i;\eta\rangle \leq \psi$ and so

$$\langle m(\sum \varphi_i \xi_i);\eta\rangle = \sum m(\varphi_i)\langle \xi_i;\eta\rangle = m(\langle \sum \varphi_i \xi_i;\eta\rangle) \leq m(\psi).$$

Since $\eta \in E_1'$ and $\varphi \geq \|\sum \varphi_i \xi_i\|$ are arbitrary,

$$\|m(\varphi)\| \leq m(\|\varphi\|) \leq M(\varphi) , \quad \text{for} \quad \varphi \in \mathcal{R} \otimes E.$$

The map $m: \mathcal{R} \otimes E \to E$ is therefore well defined and has finite semi-variation majorized by M. We may, then, extend m by continuity and obtain the integral

$$\int f dm = \int f(x) dm(x) \in E , \quad \text{for} \quad f \in \mathcal{L}_E^1(\mathcal{R},M).$$

Let $\eta \in E'$. From the definition, $\int \langle f;\eta\rangle dm = \langle \int f dm;\eta\rangle$ if $f \in \mathcal{R} \otimes E$; and since both sides depend continuously on f, this equation holds for all $f \in \mathcal{L}_E^1$:

<u>10.3. Proposition</u>. If $m \geq 0$ and $\eta \in E'$ then

$$\int \langle f;\eta\rangle dm = \langle \int f dm;\eta\rangle \quad \text{and}$$

$$\left|\int \langle f;\eta\rangle dm\right| \leq \|\eta\| \int \|f\| dm \leq \|\eta\| M(f) , \quad \text{for} \quad f \in \mathcal{L}_E^1 . \ \#$$

If A is a subset of E, let $co(A)$ denote its <u>convex hull</u>, the set of all finite convex linear combinations of elements of A, and $\overline{co}(A)$ its <u>closed convex hull</u>, the closure of $co(A)$. The following two results give estimates for the value of the integral.

<u>10.4. Corollary</u>. Suppose $(\mathcal{R},m)$ and M are as above and K is an $(\mathcal{R},M)$-integrable set. Then

$$\int fK\varphi dm \in \overline{co}(f(K)) \cdot \int K\varphi dm \ , \quad \text{for} \quad \varphi \in \mathcal{R}_+ .$$

<u>Proof</u>. If $\int \varphi K dm = 0$ there is nothing to prove, and so we assume that $\int K dm > 0$; in fact we multiply m by a suitable constant and arrive at $\int \varphi K dm = 1$.

If $\xi \notin \overline{co}(f(K))$ then there are an $\eta \in E'$ and an $a \in \mathbb{R}$ such that $\langle \xi;\eta \rangle > a$ and $\langle f(K);\eta \rangle < a$ (Hahn-Banach theorem). Then $\langle \int fK\varphi dm;\eta \rangle = \int \langle f;\eta \rangle K\varphi dm < a \int K\varphi dm = a$, and so $\xi \neq \int fK\varphi dm$.#

For any set $S \subset E$, the <u>convex equilibrated hull</u>, $ce(S)$, is the convex hull of $S \cup -S$. Its closure is denoted by $\overline{ce}(S)$. The fact that $ce(f(K)) = \overline{ce}(f(K)\varphi/|\varphi|)$ implies the following.

<u>10.5. Corollary</u>. If $f \in \mathcal{L}^1_E(\mathcal{R},M)$ then

$$\int f\varphi K dm \in \int |\varphi| K dm \cdot \overline{ce}(f(K)),$$

for all $\varphi \in \mathcal{R}$ and all $K \in \mathfrak{C}(\mathcal{R},M)$.#

B. <u>Compactness of the integral</u>

Recall that a linear map from one topological vector space to another is <u>compact</u> if it maps bounded sets into relatively compact sets. In studying compactness properties of the integral of a map $U: \mathcal{R} \otimes E \to G$, several different topologies on $\mathcal{R} \otimes E$ and on G are of interest. We start with the investigation of the relevant topologies on G with respect to which compactness of U can be formulated and, in order to avoid repetitions, lay down the following definition.

Definition. A topology τ on the Banach space G is said to be admissible, if (G,τ) is a topological vector space, if τ is coarser than the norm topology, and if the space of τ-continuous linear functionals on G is norming (4D).#

Examples. The norm topology on G is admissible. So is the weak topology $\sigma(G,G')$. If V is a norming subspace of G', then the V-weak topology $\sigma(G,V)$ is admissible. An admissible topology is finer (or equal to) the V-weak topology $\sigma(G,V)$, where V denotes the τ-continuous linear functionals on G.#

Below, the following compactness criterion will be needed.

10.6. Lemma. Let τ be an admissible topology on the Banach space G and let C be a subset of G.

If for every $\varepsilon > 0$ there is a τ-compact subset K of G such that C is contained in the (norm-) ε-neighborhood of K, then C is relatively τ-compact.

Proof. It will be shown that any filter $\mathcal{F}$ on C has a convergent refinement. To simplify the notation, let (a) denote the ball of radius a around zero. For every positive integer k, let K^k be a compact set such that C is contained in $K^k + (2^{-k})$. Put

$$\mathcal{F}^1 = \{(F + (2^{-1})) \cap K^1 : F \in \mathcal{F}\}.$$

Clearly $\mathcal{F}^1$ is a filter on K^1, and due to compactness, $\mathcal{F}^1$ has a refinement $\mathcal{F}_1^1$ which converges in the topology τ to an element ξ^1 of K^1. Put

$$\mathcal{F}_1 = \{(F + (2^{-1})) \cap G: F \in \mathcal{F}_1^1, G \in \mathcal{F}\}.$$

Clearly $\mathcal{F}_1$ is a filter again and is a refinement of $\mathcal{F}$. If refinements $\mathcal{F} \subset \mathcal{F}_1 \subset \ldots \subset \mathcal{F}_{k-1}$ have been defined up to the index $k - 1$, proceed as follows. Put $\mathcal{F}^k = \{(F + (2^{-k})) \cap K^k: F \in \mathcal{F}_{k-1}\}$ and let $\mathcal{F}_k^k$ be a convergent refinement with limit $\xi^k \in K^k$; and put

$$\mathcal{F}_k = \{(F + (2^{-k})) \cap G: F \in \mathcal{F}_k^k, G \in \mathcal{F}_{k-1}\}.$$

$\mathcal{F}_k$ converges to $\xi^k + (2^{-k})$ in the sense that if V is any neighborhood of zero in τ, then there is an $F \in \mathcal{F}_k$ entirely contained in $\xi^k + (2^{-k}) + V$. Indeed, if $F' \in \mathcal{F}_k^k$ is such that $F' \subset \xi^k + V$, then clearly $F = F' + (2^{-k}) \in \mathcal{F}_k$ satisfies this description.

Let a τ-continuous linear functional η of norm one be given. For any $\varepsilon > 0$ there is an $F \in \mathcal{F}_k$ such that

$$|\langle \xi^k - \xi; \eta \rangle| \leq 2^{-k} + \varepsilon \quad \text{for all} \quad \xi \in F.$$

There is a $G \in \mathcal{F}_{k+1}$ with $G \subset F$ such that

$$|\langle \xi^{k+1} - \xi; \eta \rangle| \leq 2^{-k-1} + \varepsilon \quad \text{for all} \quad \xi \in G.$$

Hence

$$|\langle \xi^k - \xi^{k+1}; \eta \rangle| \leq 2^{-k} + 2^{-k-1} + 2\varepsilon$$

for all $\varepsilon > 0$ and all τ-continuous linear functionals η of norm one, and since these are norming,

$$\|\xi^k - \xi^{k+1}\| \leq 2^{-k} + 2^{-k-1} .$$

This shows that (ξ^k) is a Cauchy sequence and so has a limit, ξ^o. Put $\mathcal{F}' = \cup\{\mathcal{F}_k : k \in \mathbb{N}\}$. Evidently, $\mathcal{F}'$ is a refinement of $\mathcal{F}$. It is left to be shown that $\mathcal{F}'$ converges to ξ^o in the topology τ.

Let $V \in \tau$ be any neighborhood of zero. There is a positive integer k such that $(2^{-k}) \subset V$. Enlarging k if necessary we may assume that $\xi^j \in \xi^o + V$ for $j \geq k$. There is a set $F \subset \mathcal{F}_k$ such that

$$F \subset \xi^k + (2^{-k}) + V \subset \xi^o + 2(2^{-k}) + V \subset \xi + 3V;$$

this shows that $\mathcal{F}'$ converges to ξ^o.#

<u>Definition</u>. Let τ be an admissible topology on the Banach space G. The linear map $U: \mathfrak{R} \otimes E \to G$ is said to be τ-<u>compact</u> if it is compact on each of the (sup-) normed spaces $\mathfrak{R}[K] \otimes E$, $K \in \mathcal{K}(\mathfrak{R})$, with respect to the topology τ on G (Cf. 4.16).

If M is any norm or seminorm on $\mathfrak{R} \otimes E$, then U is said to be (τ,M)-<u>compact</u> if it is a compact map from the M-normed space $\mathfrak{R} \otimes E$ to (G, τ).#

<u>Examples</u>. (1) If G is reflexive and U has finite semivariation, then U is weakly compact and weakly $\|U\|$-compact. Indeed, if $B \subset \mathfrak{R}[K] \otimes E$ is bounded in the uniformity of dominated uniform convergence or with respect to the norm $\|U\|$, then $U(B)$ is norm-bounded in G; and the bounded sets in G are weakly relatively compact.

(2) If U is weakly compact then it has finite semivariation. Indeed, the image of a bounded set in $\mathfrak{R}[K] \otimes E$ is then weakly bounded, hence strongly bounded.

(3) A scalar measure of finite semivariation is compact.#

For the remainder of this subsection, $U: \mathcal{R} \otimes E \to G$ is a linear map majorized by the upper norm M (and hence of finite semivariation), and τ is an admissible topology on G. The following result describes how τ-compactness of U is passed on to the integral extension of U.

<u>10.7. Proposition</u>. Let $h: X \to \overline{\mathbb{R}_+}$ be a function and denote by $(h)_E$ the set of all $(\mathcal{R},M)$-integrable E-valued functions f with $\|f\| \leq h$.

(i) If U is τ-compact and h $(\mathcal{R},M)$-integrable, then $\int (h)_E dU$ is relatively τ-compact.

(ii) If U is (M,τ)-compact and $M(h) < \infty$, then $\int (h)_E dU$ is relatively τ-compact.

<u>Proof</u>. (i) Let $\varepsilon > 0$ be given. There is a $\psi \in \mathcal{R}_+$ with $M(h - \psi) < \varepsilon$. For $f \in (h)_E$ define

$$f_\psi = f\psi/(\|f\| \vee \psi) \ , \quad f_\psi = 0 \ \text{ on } \ [\psi = 0] .$$

Evidently $\|f_\psi\| \leq \psi$. Suppose first that $f = \varphi \in \mathcal{R}_E$, and consider the analogue of φ_ψ on the spectrum of $\mathcal{R}$,

$$\hat{\varphi}_{\hat{\psi}} = \hat{\varphi}\,\hat{\psi}/(\|\varphi\|^{\wedge} \vee \hat{\psi}) .$$

It is clearly continuous, hence uniformly continuous (the closure of $[\hat{\psi} \neq 0]$ is compact), and so is, then, φ_ψ . Therefore φ_ψ belongs to $\overline{\mathcal{R}}_E$ (5.8) and so is integrable (7.17). Moreover, $\varphi_\psi \in (\psi)_E \cap \overline{\mathcal{R}}_E = (\psi)_E^{\circ}$ and so $U(\varphi_\psi)$ belongs to the relatively τ-compact set $K = U((\psi)_E^{\circ})$.

Now, let $f \in (h)_E$ be arbitrary, and select $\varphi \in \overline{\mathcal{R}}_E$ so that $M(f - \varphi) \leq \varepsilon/2$. The chain of inequalities

$$\|f_\psi - \varphi_\psi\| = \|\psi f/(\|f\| \vee \psi) - \psi\varphi/(\|\varphi\| \vee \psi)\|$$

$$\leq \psi\|f - \varphi\|/(\|f\| \vee \psi) + \psi\|\varphi\| \, |(\|\varphi\| \vee \psi - \|f\| \vee \psi)|/((\|\varphi\| \vee \psi)(\|f\| \vee \psi))$$

$$\leq \|f - \varphi\| + (\|\varphi\| \vee \psi - \|f\| \vee \psi) \leq \|f - \varphi\| + \|\varphi\| - \|f\| \leq 2\|f - \varphi\|$$

shows that f_ψ is integrable and, indeed, lies in an ε-neighborhood of $(\psi)_E^o$. From

$$\|f - f_\psi\| = \|f\|(1 - \psi/(\|f\| \vee \psi)) = (\|f\| \vee \psi - \psi)\|f\|/(\|f\| \vee \psi) \leq h \vee \psi - \psi \leq |h - \psi|$$

we obtain $M(f - f_\psi) < \varepsilon$, so that f lies actually in a (2ε)-neighborhood of $(\psi)_E^o$. Therefore $\int f dU$ lies in a (2ε)-neighborhood of the relatively τ-compact set K. As f was arbitrary in $(h)_E$, this shows that, for every $\varepsilon > 0$, $\int (h)_E dU$ is contained in the ε-neighborhood of some τ-compact set K, and so is relatively compact, by lemma 10.6. This finishes the proof of (i).

(ii) is much simpler to prove: If $M(h) = a < \infty$, then $(h)_E$ is contained in the M-closure of the ball B in $\mathcal{R} \otimes E$ of M-radius a. Hence $\int (h)_E dU$ is contained in the norm closure of $\int B dU$, which is by assumption relatively τ-compact.#

<u>10.8. Proposition</u>. If U is (τ,M)-compact on $\mathcal{R} \otimes E$ then $f \to \int f dU$ also is (τ,M)-compact on $\mathcal{L}_E^1(\mathcal{R},M)$.

<u>Proof</u>. Let C be the image under U of the M-unit ball of $\mathcal{R} \otimes E$. It is relatively compact, and the image of every function in the unit ball of $\mathcal{L}_E^1$ lies in the norm closure of C, which is compact.#

<u>10.9. Proposition</u>. Let m be a scalar measure majorized by the upper norm M, and let $g: X \to E$ be an integrable function. The map

$$f \to \int fg\,dm$$

from the space $UC(\mathcal{R})$ of all $\mathcal{R}$-uniformly continuous real-valued functions (with the topology of uniform convergence) to E is norm compact.

Proof. First, assume that g belongs to $\mathcal{R} \otimes E$, $g = \sum_{i=1}^{n} \varphi_i \xi_i$ say. We assume that $\int \varphi_i dm = 1$ for all indices $1 \leq i \leq n$. Then $fg \in \bar{\mathcal{R}} \otimes E$, and if $|f| \leq 1$, then

$$\int fg\,dm = \sum \left(\int f\varphi_i dm\right)\xi_i$$

belongs to the convex equilibrated hull $ce(\xi_1, \ldots, \xi_n)$, which is compact.

If g is arbitrary integrable and $\varepsilon > 0$ is given, then there is a $\varphi \in \mathcal{R} \otimes E$ with $M(\varphi - g) < \varepsilon$. Consequently $M(f\varphi - fg) < \varepsilon$ for $f \in UC(\mathcal{R})$ with $|f| \leq 1$, and so fg is integrable. Moreover, $\left\|\int fg\,dm - \int f\varphi\,dm\right\| < \varepsilon$, which shows, together with the above, that $\int UC_1(\mathcal{R}) g\,dm$ is contained in the ε-neighborhood of a norm compact set. From lemma 10.6, $\int UC_1(\mathcal{R}) g\,dm$ is relatively norm compact.#

C. The integral of a vector measure

Let E, F, G be three Banach spaces with norms $\|\ \|$, and let there be a bilinear map from $E \times F$ to G of norm one, denoted by juxtaposition: $\|\xi\eta\| \leq \|\xi\|\,\|\eta\|$ for $\xi \in E$, $\eta \in F$, $\xi\eta \in G$. Also, let $m\colon \mathcal{R} \to F$ be a measure majorized by the upper norm M. One might define the integral $\int \varphi\,dm$ for functions $\varphi = \sum \varphi_i \xi_i$ in $\mathcal{R} \otimes E$ by

(1) $$\int \varphi\,dm = \sum \xi_i m(\varphi_i) \in G.$$

It is not easy to see that $\int \varphi\,dm$ does not depend on the particular choice of the representation of φ in the form $\sum \varphi_i \xi_i$ $(\varphi_i \in \mathcal{R},\ \xi_i \in E)$.

If m has finite variation $|m|$, though, and if $|m|$ is majorized by a weak upper gauge $(\mathcal{R},M)$, then the situation is satisfactory; and so it will be assumed for the remainder of this section that this is the case.

<u>10.10. Lemma</u>. $\int d|m|$ is the variation of $\int dm$ on $\mathcal{L}^1(\mathcal{R},M)$.

<u>Proof</u>. Clearly $\int d|m|$ majorizes $\int dm$ on $\mathcal{L}^1$ and so $|\int dm| \leq \int d|m|$. On the other hand, $\int d|m| = |\int dm|$ on $\mathcal{R}_+$, and by continuity on $\mathcal{L}^1$.#

<u>10.11. Proposition</u>. There is a unique continuous linear map $\int dm$ from $\mathcal{L}^1_E(\mathcal{R},M)$ to G, given by (1) on $\mathcal{R} \otimes E$. It satisfies

$$\|\int f dm\| \leq \int \|f\| d|m| \leq M(f), \quad \text{for } f \in \mathcal{L}^1_E(\mathcal{R},M).$$

<u>Proof</u>. We use the fact that the E-valued integrable step functions $\mathcal{S}(\mathcal{C}(\mathcal{R},M)) \otimes E$ are dense in $\mathcal{L}^1_E$ (8.4). For such a function, $s = \sum A_i \xi_i$ say, define

$$\int s dm = \sum \xi_i \int A_i dm \in G.$$

As in 1.1, this definition is seen not to depend on the particular representation of s chosen; indeed, it may be assumed that the sets A_i are mutually disjoint. Therefore,

$$\|\int s dm\| \leq \sum \|\xi_i\| \, \|\int A_i dm\| \leq \sum \|\xi_i\| \int A_i d|m| = \int \|\varphi\| d|m| \leq M(s).$$

Hence $\int dm$ is continuous on $\mathcal{S}(\mathcal{C}(\mathcal{R},M)) \otimes E$, and has a unique extension $\int dm$ to $\mathcal{L}^1_E$. If $\xi \in E$, $\varphi \in \mathcal{R}$, and (s_n) is a sequence in $\mathcal{S}(\mathcal{C}(\mathcal{R},M))$ which converges to φ in mean (8.4), then (ξs_n) converges to $\xi\varphi$; and so $\int \xi\varphi dm = \xi m(\varphi)$. Linearity yields (1). Since $\mathcal{R} \otimes E$ is dense, $\int dm$ as defined here is the unique continuous linear map from $\mathcal{L}^1_E$ to G which is given by (1) on $\mathcal{R} \otimes E$.#

Note that we have made essential use of the fact that there is an abundance of integrable sets, which was ensured by $(\mathcal{R},M)$ being a weak upper gauge. The next result describes the variation of a measure with values in E' in terms of its integral on E-valued functions.

<u>10.12. Proposition</u>. If $F = E'$, $G = \mathbb{R}$, and if the map $E \times F \to G$ is the duality, then

(2) $$\int |g| d|m| = \sup\{\int f dm \colon f \in \mathcal{L}^1_E ,\ \|f\| \leq g\},$$

for $g \in \mathcal{L}^1_+(\mathcal{R},M)$, and for $\psi \in \mathcal{R}_+$,

(3) $$|m|(\psi) = \sup\{m(\varphi)\colon \varphi \in \mathcal{R} \otimes E,\ \|\varphi\| \leq \psi\}.$$

<u>Proof</u>. The inequalities $\geq$ follow from 10.11. Let $\psi \in \mathcal{R}_+$ and suppose that $m(\psi) > a \in \mathbb{R}$. There are functions $\varphi_1,\dots,\varphi_n \in \mathcal{R}_+$ with sum ψ such that $\sum \|m(\varphi_i)\| > a$. Given any $\varepsilon > 0$, find unit vectors $\xi_1,\dots,\xi_n$ in E such that $\langle \xi_i;m(\varphi_i)\rangle > \|m(\varphi_i)\| - \varepsilon/n$. Then $\varphi = \sum \varphi_i \xi_i \in \mathcal{R} \otimes E$ has norm not exceeding ψ and $\int \varphi dm = \sum \langle \xi_i;m(\varphi_i)\rangle > a - \varepsilon$. This shows the reverse inequality in (3). For (2) one proceeds analogously, replacing $\mathcal{R}$ by $\mathcal{L}^1$.#

<u>10.13. Corollary</u>. Equation (3) holds for any measure $m\colon \mathcal{R} \to E'$ of finite variation.

<u>Proof</u>. The measure m has property (3) if and only if its Gelfand transform $\hat{m}$ does. The latter has finite variation and is B-continuous (4.11). It is therefore majorized by an upper integral, namely by $\hat{m}^B$ (9.1), and so enjoys property (3) by the preceding proposition.#

11. Extension of linear maps.

This section contains the integration theory of linear maps $U: \mathcal{R} \otimes E \to G$ that do not have finite variation but merely finite semi-variation. To retain the all-important Lebesgue continuity property, we shall have to assume weak *-continuity.

The important result of the following subsection will be used below to show that a weakly compact linear map U can be majorized by a weak upper gauge. With this done, the whole integration theory of U is already contained in sections 8 and 10 above.

A. The theorem of Vitali-Hahn-Saks.

Let $m: \mathcal{A} \to F$, $n: \mathcal{A} \to \mathbb{R}_+$ be σ-additive measures on a σ-algebra $\mathcal{A}$. We say that m is absolutely continuous with respect to n, denoted $m \ll n$, if $n(A) \to 0$ implies $m(A) \to 0$. In other words,

(*) for all $\varepsilon > 0$ there is a $\delta > 0$ such that $A \in \mathcal{A}$ and $n(A) \leq \delta$ imply $\|m(A)\| < \varepsilon$.

This coincides with our earlier definition (3D) if m has finite variation (see 17.3) and extends it. A family $\mathcal{M}$ of measures $m: \mathcal{A} \to F$ is said to be uniformly absolutely continuous with respect to n if (*) holds for every $m \in \mathcal{M}$ and the δ can be chosen independently of m.

Let us interpret absolute continuity in topological terms. With the positive measure n there is associated a metric d on $\mathcal{A}$ given by

$$d(A,B) = n(A \triangle B) = n^*(|A - B|) \qquad , A,B \in \mathcal{A}.$$

11.1. Lemma. Let $\mathcal{M}$ be a family of measures on $\mathcal{A}$, n a positive measure.

(i) Equivalent are:

(a) $\mathcal{M}$ is uniformly absolutely continuous with respect to n.

(b) $\mathcal{M}$ is a uniformly equicontinuous family of functions on $\mathcal{Q}$.

(c) $\mathcal{M}$ is equicontinuous at some point $C \in \mathcal{Q}$.

(ii) When $\mathcal{M}$ consists of scalar measures these conditions are equivalent to:

(d) $\{|m| : m \in \mathcal{M}\}$ is a uniformly equicontinuous family of functions on $\mathcal{Q}$.

<u>Proof</u>. (a) $\Rightarrow$ (b): For $\varepsilon > 0$, let $\delta > 0$ be such that $\|m(A)\| \leq \varepsilon$ for all $m \in \mathcal{M}$ whenever $n(A) < \delta$. If $C, A \in \mathcal{Q}$ are such that $d(C,A) = n(|C - A|) < \delta$, then $n(C \setminus A) < \delta$ and $n(A \setminus C) < \delta$ and thus

$$\|m(C) - m(A)\| = \|m(C \setminus A) - m(A \setminus C)\| \leq \|m(C \setminus A)\| + \|m(A \setminus C)\| < \varepsilon,$$

for all $m \in \mathcal{M}$. That is, $\mathcal{M}$ is uniformly equicontinuous.

(b) $\Rightarrow$ (c) is obvious.

(c) $\Rightarrow$ (a): Suppose $\mathcal{M}$ is equicontinuous at $C \in \mathcal{Q}$. For $\varepsilon > 0$, let $\delta > 0$ be such that $\|m(C) - m(A)\| < \varepsilon$ for all $m \in \mathcal{M}$ whenever $n(|C - A|) < \delta$. Suppose $n(E) < \delta$. Then both $C \cup E$ and $C \setminus E$ have distance less than δ from C, and so $\|m(C) - m(C \cup E)\| < \varepsilon$ and $\|m(C) - m(C \quad E)\| < \varepsilon$. Combining these two inequalities, we get

$$\|m(E)\| \leq \|m(E \setminus C)\| + \|m(E \cap C)\| < 2\varepsilon.$$

For (ii), observe that

$$|m|(E) = \sup\{m(A) - m(B) : A, B \in \mathcal{Q},\ A + B = E\} \leq 2\varepsilon$$

if $n(E) < \delta$, where δ is so that $n(A) < \delta$ implies $\|m(A)\| \leq \varepsilon$. This proof actually shows more:

11.2. Corollary. If the measures of the family $\mathcal{M}$ all have oscillation less than ε on a ball $B_\delta(C)$ and are scalar, then the family $\mathcal{M} \cup \{|m| : m \in \mathcal{M}\}$ has oscillation less than 4ε on the δ-ball around any other point of $\mathcal{Q}$ (in particular around $\emptyset \in \mathcal{Q}$).

11.3. Theorem. (Vitali-Hahn-Saks). Let $\mathcal{Q}$ be a σ-algebra and (m_k) a sequence of σ-additive scalar measures. Suppose $m_\infty(A) = \lim\{m_k(A): k = 1,2,\ldots\}$ exists at all points A of $\mathcal{Q}$.

(i) If (A_n) is any sequence of sets in $\mathcal{Q}$ such that $\lim\{|m_k|(A_n): n = 1,2,\ldots\} = 0$ for all k then

$$\lim\{|m_k|(A_n): n = 1,2,\ldots\} = 0 \text{ uniformly for } k = 1,2,\ldots,\infty.$$

In particular, the family $\{m_k, |m_k| : k = 1,2,\ldots,\infty\}$ is uniformly σ-additive; i.e., $\mathcal{Q} \ni A_n \downarrow 0$ implies $|m_k(A_n)| + |m_k|(A_n) \to 0$ as $n \to \infty$, uniformly in k.

(ii) If each m_k is absolutely continuous with respect to the σ-additive positive measure n then the family $\{m_k, |m_k| : k = 1,2,\ldots,\infty\}$ is uniformly absolutely continuous with respect to n.

Proof. Recall that a scalar σ-additive measure on $\mathcal{Q}$ has finite variation, so the statement makes sense (6.15). We start with the proof of (ii). Under the metric $d(A,B) = n(|A - B|) = n^*(|A - B|)$, $\mathcal{Q}$ is a complete pseudometric space. Indeed, a Cauchy sequence contains an n^*-a.e. convergent subsequence (7.8) whose limit equals a.e. a function from $\mathcal{Q}$. Let $\varepsilon > 0$ be given. By the first lemma above, the sets

$$F_n = \{A \in \mathcal{Q}: \sup_{k \geq n} |(m_k - m_n)(A)| \leq \varepsilon/8\}, \quad n = 1, 2, \ldots$$

are closed in $\mathcal{Q}$. Since (m_k) converges pointwise, the F_n have union $\mathcal{Q}$.

By the Baire category theorem, at least one of them has non-void interior, F_n say. There is thus a point $C \in F_n$ and a $\delta > 0$ such that the ball $B_\delta(C)$ with radius δ around C lies inside F_n. By choosing δ sufficiently small, we can achieve that the oscillation of m_n on $B_\delta(C)$ is smaller than $\varepsilon/8$. Then clearly the family $\{m_k: n \leq k \leq \infty\}$ has oscillation smaller than $\varepsilon/4$ on $B_\delta(C)$. Choosing $\delta > 0$ even smaller we may make the oscillation of $\{m_k: 1 \leq k < n\}$ also smaller than $\varepsilon/4$ on $B_\delta(C)$. The corollary shows that $|m_k(A)| < \varepsilon$ and $|m_k|(A) < \varepsilon$ for $1 \leq k \leq \infty$ and all $A \in \mathfrak{a}$ satisfying $n(A) < \delta$.

(i): Let $a_k = |m_k|(1) \vee 1$ and put

$$n = \sum 2^{-k} a_k^{-1} |m_k| \ .$$

Clearly $m_k \ll n \geq 0$ for all k. If for all k $|m_k|(A_n) \to 0$ as $n \to \infty$ then $n(A_n) \to 0$ and by (ii), $|m_k|(A_n) \to 0$ uniformly in k. In particular, if $\mathfrak{a} \ni A_n \downarrow 0$ then $n(A_n) \to 0$ and $|m_k|(A_n) \to 0$ uniformly in k.#

B. <u>The weak integral.</u>

The data considered in the rest of the section are as follows. $\mathcal{R}$ is a dominated integration lattice on X, E and G are two Banach spaces with norms $\| \ \|$, and

$$U: \mathcal{R} \otimes E \to G$$

is a map of finite semivariation $\|U\|: \mathcal{R} \to \mathbb{R}_+$. If $E = \mathbb{R}$ then U is simply a measure of finite semivariation. In addition, it is assumed that V is a norming subspace of G' and that U is V-weakly $*$-continuous, where $* = S$ or B (see 4D).

To apply the results of section 10, an upper norm majorizing U is needed. We define

$$U^*(f) = \sup\{|Uv|^*(f): v \in V_1\} \qquad , \text{ for } f: X \to \overline{\mathbb{R}}_+.$$

Recall that the measure $Uv: \mathcal{R} \to E'$ is defined by $(Uv)(\varphi)(\xi) = \langle U(\varphi\xi);v\rangle$ $(\varphi \in \mathcal{R},\ \xi \in E)$, and is $*$-continuous and of finite variation

(1) $$|Uv| \leq \|U\| \qquad \text{(for } v \in V_1\text{; Cf. 4.14).}$$

<u>11.4. Lemma</u>. U^* is an upper S-norm and coincides with $\|U\|$ on $\mathcal{R}_+$. If $* = B$ then $(\mathcal{R}, U^B)$ is B-continuous.

<u>Proof</u>. As the supremum of upper S-norms, U^* is an upper S-norm (7.21). The inequality $U^* \leq \|U\|$ on $\mathcal{R}_+$ is evident from (1). To see the reverse inequality, choose a $\varphi \in \mathcal{R}_+$ and then a $\psi = \sum \varphi_i \xi_i \in \mathcal{R} \otimes E$ such that $\|\psi\| \leq \varphi$. From 10.12,

$$\begin{aligned}\|U\psi\| &= \sup\{\langle \textstyle\sum U(\varphi_i\xi_i);v\rangle: v \in V_1\} = \sup\{\textstyle\sum \langle \xi_i; Uv(\varphi_i)\rangle: \ldots\}\\ &= \sup\{\textstyle\int \sum \varphi_i\xi_i\, d(Uv): \ldots\} \leq \sup\{\textstyle\int \|\psi\| d|Uv|: \ldots\} = U^*(\|\psi\|) \leq U^*(\varphi).\end{aligned}$$

Taking the supremum over all $\psi \in \mathcal{R} \otimes E$ with $\|\psi\| \leq \varphi$ yields $\|U\|(\varphi) \leq U^*(\varphi)$.#

Next, $\mathcal{L}_E^{1V}$ is defined as the set of functions f in

$$\cap\ \{\mathcal{L}_E^1(\mathcal{R}, |Uv|^*): v \in V_1\}$$

that satisfy $U^*(f) < \infty$. It is the space of V-<u>weakly</u> U-<u>integrable</u> functions. As the intersection of full vector spaces, $\mathcal{L}_E^{1V}$ is a full vector space (6A). Recall that $\mathcal{R}_E^*$ denotes the dominated Baire functions (if $* = S$, see 6A) or Borel functions (if $* = B$, see 8C). Clearly

$$\mathcal{R}_E^* \subset \mathcal{L}_E^{1V} .$$

For every $f \in \mathcal{L}_E^{1V}$ we define a linear map

$$\int^V fdU \in V' \quad \text{by} \quad \langle \int^V fdU; v\rangle = \int fd(Uv) \qquad (v \in V).$$

<u>11.5. Proposition</u>. (i) For $f \in \mathcal{L}_E^{1V}$, $\int^V fdU$ is an element of V' and satisfies

$$\|\int^V fdU\| \leq U^*(f).$$

(ii) If (f_n) is a sequence of V-weakly integrable functions, converging U^*-a.e. to a function f and majorized by a function g with $U^*(g) < \infty$, then f is V-weakly integrable and

$$\lim\langle \int^V f_n dU; v\rangle = \langle \int fdU; v\rangle \qquad , \text{ for all } v \in V.$$

(iii) Suppose U maps the bounded subsets of $\mathcal{R}[K] \otimes E$ $(K \in \mathcal{K}(\mathcal{R}))$ into subsets of G that are sequentially $\sigma(G,V)$-complete. (This is the case, for instance, if U is V-weakly compact (10.B).) Then

$$\int^V \mathcal{R}_E^S dU \subset G.$$

(iv) If $V = G'$ and U is weakly $\|U\|$-compact (in particular if G is reflexive) then

$$\int^W \mathcal{L}_E^{1W} dU \subset G .$$

<u>Remark</u>. In general, a function $f \in \mathcal{L}_E^{1V}$ is called V-Pettis integrable provided

$$\int^V fdU \in G .$$

If $V = G'$, it is customary to write

$$\mathcal{L}_E^{1G'} = \mathcal{L}_E^{1w} \text{ and } \int^V fdU = \int^w fdU$$

and to call the elements f weakly integrable, Pettis-integrable if $\int^w fdU \in G$.

Proof of 11.5. From $|\langle\int^V fdU;v\rangle| \leq \int \|f\| d|Uv| \leq U^*(f)\cdot\|v\|$ we see that $\int^V fdU$ is a linear functional on V of norm not exceeding $U^*(f)$. If f_n, f, g are as in (ii), then (f_n) converges to f in $|Uv|^*$-mean for all $v \in V$, and so

$$\langle\int^V f_n dU;v\rangle = \int f_n d(Uv) \to \int fd(Uv) = \langle\int^V fdU;v\rangle .$$

Since $\|f\| \leq g$, $f \in \mathcal{L}_E^{1V}$.

(iii): Let $K \in \mathcal{K}(\mathcal{R})$, $B = \{\varphi \in \mathcal{R}[K] \otimes E: \|\varphi\| \leq 1\}$ and C a sequentially $\sigma(G,V)$-complete subset of G with $U(B) \subset C$. Consider the set $\mathcal{F} \subset (\mathcal{R}[K] \otimes E)^S$ of functions f with $\int^V fdU \in C$. If (f_n) is a dominated sequence in $\mathcal{F}$ with pointwise limit f, then $\int^V f_n dU$ is a $\sigma(G,V)$-Cauchy sequence in C. By the nature of C, its limit $\int^V fdU$ also belongs to $C \subset G$. Hence $\mathcal{F}$ is S-closed and so equals $(\mathcal{R}[K] \otimes E)^S$. Since $\mathcal{R}_E^S = \cup \{(\mathcal{R}[K] \otimes E)^S: K \in \mathcal{K}(\mathcal{R})\}$, $\int^V \mathcal{R}_E^S dU \subset G$.

(iv): Suppose the image C of the $\|U\|$-unit ball B of $\mathcal{R} \otimes E$ in G is relatively weakly compact. If $f \in \mathcal{L}_E^{1w}$ with $U^*(f) \leq 1$, then for every $v \in G'$ there is a sequence $(\varphi_n) \in B$ with $\langle\int^w fdU;v\rangle = \lim\langle U(\varphi_n);v\rangle$. Hence $\int^w fdU$ belongs to the weak closure $\bar{C} \subset G$ of C.#

C. Weakly compact linear maps.

11.6. Theorem. Suppose $U: \mathcal{R} \otimes E \to G$ is a weakly $*$-continuous map of finite semivariation.

(i) Then $(\mathcal{R},U^*)$ is an inner regular $*$-continuous weak upper gauge majorizing U provided

(a) U is weakly compact <u>or</u>

(b) For every $K \in \mathcal{K}(\mathcal{R})$, U maps bounded subsets of $\mathcal{R}[K] \otimes E$ into weakly sequentially complete subsets of G <u>or</u>

(c) U has a weakly S-continuous extension to $\mathcal{R}_E^S$. (This extension must equal $\int^w dU$.)

(ii) $(\mathcal{R},U^*)$ is an inner regular, $*$-continuous upper gauge provided

(a) G is reflexive <u>or</u>

(b) U is weakly $\|U\|$-compact <u>or</u>

(c) $\int^w dU$ maps $\mathcal{L}_E^{1w}$ into G.

(iii) In both cases, U is (strongly) $*$-continuous: If $\mathcal{R}$ $\Phi\downarrow *0$ then $\lim\{\|U\|(\varphi): \varphi \in \Phi\} = 0$.

We begin the proof with a simple characterization of (weak) upper gauges.

<u>11.7. Lemma</u>. Let M be an upper S-norm finite on $\mathcal{R}_+$.

(i) Equivalent are

(α) $(\mathcal{R},M)$ is a weak upper gauge.

(β) If $\varphi, \varphi_n \in \mathcal{R}_+$ are such that $\sum_{n=1}^{\infty} \varphi_n \le \varphi$ then $M(\varphi_n) \to 0$ as $n \to \infty$.

(γ) If (ψ_n) is an increasing sequence in $\mathcal{R}_+$ with $h = \sup \psi_n \le \varphi$ for some $\varphi \in \mathcal{R}_+$, then $M(h - \psi_n) \to 0$.

(ii) $(\mathcal{R},M)$ is an upper gauge if and only if for every sequence (φ_n) in $\mathcal{R}_+$, $M(\sum \varphi_n) < \infty$ implies $M(\varphi_n) \to 0$.

<u>Proof</u>. $(\alpha) \Rightarrow (\beta)$ is clear. Assume (β) and let $(\psi_n), \varphi$ be as in (γ). If $M(h - \psi_n) \not\to 0$, then there is an $a > 0$ with $M(h - \psi_n) > a$ for all n. Set $\psi_{n(1)} = \psi_1$. Since $M(h - \psi_1) = \sup_n M(\psi_n - \psi_{n(1)}) > a$, there is an

$n(2) \geq n(1)$ such that $M(\psi_{n(2)} - \psi_{n(1)}) > a$. Then there is an $n(3)$ such that $M(\psi_{n(3)} - \psi_{n(2)}) > a$, etc. We arrive at a sequence $\varphi_k = \psi_{n(k+1)} - \psi_{n(k)}$ of sum smaller than φ and $M(\varphi_k) > a$ for all k. This contradicts (β).

$(\gamma) \Rightarrow (\alpha)$: Let $(\varphi_n) \in \mathcal{R}$ be increasing and so that $h = \sup \varphi_n$ is majorized by an $(\mathcal{R},M)$-integrable function g. We have to show that $M(h - \varphi_n) \to 0$. Let $\varepsilon > 0$. There is a $\varphi \in \mathcal{R}_+$ with $M(g - \varphi) < \varepsilon$. Then $\psi_n = \varphi_n \wedge \varphi$ increases to $h \wedge \varphi \leq \varphi$. Since $h - h \wedge \varphi \leq g - \varphi \leq |g - \varphi|$, we get from (γ)

$$\inf M(h - \varphi_n) \leq \inf M(h - \psi_n) \leq M(h - h \wedge \varphi) + \inf M(h \wedge \varphi_n - \psi_n) \leq M(g - \varphi) < \varepsilon.$$

The proof of (ii) is simple and is left to the reader.#

<u>Proof</u> of the theorem. From the proposition (a) $\Rightarrow$ (b) $\Rightarrow$ (c) in (i) and in (ii). We need to show only that $(\mathcal{R},U^*)$ is a weak upper gauge if (ic) holds, an upper gauge if (iic) holds.

To do that we consider a sequence (φ_n) in $\mathcal{R}_+$ whose sum is majorized by a function h. In case (i) we take $h \in \mathcal{R}_+$, in case (ii) merely $U^*(h) < \infty$. It has to be shown that $U^*(\varphi_n) \to 0$. We reason by the absurd, and assume that there is an $a > 0$ such that $U^*(\varphi_n) > \infty$ for all n.

By 11.4 and the definition of $\|U\|$, there are functions $\psi_n \in \mathcal{R} \otimes E$ with $\|\psi_n\| \leq \varphi_n$ and $\|U(\psi_n)\| > a$ for all n.

Let $\mathcal{A}$ denote the σ-algebra of all subsets of $\mathbb{N}$, and define

$$m(A) = \int^{w} \sum \{\psi_n : n \in A\} dU \in G \qquad (A \in \mathcal{A}).$$

This makes sense since $\psi_A = \sum \{\psi_n : n \in A\}$ belongs to $\mathcal{R}_E^S$ in case (i) and to $\mathcal{L}_E^{1w}$ in case (ii), respectively. Clearly, m is a measure on $\mathcal{A}$. For any $v \in G'$, the scalar measure

$$\langle m;v\rangle : A \to \langle m(A);v\rangle = \int \Sigma\,\{\psi_n : n \in A\}d(Uv)$$

is σ-additive. For if $\mathfrak{A}$ $A_k \downarrow 0$ then there is a $k(N)$ such that $\{1,\ldots,N\} \cap A_{k(N)} = \emptyset$, for any $N \in \mathbb{N}$; and so

$$\inf_k |\langle m;v\rangle(A_k)| \leq \inf_N \int \Sigma\,\{\varphi_n : n \geq N\}d|Uv| = 0.$$

It follows that m takes its values in the separable subspace G^s of G that is spanned by the vectors $m(\{n\}) = U(\psi_n)$, $n = 1,2,\ldots$. For if $v \in G'$ annihilates G^s then $\langle m(A);v\rangle = \Sigma\,\{\int \psi_n d(Uv) : n \in A\} = 0$.

Let us fix a countable dense subset H of G^s. Since $\|m(\{n\})\| > a$ there are unit vectors $v_n \in G_1^{s'} \subset G_1'$ such that

$$(*) \qquad \langle m(\{n\});v_n\rangle = \langle m;v_n\rangle(\{n\}) > a$$

for all n. By the usual diagonal process, we may extract a subsequence $(v_{n(k)})$ such that $\langle \xi, v_{n(k)}\rangle$ converges for all $\xi \in H$. Since $\{v_{n(k)}\}$ is uniformly bounded, the sequence $\langle \xi;v_{n(k)}\rangle$ converges for all $\xi \in G^s$. Consequently, the sequence $(\langle m;v_{n(k)}\rangle)$ of scalar measures converges at every set $A \in \mathfrak{A}$. Now,

$$|\langle m;v_{n(k)}\rangle|(\{n\}) = |\langle m;v_{n(k)}\rangle(\{n\})| = |\int \psi_n d(Uv_{n(k)})|$$

converges to zero as $n \to \infty$, for every k. By the theorem of Vitali-Hahn-Saks, the convergence is uniform in k. This contradicts (*) and shows that $U^*(\varphi_n) \to 0$, after all.

If U is weakly B-continuous, then $(\mathcal{R}, U^B)$ is B-continuous from 11.4.

The fact that $(\mathcal{R},U^*)$ is inner regular follows from 8.22. Indeed, if $A \in \mathcal{C}(\mathcal{R},U^*)$ has $U^*(A) > 0$, then there is a $v \in G'$ with $|Uv|^*(A) > 0$. Consequently, there is a subset K of A in $\mathcal{K}^*(\mathcal{R})$ with $|Uv|^*(K) > 0$ (9.4) and a fortiori $U^*(K) > 0$.

The last statement (iii) is obvious (8.4, 8.15).#

Remarks. The fact that $(\mathcal{R},U^*)$ is a (weak) upper gauge implies that the dominated Baire or Borel functions are Bochner-integrable: $\mathcal{R}_E^* \subset \mathcal{L}_E^1(\mathcal{R},U^*)$; and that Lebesgue's dominated convergence theorem holds in $\mathcal{L}_E^1(\mathcal{R},U^*)$. Conversely, the theorem shows that either property of $\mathcal{L}_E^1(\mathcal{R},U^*)$ entrains that $(\mathcal{R},U^*)$ is a weak upper gauge. It can be shown that conditions (ia) and (iia), which seem stronger at first, are also necessary: at least if E is reflexive then U is weakly compact whenever U^* is a weak upper gauge.

The theorem implies the Orlicz-Pettis theorem to the effect that if $m: \mathcal{F} \to G$ weakly S-continuous measure of finite semivariation defined on a δ-ring or full integration lattice $\mathcal{F}$, then m is S-continuous. Indeed, if $\mathcal{F}$ is a full integration lattice, this follows from our theorem with $E = \mathbb{R}$: If $\mathcal{F} \ni \varphi_n \downarrow 0$ then $m^*(\varphi_n) \to 0$ and so $\|m(\varphi_n)\| \to 0$. If $\mathcal{F}$ is a δ-ring, then m can be extended by continuity to $\overline{\mathcal{S}(\mathcal{F})}$, the closure of $\mathcal{S}(\mathcal{F})$ in dominated uniform convergence. Since $\overline{\mathcal{S}(\mathcal{F})}$ is full (6.9), the result follows also in this case.

An application. Let $\mathcal{R}$ be a dominated integration lattice on X. Then $\mathcal{R}$ is a topological vector space in the topology of dominated uniform convergence and $I_0(\mathcal{R})$ is its dual (4.18). We equip $I_0(\mathcal{R})$ with the strong topology of uniform convergence on bounded sets. Consider a weakly compact subset

$\mathcal{M} \subset I_0(\mathcal{R})$. That is, $\mathcal{M}$ is $\sigma(I_0(\mathcal{R}), I_0(\mathcal{R})')$-compact. We claim that $\mathcal{M}$ is uniformly additive. That is to say, if (φ_n) is a sequence in $\mathcal{R}_+$ whose sum is majorized by a function $\varphi \in \mathcal{R}_+$, then $\lim m(\varphi_n) \to 0$ uniformly for $m \in \mathcal{M}$.

To show this, we may go to the Gelfand-Bauer transforms $\hat{m}$: we may assume that X is locally compact in the $\mathcal{R}$-topology, so that all the $m \in \mathcal{M}$ are B-continuous.

Now, we look at the map $U: \mathcal{R}^S \to C(\mathcal{M})$ defined by $U(\varphi)(m) = \int \varphi dm$. Evidently, $U(\varphi)$ is a continuous function on the compact space $\mathcal{M}$. Also, U has finite semivariation. For if $K \in \mathcal{K}(\mathcal{R})$, then

$$\begin{aligned} \|U\|(K) &= \sup\{\|U(\varphi)\|: |\varphi| \leq K\} \\ &= \sup\{\int \varphi dm: |\varphi| \leq K, m \in \mathcal{M}\} \end{aligned}$$

is evidently finite. By the theorem, $(\mathcal{R}, U^B)$ is a weak upper gauge, and so $\mathcal{M}$ is uniformly additive.

It can be shown that the converse is also true: if $1 \in \mathcal{R}_\uparrow^S$ and $\mathcal{M}$ is uniformly additive then $\mathcal{M}$ is weakly compact.

D. Supplements.

11.8. Suppose $m: \mathcal{R} \to F$ is a weakly compact weakly $*$-continuous measure, and $\mathcal{R}$ contains the constants. A function f is then $(\mathcal{R}, \|m\|^*)$-integrable if and only if there exists a sequence (φ_n) in $\mathcal{R}$ that converges pointwise $\|m\|^*$-a.e. to f in such a way that $\int K\varphi_n dm$ converges in norm, for all Baire sets K. (This is the standard definition of integrability for f.)

11.9. If $U: \mathcal{R} \otimes E \to G$ is a weakly compact and weakly $*$-continuous linear map, then $f \to \int f dU$ is a weakly compact linear map from $\mathcal{L}^1_{E\rho}$, the space of bounded, $\mathcal{L}^1$-dominated functions in $\mathcal{L}^1_E$ equipped with the topology of $\mathcal{L}^1$-dominated uniform convergence, to G. If U is weakly $\|U\|$-compact, then $f \to \int f dU$ is a weakly compact map from the seminormed space $\mathcal{L}^1_E(\mathcal{R}, \|U\|^*)$ to G.

11.10. (Riesz Representation of weakly compact linear maps). To every weakly compact and weakly $*$-continuous linear map $U: \overline{\mathcal{R}}_E \to F$ there correspond a δ-ring S containing $\mathcal{C}(\mathcal{R}^*)$ and an inner regular, σ-additive (both in the operator norm) measure m having values in the space $L_0(E,F)$ of weakly compact linear maps and having the following property: For every $A \in S$, the elements $\sum T(B_i)\xi_i$ form a relatively weakly compact set in F, where $\sum B_i\xi_i$ ranges over all integrable step functions satisfying $\| \sum B_i\xi_i \| \le A$. The map $U \to m|\mathcal{C}(\mathcal{R}^*)$ is a linear isomorphism onto the set of measures of this description.

11.11. Suppose $\mathcal{R}$ contains the constants and let $L(\overline{\mathcal{R}}_E, F)$ denote the vector space of linear maps $U: \mathcal{R}_E \to F$ of finite semivariation. Then $U \to \|U\|(1)$ is a norm on $L(\overline{\mathcal{R}}_E, F)$ under which this space is complete; and the weakly $*$-continuous $(* = S, B)$ maps form a closed subspace. So do the τ-compact maps.

LITERATURE: [11], [14].

§3. Various new upper gauges

We shall discuss four methods for constructing new upper gauges from given ones. The upper integrals, m^*, and (weak) upper gauges, $\|U\|^*$, that we have met are often too big for special purposes. Recall that the smaller an upper gauge the more integrable functions there are. Many more upper gauges will appear naturally in chapter IV.

12. The p-norms $(1 \leq p < \infty)$

A. Convexity theorems

We fix a set X and an upper norm M on X.

12.1. Theorem. Let $F\colon \mathbb{R}_+^n \to \mathbb{R}_+$ be a continuous map satisfying the following conditions:

A) $t_i > 0$ for $1 \leq i \leq n$ implies $F(t_1,\ldots,t_n) > 0$;

B) $F(rt_1,\ldots,rt_n) = rF(t_1,\ldots,t_n)$ for $0 < r \in \mathbb{R}$ and $\underline{t} = (t_1,\ldots,t_n) \in \mathbb{R}_+^n$;

C) The set $K = \{\underline{t} \in \mathbb{R}_+^n : F(\underline{t}) \geq 1\}$ is convex.

If $f_i\colon X \to \mathbb{R}_+$ are such that $M(f_i) < \infty$ for $1 \leq i \leq n$, then

$$M(F(f_1,\ldots,f_n)) \leq F(M(f_1),\ldots,M(f_n)).$$

Proof. The inner product of two n-tuples $\underline{t},\underline{r}$ in ^{n} will be denoted by $\langle \underline{t};\underline{r}\rangle$. The half-space $H(\underline{r},c)$ determined by the n-tuple $\underline{r}$ and the constant c is the set of all n-tuples $\underline{t}$ satisfying the inequality $\langle \underline{t};\underline{r}\rangle \geq c$. Since K is convex and closed, it is the intersection of all half-spaces containing it. Let $\mathcal{H}$ denote the set of all such half-spaces. Let $H(\underline{r},c) \in \mathcal{H}$ and $\underline{t} \in K$. From A) and B), $s\underline{t} \in K$ for all $s \geq 1$ and thus $s\underline{t} \in H(\underline{r},c)$; i.e., $s\langle \underline{r},\underline{t}\rangle \geq c$ for all large s. This shows that $\langle \underline{r},\underline{t}\rangle \geq 0$ for all $\underline{t} \in K$, and so $H(\underline{r},0) \in \mathcal{H}$. Furthermore, since there is a point of K on each of the positive axes of $\mathbb{R}^n$, each $r_i \geq 0$. Since $H(\underline{r},c) \supset H(\underline{r},0)$ for $c < 0$, K is actually the intersection of the positive half-spaces $\mathcal{H}_+ \subset \mathcal{H}$; i.e., of those $H(\underline{r},c) \in \mathcal{H}$ with $\underline{r} \geq 0$ and $c \geq 0$.

Now, if $0 \leq \underline{t} \in \mathbb{R}^n$ and $H(\underline{r},c) \in \mathcal{H}_+$, then $\langle \underline{r};\underline{t}\rangle \geq F(\underline{t})c$; this is clear if $F(\underline{t}) = 0$, and if $F(\underline{t}) \neq 0$ then $\underline{t}(F(\underline{t}))^{-1} \in K$ and so $\langle \underline{r}; (F(\underline{t}))^{-1}\underline{t}\rangle \geq c$. If $f_1,\ldots,f_n \in \mathbb{R}_+^X$ then

$$\sum r_i f_i(x) \geq cF(f_1(x),\ldots,f_n(x)) \quad , \quad (x \in X),$$

and, using sublinearity,

$$\sum r_i M(f_i) \geq cM(F(f_1,\ldots,f_n)).$$

This is true for all $H(\underline{r},c) \in \mathcal{H}_+$ and shows that, if $M(F(f_1,\ldots,f_n)) \neq 0$, then the n-tuple with components $(M(F(f_1,\ldots,f_n)))^{-1}M(f_i)$ is in K; hence in this instance

$$M(F(f_1,\ldots,f_n)) \leq F(M(f_1),\ldots,M(f_n)),$$

as desired. If $M(F(f_1,\ldots,f_n)) = 0$, there is nothing to prove.#

<u>12.2. Corollary</u>. Suppose N is an upper S-norm and let $F\colon \overline{\mathbb{R}_+^n} \to \overline{\mathbb{R}_+}$ be a continuous map satisfying the following:

A_1) $0 < t_i < \infty$ $(i = 1,\ldots,n)$ implies $0 < F(t_1,\ldots,t_n) < \infty$.

A_2) If $t_i = \infty$ for one i in $\{1,\ldots,n\}$ then $F(t_1,\ldots,t_n) = \infty$.

B) $F(rt_1,\ldots,rt_n) = rF(t_1,\ldots,t_n)$ for $0 < r \in \mathbb{R}$.

C) The set $K = \{(t_1,\ldots,t_n) \in \mathbb{R}_+^n : F(t_1,\ldots,t_n) \geq 1\}$ is convex in $\mathbb{R}^n$.

Also, let f_i $(i = 1,\ldots,n)$ be n positive functions, defined M-almost everywhere. Then

$$M(F(f_1,\ldots,f_n)) \leq F(M(f_1),\ldots,M(f_n)).$$

Proof. The union A of the sets A_i where f_i is not defined $(i = 1,\ldots,n)$ is negligible, and so $F(f_1,\ldots,f_n)$ is almost everywhere defined. The inequality is true if one of the $M(f_i)$, $i \in \{1,\ldots,n\}$, is infinite, due to A_2). If all the $M(f_i)$ are finite, the f_i may be replaced by everywhere defined and finite functions, without altering either side of the inequality. This reduces the problem to the one already answered in the theorem above.#

12.3. Corollary. (Hölder's inequality). Let M be as above, let $0 < a < 1$, and let f, g be two M-almost everywhere defined positive numerical functions on X. Then

$$M(f^a g^{1-a}) \leq (M(f))^a (M(g))^{1-a} .$$

Proof. We have to show that the function $F: (t,s) \to t^a s^{1-a}$ on $\mathbb{R}^2_+$ has the properties A_1), A_2), B), and C), and is continuous. With the definition $\infty^a = \infty$, F is continuous and satisfies A_1, A_2), and B). To prove C), let (t_1,s_1) and (t_2,s_2) be in K and consider the points $(T(r),S(r)) = (rt_1 + (1-r)t_2,\ rs_1 + (1-r)s_2) = r(t_1,s_1) + (1-r)(t_2,s_2)$ $(0 \leq r \leq 1)$ on the segment joining (t_1,s_1) with (t_2,s_2). We have to show that $f(r) = T(r)^a S(r)^{1-a} \geq 1$ for $0 \leq r \leq 1$. The inequality is true for $r = 0, 1$. Now, the derivative $f'(r)$ equals

$$f'(r) = f(r)(a(t_1 - t_2)T(r)^{-1} + (1-a)(s_1 - s_2)S(r)^{-1})$$

and is either zero for $0 < r < 1$, in which case we are finished; or it vanishes nowhere in $(0,1)$, in which case $f(r)$ lies between $f(0)$ and $f(1)$, and $f(r) \geq 1$ throughout again; or it vanishes in exactly one point r_0 in $(0,1)$, given by the linear equation

$$a(t_1 - t_2)S(r_0) + (1-a)(s_1 - s_2)T(r_0) = 0.$$

(Note that $T(r) > 0 < S(r)$ in $[0,1]$ and hence $f(r) > 0$.) At r_0, we find $f''(r_0) = -f(r_0)(a(t_1 - t_2)^2 T(r)^{-2} + (1-a)(s_1 - s_2)^2 S(r)^{-2}) \leq 0$. Hence f has a maximum in r_0, and in this case also $f(r) \geq 1$ in $[0,1]$.#

12.4. Corollary. (Minkowski's inequality). Let $1 \leq p < \infty$ and let f,g be two positive numerical functions on X, M-a.e. defined. Then

$$M((f+g)^p)^{1/p} \leq (M(f^p))^{1/p} + (M(g^p))^{1/p} .$$

Proof. We will show that the continuous function $F\colon (t,s) \to (t^{1/p} + s^{1/p})^p$ satisfies the conditions of 12.2; from this the claim follows immediately.

There is only C) to prove, the rest being obvious. Resume the notation of the proof of 12.3. Assume that (t_1, s_1) and (t_2, s_2) are in K and consider the point $(T(r), S(r))$ on the segment joining these points. Put

$$f(r) = \{(T(r))^{1/p} + (S(r))^{1/p}\}^p.$$

An easy calculation gives

$$f''(r) = p^{-1}(p^{-1} - 1)\{(t_1 - t_2)^2 (T(r))^{(1/p)-2} + (s_1 - s_2)^2 (S(r))^{(1/p)-2}\} \leq 0$$

in $(0,1)$, and so f is a concave function. Hence $f(r) \geq 1$ for $0 < r < 1$.#

Let M be an upper norm on X. For $1 \leq p < \infty$, define

$$M_p(f) = M(f^p)^{1/p}, \quad \text{for} \quad f\colon X \to \overline{\mathbb{R}}_+ .$$

Evidently M_p is positively homogeneous, increasing and, from Minkowski's inequality, also subadditive; it is therefore an upper norm. If f is a function on X having values in a Banach space, we put, as usual

$$M_p(f) = M_p(\|f\|).$$

The next three results describe certain simple properties of these p-norms.

<u>12.5. Corollary</u>. If M is any upper norm on X, then M_p is also an upper norm, for $1 \leq p < \infty$, and furthermore

$$(M_p)_q = M_{pq}, \quad \text{for} \quad 1 \leq p,q < \infty.\#$$

<u>12.6. Corollary</u>. Let $1 \leq p,q,r < \infty$ satisfy $1/p + 1/q = 1/r$. For any two positive numerical functions f,g (M-almost everywhere defined if M is an upper S-norm) we have

$$M_r(fg) \leq M_p(f)M_q(g).$$

<u>Proof</u>. We apply Hölder's inequality with $a = r/p$, to f^p and to g^q. Since $1 - a = r/q$,

$$M_r(fg) = (M((f^p)^a(g^q)^{1-a}))^{1/r} \leq (M(f^p))^{1/p}(M(g^q))^{1/q} = M_p(f)M_q(g),$$

as desired.#

<u>12.7. Corollary</u>. Let M be an upper S-norm and let f be a function M-a.e. defined and with values either in the extended reals or in a

Banach space. The set I of real numbers $1/p$ $(p \geq 1)$ such that $M_p(f) < \infty$ is either empty or is an interval. If $M_p(f) = 0$ for some $p \geq 1$, then $M_p(f) = 0$ for all $p \geq 1$, and $I = [1,\infty)$. The function $1/p \to M_p(f)$ is logarithmically convex on I and hence is continuous in the interior of I.

Proof. Suppose that f is not negligible, let $1/p$ and $1/q$ be in I, and let $1/s = \theta/p + (1-\theta)/q$ $(0 < \theta < 1)$ be a point in between. If $a = s\theta/p$ then $1 - a = (1-\theta)s/q$, and therefore $0 < a < 1$. A simple calculation gives $s = ap + (1-a)q$. The following sequence of inequalities is a consequence of Hölder's inequality applied to the functions $|f|^p$ and $|f|^q$:

$$M(|f|^s) = M(|f|^{pa}|f|^{q(1-a)}) \leq M(|f|^p)^a M(|f|^q)^{1-a} ;$$

$$M_s(f) \leq (M_p(f))^\theta (M_q(f))^{1-\theta} ;$$

$$\log M_s(f) \leq \theta \log M_p(f) + (1-\theta) \log M_q(f) .$$

The last inequality shows that $1/s \in I$ and proves the desired logarithmic convexity of $1/s \to M_s(f)$. The other statements of the corollary are obvious.#

B. p-integrable functions

Throughout this section it will be assumed that $(\mathcal{R},M)$ is an upper gauge on the dominated integration lattice $\mathcal{R}$.

12.8. Theorem. Let $1 \leq p,q < \infty$.

(i) Then $(\mathcal{R},M_p)$ is an upper gauge, and is B-continuous or strong if $(\mathcal{R},M)$ is.

(ii) A function f, defined M-a.e. and with values in a Banach space E or in $\overline{\mathbb{R}}$, is $(\mathcal{R},M_p)$-integrable if and only if the function $f\|f\|^{(p/q)-1}$ is $(\mathcal{R},M_q)$-integrable.

(iii) The $(\mathcal{R},M_p)$-integrable sets and the $(\mathcal{R},M_q)$-integrable sets coincide.

Proof. (i) If $\varphi \in \mathcal{R}_+$ then φ^p is $\mathcal{R}$-uniformly continuous, hence in $\overline{\mathcal{R}}$ (5.6), and so $M_p(\varphi) < \infty$: M_p is finite on $\mathcal{R}_+$. The same argument shows that φ^p is $(\mathcal{R},M)$-integrable (4.7).

If (φ_n) is an increasing sequence in $\mathcal{R}_+$ with $\sup M_p(\varphi_n) < \infty$, then

$$\lim\{M(\varphi_n^p - \varphi_m^p) : n \geq m \to \infty\} = 0$$

from the monotone convergence theorem (8.9). The elementary inequality

$$\varphi_n^p - \varphi_m^p \geq (\varphi_n - \varphi_m)^p \qquad (n \geq m,\ p \geq 1)$$

gives $\lim\{M_p(\varphi_n - \varphi_m) : n \geq m \to \infty\} = 0$, so that $(\mathcal{R},M_p)$ is an upper gauge. It is obvious that $(\mathcal{R},M_p)$ is B-continuous if $(\mathcal{R},M)$ is (8.14). It will be shown at the end of the proof that strongness is inherited.

For the second statement, let $f \in \mathcal{L}_E^1(\mathcal{R},M_p)$. From 7.12, there is a sequence (φ_n) in $\mathcal{R} \otimes E$ such that

$$\sum M_p(\varphi_n) < \infty \quad \text{and} \quad f = \sum \varphi_n$$

M-a.e. and in M_p-mean. Note that

$$M((\sum \|\varphi_n\|)^p) \leq (\sum M_p(\varphi_n))^p < \infty.$$

Put $\psi_n = \|\varphi_1 + \ldots + \varphi_n\|^{(p/q)-1}(\varphi_1 + \ldots + \varphi_n)$, for $n = 1, 2, \ldots$.

As $\psi_n \in \overline{\mathcal{R}}_E$, it is $(\mathcal{R},M_q)$-integrable (7.17). Moreover,

$$\|\psi_n\|^q = \|\varphi_1 + \ldots + \varphi_n\|^p \leq (\sum^{\infty} \|\varphi_n\|)^p = h^q ,$$

where the function $h: X \to \overline{\mathbb{R}_+}$ so defined satisfies $M_q(h) < \infty$. Since (ψ_n) converges M-a.e. to $f\|f\|^{(p/q)-1}$ and is majorized by h, it follows from Lebesgue's theorem (8.12) that $f\|f\|^{(p/q)-1}$ is $(\mathcal{R},M_q)$-integrable. A simple calculation with exponents shows that this is also sufficient for f to be $(\mathcal{R},M_p)$-integrable.

The last statement is now obvious. To finish the proof, suppose $(\mathcal{R},M)$ is a strong upper gauge. If $0 \leq f \leq g$ in $\mathcal{L}^1(\mathcal{R},M_p)$ and $M_p(f) = M_p(g)$, then $0 \leq f^p \leq g^p$ in $\mathcal{L}^1(\mathcal{R},M)$ and $M(f^p) = M(g^p)$; hence $f \doteq g$ M-a.e. and so $M_p(g - f) = 0$.#

When $q = 1$, the theorem states that $f \geq 0$ is $(\mathcal{R},M_p)$-integrable if and only if f^p is $(\mathcal{R},M)$-integrable; and so the elements of $\mathcal{L}^1_E(\mathcal{R},M_p)$ are called p^{th} power integrable. We adopt the usual notation and write

$$\mathcal{L}^p_E(\mathcal{R},M) \quad \text{for} \quad \mathcal{L}^1_E(\mathcal{R},M_p).$$

If $\mathcal{R}$ is not dominated, M_p need not be finite on $\mathcal{R}_+$. In this instance, $\mathcal{L}^p_E(\mathcal{R},M)$ is defined to be the closure of $\mathcal{R}_0 \otimes E$ under M_p. For $p = 1$ this changes nothing (8.4).

C. Supplement

12.9. $(\mathcal{R},M)$ and $(\mathcal{R},M_p)$ are equivalent upper gauges, $1 \leq p < \infty$.

12.10. Suppose $(\mathcal{R},M)$ is an upper gauge such that $r = M(1)$ is finite, and let $f: X \to \overline{\mathbb{R}}_+$. Then $p \to r^{-1/p} M_p(f)$ is an increasing function of p; and for $1 \leq p \leq q < \infty$, $\mathcal{L}^q_E(\mathcal{R},M) \subset \mathcal{L}^p_E(\mathcal{R},M)$ and the inclusion is a continuous map with norm $r^{(1/p + 1/q)}$.

12.11. Suppose $f: X \to E$ is bounded. If $1 \leq p \leq q < \infty$ and $M_p(f) < \infty$ then $M_q(f) < \infty$.

LITERATURE: The convexity theorems are proved as in [5], ch. IV.

13. Essential upper gauges

In this section, $(\mathcal{R},M)$ is a fixed weak upper gauge on X.

A. Definition. A subset K of X is said to be $(\mathcal{R},M)$-finite if it is contained in an $(\mathcal{R},M)$-integrable set. K is called σ-finite (with respect to $(\mathcal{R},M)$) if it is contained in a countable union of $(\mathcal{R},M)$-integrable sets. Finally, $(\mathcal{R},M)$ is said to be finite (σ-finite) if X is an $(\mathcal{R},M)$-finite (σ-finite) set.#

A σ-finite set is contained in the union of countably many dominated Baire sets and a negligible set (8.7).

Definition. The essential upper gauge $(\mathcal{R},\overline{M})$ associated with $(\mathcal{R},M)$ is defined by

$$\overline{M}(f) = \sup M(Af) \ , \quad \text{for} \quad f: X \to \overline{\mathbb{R}}_+ \ ,$$

where A ranges over all sets that are σ-finite with respect to $(\mathcal{R},M)$; and $(\mathcal{R},M)$ is said to be essential if it equals $(\mathcal{R},\overline{M})$.#

Let $f \in \overline{\mathbb{R}}_+^X$ and let (A_n) be a sequence of σ-finite sets such that

$$\overline{M}(f) = \sup M(fA_n).$$

The A_n may be chosen to be increasing. Then $A = \cup A_n$ is also σ-finite, and, using the S-continuity of M, $\overline{M}(f) = M(fA)$.

Suppose $\{f_n\}$ is a countable number of positive numerical functions on X. For each n, let A_n be a σ-finite set for which $\overline{M}(f_n) = M(fA_n)$. Then $A = \cup A_n$ is σ-finite, and so the following result is proved.

13.1. Lemma. Let $\{f_n\}$ be a countable collection of positive numerical functions. Then there is a σ-finite set A such that

$$\overline{M}(f_n) = M(Af_n) , \quad \text{for } n = 1, 2, \ldots \quad .\#$$

Using this, it is seen immediately that $\overline{M}$ is an upper S-norm on X and is smaller than M. Since every $(\mathcal{R},M)$-integrable function f vanishes off a σ-finite set (8.7),

$$M(f) = \overline{M}(f) , \quad \text{for } f \in \mathcal{L}^1(\mathcal{R},M).$$

This will show that $(\mathcal{R},\overline{M})$ is a weak upper gauge. For if (φ_n) is an increasing sequence in $\mathcal{R}_+$ whose supremum f is majorized by $h \in \mathcal{L}^1(\mathcal{R},\overline{M})$ then there is a sequence (ψ_n) in $\mathcal{R}_+$ so that $\overline{M}(f - \psi_n) \to 0$. Choose A as in lemma 13.1, for the countable collection $\{f, h, \varphi_1, \ldots, \psi_1, \ldots\}$. Clearly Ah is $(\mathcal{R},M)$-integrable, with $(A\psi_n)$ approximating it in M-mean, and it majorizes $Af = \sup(A\varphi_n)$. Hence $\lim \overline{M}(f - \varphi_n) = \lim M(A(f - \varphi_n)) = 0$. If h merely satisfies $\overline{M}(h) < \infty$, then the same argument applies and shows that $(\mathcal{R},\overline{M})$ is an upper gauge if $(\mathcal{R},M)$ is.

Suppose $(\mathcal{R},M)$ is a strong upper gauge, and let $0 \leq f \leq g$ in $\mathcal{L}^1(\mathcal{R},\overline{M})$ with $\overline{M}(g) = \overline{M}(f)$. Choose a σ-finite set A so that

$$\overline{M}(g) = M(gA), \ \overline{M}(f) = M(fA), \text{ and } \overline{M}(g - f) = M(A(g - f)).$$

It may be assumed that A is the union of an increasing sequence of $(\mathcal{R},M)$-integrable sets. From Lebesgue's theorem, fA, gA, and $(g - f)A$ are integrable. As $(\mathcal{R},M)$ satisfies (SUG), $\overline{M}(g - f) = M(A(g - f)) = 0$.

Recall that an $(\mathcal{R},M)$-adequate cover of integrable sets is a family $C \subset \mathcal{C}(\mathcal{R},M)$ so that every integrable set can be covered a.e. by countably many sets in C. Evidently any σ-finite set can also be covered a.e. by a countable subfamily of C. Remember that an $(\mathcal{R},M)$-dense family is an adequate cover.

<u>13.2. Lemma</u>. If C is an $(\mathcal{R},M)$-adequate cover, in particular if it is a dense family, then

$$\bar{M}(f) = \sup\{M(fK)\} , \quad f: X \to \overline{\mathbb{R}}_+ ,$$

where K ranges over all finite unions of sets in C.

<u>Proof</u>. There is a σ-finite set A such that $\bar{M}(f) = M(fA)$, and then there is an increasing sequence (K_n) of finite unions of sets in C converging a.e. to A. From the S-continuity of M,
$\bar{M}(f) = M(fA) = \sup M(fK_n)$.#

The following formulae follow directly from the lemma and from 8.7 and 8.4. For $f: X \to \overline{\mathbb{R}}_+$

13.3. $$\bar{M}(f) = \sup\{M(fK): K \in \mathfrak{K}(\mathcal{R}^S)\} , \quad \text{and}$$

13.4. $$\bar{M}(f) = \sup\{M(f\varphi): \varphi \in \mathcal{R}_+, \varphi \leq 1\} .$$

From 13.4, a set or function f on X is $\bar{M}$-negligible if and only if fK is M-negligible for all dominated sets K. For this reason the $\bar{M}$-negligible functions and sets are called <u>locally</u> M-<u>negligible</u>, and one talks about properties true <u>locally almost everywhere</u> (loc. a.e.).

<u>13.5. Corollary</u>. $(\mathcal{R},\bar{\bar{M}}) = (\mathcal{R},\bar{M})$; that is, $(\mathcal{R},\bar{M})$ is essential.

<u>Proof</u>. For $f: X \to \overline{\mathbb{R}}_+$, $\bar{\bar{M}}(f) = \sup\{\bar{M}(f\varphi): 1 \geq \varphi \in \mathcal{R}_+\}$
$= \sup\{M(f\varphi): 1 \geq \varphi \in \mathcal{R}_+\} = \bar{M}(f)$.#

We summarize:

<u>13.6. Proposition</u>. If $(\mathcal{R},M)$ is a (weak, strong) upper gauge or an upper integral, then so is $(\mathcal{R},\overline{M})$, and the latter is essential.

$(\mathcal{R},\overline{M})$ is smaller than $(\mathcal{R},M)$ and coincides with $(\mathcal{R},M)$ on all functions of σ-finite carrier, in particular on the $(\mathcal{R},M)$-integrable functions. If $(\mathcal{R},M)$ is B-continuous then so is $(\mathcal{R},\overline{M})$. Finally, if $(\mathcal{R},M)$ is an upper gauge and $1 \leq p < \infty$, then $\overline{M}_p = (M_p)^-$.

<u>Proof</u>. Only the last two statements have not yet been shown. Assume first that $(\mathcal{R},M)$ is B-continuous and let $\mathcal{R}_+ \supset \Phi \uparrow^B h$. If $a \in$ is such that $\overline{M}(h) > a$, then there is a $\psi \in \mathcal{R}_+$ such that $\psi \leq 1$ and $M(h\psi) > a$ (10.4). Since the functions $\varphi\psi$ $(\varphi \in \Phi)$ are in $\overline{\mathcal{R}}_+ \subset \mathcal{R}_\uparrow^B$ (5.6), it follows from 8.14 that

$$\sup \overline{M}(\Phi) \geq \sup\{M(\varphi\psi) : \varphi \in \Phi\} = M(h\psi) > a,$$

so that $\sup \overline{M}(\Phi) = \overline{M}(h)$: $(\mathcal{R},\overline{M})$ is B-continuous.

The last statement follows from 13.1 and the observation that the σ-finite sets for $(\mathcal{R},M)$ coincide with those for $(\mathcal{R},M_p)$.#

The following result is a direct consequence of this proof.

<u>13.7. Corollary</u>. If $(\mathcal{R},M)$ is B-continuous, then

$$\overline{M}(h) = M(h) \ , \quad \text{for } h \in \mathcal{R}^B_{\uparrow +} \ .\#$$

If $(\mathcal{R},m)$ is a positive $*$-measure on X, then $(\mathcal{R},\overline{m^*})$ is called the <u>standard essential upper integral</u> associated with m. The $(\mathcal{R},\overline{m^*_p})$ are the <u>standard essential</u> p-<u>norms</u> $(1 \leq p < \infty)$.

B. Supplements

13.8. $(\mathcal{R},M)$ and $(\mathcal{R},\bar{M})$ are equivalent upper gauges, and are essentially equal (7C); hence $L^1_E(\mathcal{R},M) = L^1_E(\mathcal{R},\bar{M})$.

13.9*. If $\mathcal{R}$ is countably generated, then every essential upper gauge $(\mathcal{R},M)$ is σ-finite.

13.10*. The map $m \to \bar{m}^*$ from $\bar{M}^*_{\ddagger}(\mathcal{R})$ to the set of upper integrals on $(X,\mathcal{R})$ is positively homogeneous, additive, and preserves suprema. If $m,n \in M^*_{\ddagger}(\mathcal{R})$, then $m \ll n$ if and only if an $\bar{n}^*$-negligible set is $\bar{m}^*$-negligible (cf. 6.7).

13.11. (Riesz' representation theorem). Let m be an F-valued Radon measure of finite variation on the locally compact space X. There is then an F-valued σ-additive content m' of finite variation on the δ-ring of dominated Borel sets, $\mathcal{C} = \mathcal{C}(C_{oo}(X)^B)$, which has the following properties:

(i) it is inner regular: for $A \in \mathcal{C}$, $m'(A) = \lim(m'(K))$, where K ranges over the compact subsets of A and the limit is taken as K increases.

(ii) it is outer regular: for $A \in \mathcal{C}$, $m'(A) = \lim(m'(U))$, where U ranges over the open sets containing A and the limit is taken as U decreases.

(iii) If $\tilde{m}'$ denotes the extension by linearity to $\mathcal{S}(\mathcal{C})$, then $\tilde{\bar{m}}'^{S} = \vec{m}^B$; in particular, an $\vec{m}^B$-integrable function f is $\tilde{\bar{m}}'$-integrable and $\int f dm = \int f d\tilde{m}'$.

This theorem plays an important role in the standard treatment of integration theory. In Daniell's theory it is but an observation. A similar result can be formulated if m is not of finite variation but weakly compact instead (11.15).

13.12*. A measure m of finite semivariation on $\mathcal{R}$ is said to be bounded if

$$|||m||| = \sup\{\,||m||(\varphi):\ 1 \geq \varphi \in \mathcal{R}_+\} < \infty,$$

and totally bounded if $\mathcal{R}$ contains the constants. A totally bounded measure is bounded; and $|||\cdot|||$ is a norm on the vector space $M_b(\mathcal{R};F)$ of F-valued bounded measures on $\mathcal{R}$, under which $M_b(\mathcal{R};F)$ is a Banach space. A scalar *-measure m of finite variation is bounded if and only if $\overline{|m|}^*$ is finite (13A), and in this instance $|||m||| = \overline{|m|}^*(1)$.

13.13. A positive *-measure m on $\mathcal{R}$ is said to be σ-bounded if there is a sequence (φ_n) in $\mathcal{R}$, $0 \leq \varphi_n \leq 1$, such that $m(\varphi) = \lim(m(\varphi\varphi_n))$ for all $\varphi \in \mathcal{R}$. This holds if and only if $\overline{|m|}^*$ is σ-finite. If $1 \in \mathcal{R}^S_{\uparrow}$, then any measure in $M^S_+(\mathcal{R})$ is σ-bounded.

LITERATURE: [5], ch. IV.

14. Smooth upper gauges

In this section, $(\mathcal{R},M)$ will be a fixed strong upper gauge on X.

A. Definition. The associated smooth upper gauge, $(\mathcal{R},M^*)$, is defined by

$$M^*(f) = \inf\{M(h): f \le h \in \mathcal{L}^1(\mathcal{R},M)\}, \quad \text{for} \quad f: X \to \overline{\mathbb{R}}_+,$$

where $M^*(f) = \infty$ if f fails to be majorized by an $(\mathcal{R},M)$-integrable function. $(\mathcal{R},M)$ is said to be smooth if it equals $(\mathcal{R},M^*)$.#

From the definition of m^* and 9.3, $(\mathcal{R},m^*)$ is seen to be smooth for any positive *-measure m on $\mathcal{R}$ (* = S or B).

14.1. Lemma. For any $f \in \overline{\mathbb{R}}_+^X$ with $M^*(f) < \infty$, there is an $f' \in \mathcal{L}^1(\mathcal{R},M)$ such that

$$f \le f' \quad \text{and} \quad M^*(f) = M(f'). \tag{1}$$

Furthermore, any two functions f', f_1' satisfying (1) coincide M-almost everywhere.

Proof. Let (h_n) be a sequence in $\mathcal{L}^1(\mathcal{R},M)$ such that $h_n \ge f$ for $n = 1, 2, \ldots$ and $\inf M(h_n) = M^*(f)$. Clearly, $f' = \inf\{f_n\}$ meets the description in the statement. If f_1' is another function satisfying (1), then $0 \le g = f' \wedge f_1' \le f'$ does so, too. From (SUG) and $M(g) = M(f')$, $g \doteq f'$ (M). Similarly, $g \doteq f_1'$ (M), and so $f' \doteq f_1'$ (M), as desired.#

We call a function f' satisfying (1) a least $(\mathcal{R},M)$-integrable majorant for f.

We shall now show that $(\mathcal{R},M^*)$ is a strong upper gauge. Positive homogeneity and isotony of M^* are evident. For the subadditivity, let

$f,g \in \overline{\mathbb{R}}_+^X$. If $M^*(f) = \infty$ or $M^*(g) = \infty$ there is nothing to prove. If both numbers are finite then there are least integrable majorants f', g' for f, g, respectively. Clearly $f' + g'$ majorizes $f + g$, and so $M^*(f+g) \leq M(f+g) \leq M(f) + M(g) = M^*(f) + M^*(g)$.

For the S-continuity, let (f_n) be an increasing sequence in $\overline{\mathbb{R}}_+^X$ with supremum f. Assume that $\sup M^*(f_n) < \infty$, as there is nothing to prove otherwise. There is a least integrable majorant f'_n for f_n, for $n = 1, 2, \ldots$. Replacing, if necessary, f'_n by $\inf\{f'_m : m \geq n\}$ we may assume that (f'_n) is increasing. $f' = \sup\{f'_n\}$ majorizes f and is integrable, since $\sup M(f'_n) < \infty$ (8.9). Hence $\sup M^*(f_n) = \sup M(f'_n) = M(f') \geq M^*(f)$: S-continuity is established.

M^* is obviously bigger than M and coincides with M on $\mathcal{L}^1(\mathcal{R},M)$. Hence (UG) holds, and $L^1_E(\mathcal{R},M) = L^1_E(\mathcal{R},M^*)$ (7.15), which in turn implies (SUG). The first half of the next proposition has now been proved.

<u>14.2. Proposition</u>. (i) If $(\mathcal{R},M)$ is a strong upper gauge then so is $(\mathcal{R},M^*)$. M^* is bigger than M and coincides with it on $\mathcal{L}^1(\mathcal{R},M)$. $(\mathcal{R},M)$ and $(\mathcal{R},M^*)$ have the same integrable functions and sets.

(ii) If $(\mathcal{R},M)$ is an upper integral then so is $(\mathcal{R},M^*)$; and if $(\mathcal{R},M)$ is B-continuous, then so is $(\mathcal{R},M^*)$.

(iii) $(\mathcal{R},M^*)$ is smooth; and if $(\mathcal{R},M)$ is smooth so is $(\mathcal{R},\overline{M})$. Furthermore, $(\mathcal{R}, (M_p)^*) = (\mathcal{R}, (M^*)_p)$ for $1 \leq p < \infty$.

<u>Proof</u>. The last statement follows from the characterization 12.8 of the p-integrable functions. The fact that the $(\mathcal{R},M)$-integrable and the $(\mathcal{R},M^*)$-integrable functions coincide leads to the smoothness $(M^* = M^{**})$ of $(\mathcal{R},M^*)$.

Next, suppose $(\mathcal{R},M)$ is B-continuous and let $\mathcal{R}_+ \supset \Phi\uparrow^B h$. If $M(h) < \infty$ then h is $(\mathcal{R},M)$-integrable (8.15), and so $M^*(h) = M(h) = \sup M(\Phi) = \sup M^*(\Phi)$. If $M(h) = \infty$ then $\sup M^*(\Phi) = \sup M(\Phi) = \infty$, as well. Hence $(\mathcal{R},M^*)$ is B-continuous.

Finally, assume $(\mathcal{R},M)$ is smooth, and let $f: X \to \overline{\mathbb{R}}_+$ with $\overline{M}(f) < \infty$. For $K \in \mathscr{C}(\mathcal{R}^S)$ let g^K be a least $(\mathcal{R},M)$-integrable majorant of Kf. The corresponding classes $\dot{g}^K$ in $L^1(\mathcal{R},M)$ are increasingly directed and satisfy $\sup\{M(\dot{g}^K): K \in \mathscr{C}(\mathcal{R}^S)\} = \overline{M}(f) < \infty$; and so they have a supremum $\dot{g} \in L^1(\mathcal{R},M)$ (8.13). Let $g \in \dot{g}$. We have $g \dot{\geq} g^K$ (M) for all $K \in \mathscr{C}(\mathcal{R}^S)$ and therefore $g \dot{\geq} f$ $(\overline{M})$. Hence $g \vee f$ is $(\mathcal{R},\overline{M})$-integrable, majorizes f, and satisfies $\overline{M}(g \vee f) = M(g) = \overline{M}(f)$: $(\mathcal{R},\overline{M})$ is smooth.#

<u>14.3. Corollary</u>. Let $(\mathcal{R},M)$ be a smooth upper integral. Then

$$M(f+h) = M(f) + M(h) \quad \text{for} \quad f: X \to \overline{\mathbb{R}}_+ ,\ h \in \mathcal{L}^1_+(\mathcal{R},M).$$

If f' is a least integrable majorant of f then $f'+h$ is a least integrable majorant of $f+h$, and vice versa.

<u>Proof</u>. If $M(f) = \infty$, this is obvious. If $M(f) < \infty$, there is a least integrable majorant, g, for $f+h$. Then $g-h$ majorizes f, and the additivity of M on $\mathcal{L}^1_+$ yields

$$M(f+h) = M(g) = M(g-h) + M(h) \geq M(f) + M(h) \geq M(f+h);$$

hence $g-h$ is a least integrable majorant for f. If f' is any other such least integrable majorant, then $f'+h \doteq g$ (M).#

B. Supplements.

14.4*. $(\mathcal{R},M)$ is equivalent with $(\mathcal{R},M^*)$. The $(\mathcal{R},M)$-dense, the $(\mathcal{R},M^*)$-dense, and the $(\mathcal{R},M_p)$-dense families coincide $(1 \leq p < \infty)$.

14.5. Suppose $\mathcal{L}$ is a full integration lattice and $f \to \int fdm$ a positive measure on $\mathcal{L}$ such that the monotone convergence theorem holds: whenever (φ_n) is an increasing sequence in $\mathcal{L}_+$ with $\sup \int \varphi_n dm < \infty$, then $\sup(\varphi_n) \in \mathcal{L}$ and $\int \sup(\varphi_n)dm = \sup \int \varphi_n dm$. Then there is an upper integral $(\mathcal{L},M)$ for $\int \cdot dm$ such that $\mathcal{L} = \mathcal{L}^1(\mathcal{L},M)$.

(This shows that an integration theory in which the monotone convergence holds comes from an upper gauge: Daniell's theory is of a sufficiently general nature.)

15. The upper integral AM

A. Definition. Consider a fixed smooth upper integral $(\mathcal{R},M)$ on X and a fixed subset A of X. From these data, the upper integral $(\mathcal{R},AM)$ is defined by

$$(AM)(f) = M(Af) \ , \quad \text{for} \quad f\colon X \to \overline{\mathbb{R}}_+ \ .$$

15.1. Proposition. $(\mathcal{R},AM)$ is a smooth upper integral. If $(\mathcal{R},M)$ is B-continuous and essential, then $(\mathcal{R},AM)$ is B-continuous.

Proof. It is evident that AM is positively homogeneous, increasing, subadditive, and S-continuous. Let us show that it is additive on $\mathcal{R}_+$. Let φ,ψ be any two functions in $\mathcal{R}_0$, vanishing off the dominated set K. There is a least $(\mathcal{R},M)$-integrable majorant for AK, which may be chosen to be a set B. We claim that $B\varphi$ is a least integrable majorant for $A\varphi$. Clearly $A\varphi \leq B\varphi$. Let f be a least integrable majorant of $A\varphi$, smaller than $B\varphi$. If $M(B\varphi - f) > 0$, there is a number $a > 0$ and an integrable set L with $M(L) > 0$ such that $B\varphi > f + a$, and thus $B\varphi > A\varphi + a$ on L (8.6). The set $B - L$ is then a majorant of A, of measure $M(B - L) = M(B) - M(L) < M(AK)$; this is a contradiction. Similarly, $B\psi$ is a least integrable majorant for $A\psi$, and so

$$(AM)(\varphi+\psi) = M(A(\varphi+\psi)) = M(B(\varphi+\psi)) = M(B\varphi) + M(B\psi) =$$
$$M(A\varphi) + M(A\psi) = (AM)(\varphi) + (AM)(\psi).$$

More generally, if φ and ψ are in $\mathcal{R}_+$, they can be approximated from below by increasing sequences of dominated functions in $\mathcal{R}_+$ (8.4); the additivity of AM then follows from the S-continuity of M.

To see that AM is smooth, let $f: X \to \overline{\mathbb{R}}_+$ be such that $(AM)(f) < \infty$. There is a least $(\mathcal{R},M)$-integrable majorant f' of Af, which, since AM is smaller than M, is $(\mathcal{R},AM)$-integrable (7.15). Now,
$(AM)(f') - (AM)(f) = M(Af') - M(Af) \leq M(f') - M(Af) = 0$, and hence,
$(AM)(f') = (AM)(f)$.

The last statement in 15.1 is a consequence of the following lemma.#

<u>15.2. Lemma</u>. If $(\mathcal{R},M)$ is a B-continuous upper gauge, and g a positive numerical function, then

$$g\overline{M}: f \to \overline{M}(gf) , \quad f: X \to \overline{\mathbb{R}}_+$$

is B-continuous on $\mathcal{R}^B_{\uparrow+}$.

<u>Proof</u>. Let $\mathcal{R}_+ \supset \Phi \uparrow^B h$. For every integer $n \geq 1$ and $1 \geq \psi \in \mathcal{R}_+$,

$$\mathcal{R}^B_{\uparrow+} \supset \{(\varphi \wedge n)\psi: \varphi \in \Phi\} \uparrow^B (h \wedge n)\psi,$$

and so, $\lim\{M((h \wedge n)\psi - (\varphi \wedge n)\psi): \varphi \in \Phi\} = 0$ (8.15). Therefore,

$$\lim\{M((g \wedge k)\{(h \wedge n)\psi - (\varphi \wedge n)\psi\}): \varphi \in \Phi\} = 0$$

for all $k = 1, 2, \ldots$, and so

$$\sup\{M((g \wedge k)(\varphi \wedge n)\psi): \varphi \in \Phi\} = M((g \wedge k)(h \wedge n)\psi).$$

Taking the supremum over all integers k,n and all ψ yields

$$\sup\{\overline{M}(g\varphi): \varphi \in \Phi\} = \overline{M}(gh) , \quad \text{as desired.}$$

The case that $\mathcal{R}^B_{\uparrow} \supset H \uparrow^B h$ can be reduced to this one as in 8.15.#

The proof of 15.1 has the following corollary. If $\varphi, \psi \in \mathcal{R}_+$ and $\tilde{\varphi}, \tilde{\psi}$ are least integrable majorants of $A\varphi$, $A\psi$, respectively, then $\tilde{\varphi} + \tilde{\psi}$ is a least integrable majorant for $A(\varphi + \psi)$.

Let $m: \mathcal{R} \to F$ be a Banach-valued measure majorized by M. We may use 15.3 to define a measure $Am: \mathcal{R} \to F$ by

$$(Am)(\varphi) = \int \tilde{\varphi}\, dm \qquad \left(\varphi \in \mathcal{R}_+,\ \tilde{\varphi} \text{ a least integrable majorant of } A\varphi \right)$$

and linearity. From 15.3, Am is well defined and additive, and it is clearly majorized by AM. It is easy to see that Am has variation $|Am| = A|m| \leq |m|$. Hence if m is B-continuous so is Am. If $f \in \mathcal{L}^1_E(\mathcal{R}, M)$ then $f \in \mathcal{L}^1_E(\mathcal{R}, AM)$ and

$$\int f d(Am) = \int \tilde{f} dm$$

where $\tilde{f}$ is a least integrable majorant of Af.

B. Thick sets

Definition. The set A is called $(\mathcal{R},M)$-thick if $AM = M$ on $\mathcal{R}_+$.

15.3. Lemma. The following are equivalent:

(a) A is $(\mathcal{R},M)$-thick.

(b) $M(Af) = M(f)$ for $f \in \mathcal{L}^1_+(\mathcal{R},M)$.

(c) $M(AK) = M(K)$ for $K \in \mathscr{C}(\mathcal{R},M)$.

(d) Every $(\mathcal{R},M)$-integrable set disjoint from A is M-negligible.

(e) Every set in an $(\mathcal{R},M)$-dense family F is M-negligible if it is disjoint from A.

Proof. (a) $\Rightarrow$ (b): the equation $M(Af) = M(f)$ is true for $f \in \mathcal{R}_+$, and both sides depend continuously in M-mean on f. (b) $\Rightarrow$ (c) $\Rightarrow$ (d) $\Rightarrow$ (e) are obvious. (e) $\Rightarrow$ (a): if (a) were not true then there would be a $\varphi \in \mathcal{R}_+$ such that $M(\varphi) > M(\varphi A)$. Let f be a least integrable majorant of φA. Since $M(\varphi - f) > 0$ there is an $a > 0$ and a $K \in \mathscr{C}(\mathcal{R},M)$, with $M(K) > 0$, such that $\varphi > f + a \geq \varphi A + a$ on K. We may replace K by a set in F with the same properties. Clearly, K is disjoint from A. This contradicts (e).#

15.4. Proposition. Let $(\mathcal{R},m)$ be a $*$-measure on X, of finite scalar variation $|m|$. The subset A of X is $(\mathcal{R},\overline{m}^*)$-thick if and only if m is $*$-continuous on A; i.e., whenever $\mathcal{R}_+ \supset \Phi \uparrow^* \psi \in \mathcal{R}$ on A, then $\lim\{m(\varphi): \varphi \in \Phi\} = m(\psi)$.

Proof. We may assume that m is positive. First suppose A is $(\mathcal{R},\overline{m}^*)$-thick, and let $\mathcal{R}_+ \supset \Phi \downarrow^* 0$ on A. Then $k = \inf \Phi \in \mathcal{R}^*_\downarrow$ is $(\mathcal{R},m^*)$-integrable and $\overline{m}^*(k) = \overline{m}^*(kA) = 0$, since $k = 0$ on A.

Conversely, assume that m is *-continuous on A, and let K be a set in the $(\mathcal{R},\overline{m}^*)$-dense family $\mathcal{K}^*(\mathcal{R})$, disjoint from A. There is a family $\Phi \subset \mathcal{R}_+$ such that $\Phi\downarrow^* K$. Then $\Phi\downarrow^* 0$ on A, and so $\overline{m}^*(K) = \inf m(\Phi) = 0$.#

C. The measure induced on A

Henceforth, A is again an arbitrary subset of X, $(\mathcal{R},M)$ a smooth upper integral, and m a Banach valued measure, majorized by M.

Let $\mathcal{S} = \mathcal{R}|A$ denote the integration lattice of restrictions $\varphi|A$ of functions φ in $\mathcal{R}$ to A. Also, if f is a function on A having values in either $\overline{\mathbb{R}}_+$ or a Banach space E, let f_0 denote that extension which is zero on X - A. The equation

$$M_A(f) = (AM)(f_0) \quad (= M(Af_0)) , \quad f\colon A \to \overline{\mathbb{R}}_+ ,$$

defines an upper integral $(\mathcal{S},M_A)$ on A, having the property that

$$M_A(f|A) = M(Af) \quad \text{for} \quad f\colon X \to \overline{\mathbb{R}}_+ .$$

15.5. Proposition. $(\mathcal{S},M_A)$ is a smooth upper integral on A.

The $(\mathcal{S},M_{Ap})$-integrable functions are exactly the restrictions to A of the $(\mathcal{R},AM_p)$-integrable functions. In particular, if A is thick then

$$L_E^p(\mathcal{S},M_A) = L_E^p(\mathcal{R},AM) = L_E^p(\mathcal{R},M) , \quad \text{for} \quad 1 \le p < \infty .$$

Proof. We prove only the first part of the second statement. Due to 12.8, it suffices to consider the case $p = 1$. If $f \in \mathcal{L}_E^1(\mathcal{S},M_A)$, and if (φ_n) is a sequence in $\mathcal{R}\otimes E$ such that $f = \sum \varphi_n|A$ M_A-a.e. and in mean, then there is a sequence (K_n) of integrable sets such that $K_n\|\varphi_n\|$ is a least integrable majorant for $A\|\varphi_n\|$. Evidently f is a.e. the restriction to A of the integrable function $\sum \varphi_n K_n$.#

The <u>measure induced on</u> A is now defined by

$$m_A(\varphi) = \int \varphi_0' dm \ , \quad \text{for} \quad \varphi \in \mathcal{S},$$

where φ_0' denotes a least integrable majorant of φ_0. It is left to the reader to convince himself that this does not depend on the choice of φ_0'; that m_A is a *-measure on $(\mathcal{S},A)$ if m is (* = S or B), and that it is majorized by $(\mathcal{S},M_A)$.

D. The case of integrable A

Henceforth, $(\mathcal{R},M)$ is a fixed <u>weak</u> upper gauge on X and A is an $(\mathcal{R},M)$-integrable set. By the equation

$$(AM)(f) = M(Af) \ , \quad \text{for} \quad f: X \to \overline{\mathbb{R}_+} \ ,$$

an upper S-norm AM is defined having the following properties.

<u>15.6. Proposition</u>. A function f is $(\mathcal{R},AM)$-integrable if and only if fA is $(\mathcal{R},M)$-integrable; and the map $f \to fA$ factors to an injection, $f^{\cdot} \to (fA)^{\cdot}$, of $L^1(\mathcal{R},AM)$ into $L^1(\mathcal{R},M)$.

$(\mathcal{R},AM)$ is a weak upper gauge. If $(\mathcal{R},M)$ is an upper gauge, strong, B-continuous, essential, or an upper integral, then so is $(\mathcal{R},AM)$; moreover, $AM_p = (AM)_p$ for $1 \leq p < \infty$.

If $U: \mathcal{R} \otimes E \to G$ is a measure $(E = \mathbb{R})$ or a linear map majorized by M, then

$$(AU)(\varphi) = \int \varphi A dU , \quad \varphi \in \mathcal{R}_E ,$$

defines a measure AU majorized by AM; and for all $f \in \mathcal{L}^1_E(\mathcal{R}, AM)$,

$$\int f d(AU) = \int AfdU.$$

Proof. If $f \in \mathcal{L}^1_E(\mathcal{R}, AM)$ then there is a sequence (φ_n) in $\mathcal{R}_E$ so that $M(A(f - \varphi_n)) \to 0$. As $A\varphi_n \in \mathcal{L}^1_E(\mathcal{R}, M)$ (8.5), Af is $(\mathcal{R}, M)$-integrable. Conversely, if $Af \in \mathcal{L}^1_E(\mathcal{R}, M)$ and (φ_n) is a sequence in $\mathcal{R}_E$ converging in M-mean to Af, then $(AM)(f - \varphi_n) \leq M(Af - \varphi_n) \to 0$ and f is $(\mathcal{R}, AM)$-integrable.

AM is clearly an upper S-norm. If (φ_n) is an increasing sequence in $\mathcal{R}_+$ whose supremum f is majorized by an $(\mathcal{R}, AM)$-integrable function h, then $A\varphi_n \uparrow Af \leq Ah \in \mathcal{L}^1(\mathcal{R}, M)$, and by 8.1, $M(Af - A\varphi_n) = (AM)(f - \varphi_n) \to 0$. Hence $(\mathcal{R}, AM)$ is a weak upper gauge. The remaining statements of the second paragraph are verified along the same lines.

The last paragraph follows immediately from the first.#

Exactly as in 15C, we define a weak upper gauge $(\mathcal{S}, M_A)$ on $\mathcal{S} = \mathcal{R}|A$ by

$$M_A(f) = (AM)(f_0) = M(f_0) , \quad \text{for } f: A \to \overline{\mathbb{R}}_+ .$$

The following results are easily established by the reader.

15.7. Proposition. $(\mathcal{S}, M_A)$ is a weak upper gauge. If $(\mathcal{R}, M)$ is an upper gauge, strong, B-continuous, smooth, essential, or an upper integral, then so is $(\mathcal{S}, M_A)$; and $(M_A)_p = (M_p)_A$ for $1 \leq p < \infty$.

A function f on A is $(\mathcal{S},M_A)$-integrable if and only if it is the restriction to A of an $(\mathcal{R},M)$-integrable function.

If $U: \mathcal{R} \otimes E \to G$ is a linear map majorized by M, then

$$U_A(\varphi) = \int \varphi_0 dU \qquad (\varphi \in \mathcal{S})$$

defines a linear map $U_A: \mathcal{S} \otimes E \to G$, the "map induced on A;" and for all $f \in \mathcal{L}^1_E(\mathcal{S},M_A)$, f_0 is $(\mathcal{R},M)$-integrable and

$$\int f dU_A = \int f_0 dU.$$

<u>15.8. Application</u>. Recall that two upper gauges, $(\mathcal{R},M)$ and $(\mathcal{R},N)$, are <u>equivalent</u> if they have the same locally negligible sets (8.18); and they have then the same dominated integrable sets.

> A σ-finite strong upper gauge $(\mathcal{R},M)$ is equivalent to an upper integral $(\mathcal{R},N)$. Furthermore, if a measure $n \leq M$ is given such that $\int K dn = 0$ implies $M(K) = 0$ for all $K \in \mathcal{C}(\mathcal{R},M)$, then $(\mathcal{R},N)$ can be chosen to be an upper integral for n.

<u>Proof</u>. Let $\{K_k^\bullet : k \in \mathbb{N}\}$ be a maximal collection of non-zero classes of dominated integrable sets that are mutually disjoint, and choose a dominated set K_k in each of the $K_k^\bullet$. Suppose that for every integer k an upper integral, $(\mathcal{R},N_k)$, equivalent to $(\mathcal{R},K_k M)$ can be found. Then $\sum 2^{-k} N_k$ will clearly be an upper integral equivalent to $(\mathcal{R},M)$: We may assume that $(\mathcal{R},M)$ is finite and that $\mathcal{R}$ contains the constants. By replacing $(\mathcal{R},M)$ by $(\mathcal{R},M^*)$, it may be assumed further that $(\mathcal{R},M)$ is smooth.

Let $\mathcal{M}$ denote the set of positive scalar measures on $\mathcal{L}^1 = \mathcal{L}^1(\mathcal{R},M)$ that are majorized by M. For $m \in \mathcal{M}$, let $B^\bullet(m)$ be the supremum in L^1 of all sets $B \in \mathcal{C}(\mathcal{R},M)$ such that $\int B dm = 0$, and order $\mathcal{M}$ by putting

$$m > n \quad \text{if} \quad B^{\bullet}(m) < B^{\bullet}(n).$$

Setting $r = \inf\{M(B^{\bullet}(m)) : m \in \mathcal{M}\}$, let (m_k) be an increasing sequence in $\mathcal{M}$ such that $r = \inf M(B^{\bullet}(m_k))$. The measure $n = \sum 2^{-k} m_k$ clearly belongs to $\mathcal{M}$ and satisfies $M(B^{\bullet}(n)) = r$.

Let us show that $r = 0$, and therefore $B^{\bullet}(n) = 0$. If this is not so, then there is a non-zero linear functional, m, on L^1 majorized by the upper gauge $B(n)M$; this is due to the theorem of Hahn-Banach. m can be identified with an S-measure on $\mathcal{L}^1$, and so has finite variation (6.5). Either m_+ or m_- is nonzero and is majorized by $B(n)M$: we may assume that m is positive. Evidently, $B^{\bullet}(m/2 + n/2) < B^{\bullet}(n)$, which is a contradiction.

Finally, let N be the standard upper S-integral associated with $(\mathcal{L}^1, n)$. Then $(\mathcal{R}, N)$ is equivalent to $(\mathcal{R}, M)$. Since $n \leq M$ and since M is smooth, an M-negligible set is clearly N-negligible. Conversely, if K is an N-negligible set, then K lies in an N-negligible $\mathcal{L}^1$-Baire set, which, itself, belongs to $\mathcal{L}^1$; due to the construction of n, K is M-negligible.

E. Supplement

15.9. If $(\mathcal{R}, M)$ is any smooth upper gauge then a positive numerical function g on X is called locally majorizable if, for every dominated set $K \subset X$, $M(gK) < \infty$. For such a g,

$$gM : f \to M(gf) \quad (\text{for all} \quad f : X \to \overline{\mathbb{R}_+})$$

defines a smooth upper gauge $(\mathcal{R}_0, gM)$, which is B-continuous if $(\mathcal{R}, M)$ is B-continuous and essential. (Hint: For every countable union A of $(\mathcal{R}, M)$-integrable sets, there exists a function g' such that g'K is a least integrable majorant for gK whenever $A \supset K \in \mathfrak{G}(\mathcal{R}, M)$.)

16. Strictly localizable upper gauges

It is henceforth assumed that the upper gauges $(\mathcal{R},M)$ considered do not vanish on $\mathcal{R}_+$ and are strong.

A. Definition. Let $(\mathcal{R},M)$ be a strong upper gauge on X. An $(\mathcal{R},M)$-adequate partition of X is a collection P of mutually disjoint, integrable and non-negligible sets which constitutes an $(\mathcal{R},M)$-adequate cover; in other words, every integrable set can be covered a.e. by a countable subfamily of P.

$(\mathcal{R},M)$ is said to be strictly localizable if it has an adequate partition; and the family, possibly void, of $(\mathcal{R},M)$-adequate partitions is denoted by $\mathcal{P}(\mathcal{R},M)$.#

This notion is very important as it means that the space X can be decomposed into "the direct sum of finite spaces" (see 16.10).

Since the integrable and negligible sets are the same for $(\mathcal{R},M)$ and for $(\mathcal{R},M_p)$ (12.8), an upper gauge $(\mathcal{R},M)$ is strictly localizable if and only if $(\mathcal{R},M_p)$ is $(1 \leq p < \infty)$. For the same reason, $\mathcal{P}(\mathcal{R},M) = \mathcal{P}(\mathcal{R},M^*)$. If $(\mathcal{R},M)$ is strictly localizable, and if $(\mathcal{R},N)$ is an upper gauge with $N \leq M$, then $(\mathcal{R},N)$ is strictly localizable. In particular, $(\mathcal{R},\overline{M})$ is then strictly localizable.

B. Examples

16.1. Proposition. A σ-finite upper gauge $(\mathcal{R},M)$ is strictly localizable. There is a countable adequate partition consisting of dominated Baire sets.

Proof. Let (A_n) be a sequence of integrable sets with union X. For every integer $n \geq 1$ cover A_n a.e. by countably many dominated Baire sets B_n^k, $1 \leq k < \infty$ (8.7) and put

$$B'_n = \cup \{ B^k_m : 1 \leq m, k \leq n \} \quad \text{and} \quad C_n = B'_n - B'_{n-1} .$$

The non-negligible sets amongst the C_n clearly constitute a countable adequate partition of dominated Baire sets, since $X - \cup C_n$ is negligible.#

Definition. If P and Q are $(\mathcal{R},M)$-adequate partitions then Q is said to be a refinement of P if every set belonging to Q is entirely contained in a set belonging to P.

16.2. Lemma. If P is an $(\mathcal{R},M)$-adequate partition and F is an $(\mathcal{R},M)$-dense family, then F contains a refinement of P.

Proof. Write every set $K \in P$ as the disjoint union of countably many non-negligible sets $K_n \in F$ and a negligible set (8.21). Clearly $\{K_n : K \in P, n \in \mathbb{N}\}$ is an adequate partition, is a refinement of P, and is contained in F.#

16.3. Lemma. Suppose $(\mathcal{R},M)$ is a strong upper gauge and P is any family of mutually disjoint non-negligible sets. If A is a σ-finite set then there are at most countably many sets K in P with $M(A \cap K) > 0$.

Proof. It may be assumed that A is integrable. Let G denote the family of finite unions of sets of the form $A \cap K$, $K \in P$. G is increasingly directed with $\sup M(G) \leq M(A) < \infty$. There is a sequence (K_n) in P such that $\sup M(G) = M(\cup A \cap K_n)$. For every set $K \in P$ which does not belong to $\{K_n\}$, we have $M(\cup A \cap K_n) = M(A \cap K \cup (\cup A \cap K_n))$. From (SUG), $M(A \cap K) = 0$.#

Consider now a strong and B-continuous upper gauge $(\mathcal{R},M)$ on X.

Definition. The support of M on the set $K \subset X$ is the complement $\underline{K} = X - U_K$ of the set

$$U_K = \cup \{U \in \mathcal{U}^B(\mathfrak{R}): M(U \cap K) = 0\}.$$

$\underline{X} = X - \cup \{U \in \mathcal{U}^B: M(U) = 0\}$ is called the support of M. The support of a B-continuous measure m is the support of m^B.

U_K is in $\mathcal{U}^B(\mathfrak{R})$, and since KM is B-continuous (15.2), $M(K \cap U_K) = 0$. U_K is therefore the maximal set in $\mathcal{U}^B$ that intersects K in a negligible set. The support of $(\mathfrak{R},M)$ and of $(\mathfrak{R},\bar{M})$ coincide (13.7).

16.4. Theorem. A strong and B-continuous upper gauge is strictly localizable. There exists an adequate partition consisting of dominated sets.

Proof. Let P be a maximal family of mutually disjoint and non-negligible sets of the form $\underline{K}$, each K being dominated and integrable. The sets $\underline{K} \in P$ are also dominated and integrable (8.17). We want to show that P is an adequate partition. If U is an integrable set in $\mathcal{U}^B$, then either $U \cap \underline{K} = \emptyset$ or $M(U \cap \underline{K}) > 0$ for each set $\underline{K} \in P$. By lemma 16.3, U intersects at most countably many of the sets $\underline{K}$ in P. Hence $U - \cup \{U \cap \underline{K}: \underline{K} \in P\} = L$ is integrable. If L were not negligible, then $\underline{L}$ could be adjoined to P, in contradiction to the maximality of P. Hence L is negligible, and U can be covered adequately a.e. by a countable family in P. As every integrable set A can be covered a.e. by countably many integrable sets in $\mathcal{U}^B(\mathfrak{R})$, the same is true for A: P is an adequate cover.#

16.5. Lemma. Suppose $(\mathcal{R},M)$ is a strong upper gauge, P is an adequate partition, and P' is any family of mutually disjoint, non-negligible, and integrable sets.

Then $\text{card}(P') \leq \text{card}(P) + \aleph_0$. In particular, if P and P' are two infinite adequate partitions, then they have the same cardinality.

Proof. For $K \in P$ let P'_K denote the family of sets in P' that have non-negligible intersection with K. P'_K is at most countable (16.3), and $\cup \{P'_K: K \in P\} = P'$.#

16.6. Proposition. If $\mathcal{R}$ is first-uncountably generated, then every (non-zero) strong upper gauge $(\mathcal{R},M)$ is strictly localizable.

Proof. Let $\{\varphi_a: 0 \leq a < \aleph_1\}$ be a first-uncountable family of functions φ_a in $\mathcal{R}$ which generates $\mathcal{R}$ (1D). For each ordinal b, $0 \leq b < \aleph_1$, let $\mathcal{R}_b$ denote the integration lattice spanned by $\{\varphi_a: 0 \leq a < b\}$. $\mathcal{R}_b$ is countably generated, and hence $(\mathcal{R}_b,\overline{M})$ is σ-finite for each ordinal $b < \aleph_1$; indeed, every $(\mathcal{R}_b,M)$-integrable set is contained in the σ-finite set $\cup\{[\varphi_a \neq 0]: 0 \leq a < b\}$. Hence $(\mathcal{R}_b,\overline{M})$ is strictly localizable for all $b < \aleph_1$ (16.1).

We define now, by induction, $(\mathcal{R}_a,\overline{M})$-adequate partitions P_a (for each ordinal $a < \aleph_1$) such that $P_a \subset P_b$ for $a \leq b$. The definition of P_0 is obvious. If P_a has been defined for all $a < b$, P_b is defined to be a maximal collection of mutually disjoint non-negligible $(\mathcal{R}_b,\overline{M})$-integrable sets containing $\cup\{P_a: a < b\}$. Its existence follows from Zorn's lemma, and it is at most countable (16.3). Let $A \in \mathfrak{F}(\mathcal{R}_b,\overline{M})$, with $\overline{M}(A) > 0$. $A \setminus \cup P_b$ is $(\mathcal{R}_b,\overline{M})$-integrable and of zero measure; otherwise, it could be adjoined to P_b, which is in contradiction to the maximality of P_b. Hence P_b is, indeed, an $(\mathcal{R}_b,\overline{M})$-adequate partition.

We put $P = \bigcup\{P_a : a < \aleph_1\}$ and claim that P is an $(\mathcal{R},\overline{M})$-adequate partition. Indeed, let $A \in \mathfrak{F}(\mathcal{F},\overline{M})$. There exists a sequence (φ_n) in $\mathcal{R} = \bigcup\{\mathcal{R}_a : a < \aleph_1\}$ which converges almost-everywhere and in mean to A. Suppose φ_n belongs to $\mathcal{R}_{a(n)}$ for $0 \leq a(n) < \aleph_1$. Since $b = \sup\{a(n) : n \in \mathbb{N}\}$ is countable, it is smaller than $\aleph_1$. Hence the φ_n all belong to $\mathcal{R}_b$; A is therefore $(\mathcal{R}_b,\overline{M})$-integrable; and so A is $(\mathcal{R},\overline{M})$-adequately covered by P. Hence P is an $(\mathcal{R},\overline{M})$-adequate cover. There is an $(\mathcal{R},\overline{M})$-adequate cover consisting of dominated sets (8.7, 16.2), which is also $(\mathcal{R},M)$-adequate, automatically.#

C. <u>Supplements</u>

16.7. Any two B-continuous upper integrals $(\mathcal{R},M)$ and $(\mathcal{R},N)$ that coincide on $\mathcal{R}_+$ have the same support.

16.8. If $m,n \in M_+^B(\mathcal{R})$, then $\mathrm{supp}(m+n) = \mathrm{supp}(m) \cup \mathrm{supp}(n)$. If $M \subset M_+^B(\mathcal{R})$ and $m = \sup M$, then $\mathrm{supp}(m)$ is the $\mathcal{R}$-closure of $\bigcup\{\mathrm{supp}(n) : n \in M\}$

16.9*. Unless $(\mathcal{R},M)$ is B-continuous and the support of $(\mathcal{R},M)$ is the closure of a finite set $\{x_1,\ldots,x_n\}$ of X, every $(\mathcal{R},M)$-adequate partition has an infinite refinement.

16.10. Let $P \in \mathcal{P}(\mathcal{R},M)$, let $f \in \mathcal{L}_E^1(\mathcal{R},M)$, and let $m: \mathcal{R} \to F$ be a measure majorized by M. Then $f|K$ is $(\mathcal{R}|K,M_K)$-integrable for all $K \in P$ and

$$\int f\,dm = \sum\{\int f|K\,dm_K : K \in P\}.$$

For this behaviour, one also says that $(\mathcal{R},M)$ has the <u>direct sum property</u> if it is strictly localizable [14].

17. Bauer's theory

There is a satisfactory integration theory for linear maps or measures that satisfy both a *-continuity condition, and a condition on the variation or semi-variation. If the *-continuity is not required any more, then there is, of course, no hope to arrive at an integration theory in which Lebesgue's theorem holds. However, by going to the Gelfand-Bauer transform of a measure or linear map--which is B-continuous (4.11)--one may still make use of upper gauges in the investigation of purely additive measures:

A. The integration theories of m and $\hat{m}$

Henceforth $\mathcal{R}$ is an integration lattice with spectrum S, and E,F are fixed Banach spaces.

Recall that with every measure $m: \mathcal{R} \to F$ of finite semivariation there is associated an F-valued Radon measure $\hat{m}: C_{oo}(S) \to F$ on the spectrum $S = S(\mathcal{R})$ of $\mathcal{R}$, its Gelfand-Bauer transform, given by

$$\hat{m}(\hat{\varphi}) = m(\varphi) \qquad , \text{ for } \varphi \in \mathcal{R}_0,$$

and by extension on $C_{oo}(S)$ (5D). The map $m \to \hat{m}$ preserves semivariation and variation. If m has finite variation, then $\hat{m}$ is majorized by the upper integrals $\hat{m}^*$, $\bar{\hat{m}}^*$ ($* =$ S or B), since $\hat{m}$ is B-continuous (4.11). These can be used to investigate m, even if m is purely linear.

17.1. Theorem. (Bauer). Suppose $m: \mathcal{R} \to F$ is a measure of finite variation.

(i) Then m is *-continuous if and only if $\pi(X_0)$ is $\bar{\hat{m}}^*$-thick in the spectrum S of $\mathcal{R}$.

(ii) Via the map $f \to f \circ \pi$, $\mathcal{L}_E^p(S,\bar{\hat{m}}^*)$ is isometrically isomorphic with a dense subspace of $\mathcal{L}_E^p(\mathcal{R},\bar{m}^*)$, provided that m is *-continuous ($* =$ S or B, $1 \leq p < \infty$). Moreover, in this instance,

$$L_E^p(\mathcal{R},m^*) = L_E^p(S,\hat{m}^S) = L_E^p(S,\hat{m}^B).$$

Proof. (i) It may be assumed that $X = X_0 = \pi(X_0) \subset S$. The result follows from 15.4.

(ii) A sequence (φ_n) in $\mathcal{R}_0 \otimes E$ is m^*_p-Cauchy if and only if $(\hat{\varphi}_n)$ is $\hat{m}^*_p$-Cauchy. The last statement follows from 7.15.

17.2. Corollary. Suppose that there is a bilinear map from $E \times F$ to G, denoted by juxtaposition and of norm not exceeding one. For $\varphi = \sum \varphi_i \xi_i \in \mathcal{R} \otimes E$,

$$\int \varphi dm = \sum \xi_i m(\varphi_i) \in G \tag{1}$$

is then well-defined for every measure $m: \mathcal{R} \to F$ of finite variation and satisfies

$$\left\| \int \varphi dm \right\| \leq \int |\varphi| d|m| . \tag{2}$$

There is a unique linear map $\varphi \to \int \varphi dm$ from $\bar{\mathcal{R}}_E$ to G satisfying (1) and (2).

Proof. The map

$$\varphi \to \int \hat{\varphi} d\hat{m} = \int \varphi dm \qquad , \ \varphi \in \mathcal{R} \otimes E,$$

has all the properties in question: it is well-defined and satisfies the inequality (2), by the use of the upper integral $\hat{m}^B$ (10.11). It is linear and continuous with respect to the topology of dominated uniform convergence (4.7), and so has a unique extension to $\bar{\mathcal{R}}_E$.#

B. Rickart's decomposition

This is a generalization of the Riesz decomposition theorem 2.18 to Banach-valued measures.

17.3. Theorem. [34]. Suppose $\mathfrak{R}$ is a dominated integration lattice, $m: \mathfrak{R} \to F$ is a measure of finite variation, and A is a band in $I_0(\mathfrak{R})$.

Then m can be split uniquely into the sum, $m = m' + m^{\perp}$, of a measure m' of finite variation $|m'| \in A$ and a measure $m^{\perp}$ of finite variation $|m^{\perp}| \perp A$. Moreover, $|m| = |m^{\perp}| + |m'|$.

In particular, m can be split uniquely into the sum $m = m_{PJ} + m_{PS} + m_B$ of a purely linear measure m_{PJ}, a purely S-continuous measure m_{PS} and a B-continuous measure m_B. Moreover, $|m| = |m_{PJ}| + |m_{PS}| + |m_B|$.

Proof. The second statement is evident once the first one has been proved (3.10).

$|\hat{m}|$ may be decomposed into its part $|\hat{m}|'$ in $\hat{A}$ and its part $|\hat{m}|^{\perp}$ disjoint from $\hat{A}$ (2.18). Choose an $(S(\mathfrak{R}),\hat{m}^B)$-adequate partition, P, of relatively compact integrable sets (16.4). Every $K \in P$ decomposes into the disjoint union $K = K' \dot{\cup} K^{\perp}$ of two integrable sets $K', K^{\perp}$, so that

$$\int K' d|\hat{m}|^{\perp} = \int K^{\perp} d|\hat{m}|' = 0 \qquad (6.6).$$

We now put

$$m'(\varphi) = \sum \{ \int K'\hat{\varphi}d\hat{m}: K \in P\} \quad \text{and}$$
$$m^{\perp}(\varphi) = \sum \{ \int K^{\perp}\hat{\varphi}d\hat{m}: K \in P\} \quad , \text{ for } \varphi \in \mathfrak{R},$$

and leave to the reader the easy verification that this is the decomposition described in the statement of the theorem.#

C. Supplements

17.4. The results 17.1, 17.2, and 17.3 can be extended to measures of finite R-variation (3E).

17.5. Let $m = m_B + m_{PS} + m_{PJ}$ the unique decomposition of the positive measure m into its purely $*$-continuous parts $(* = J, S, B)$.
Then, with $m_S = m_B + m_{PS}$ and for $f: X \to \overline{\mathbb{R}}_+$

$$m_B^B(f) = \inf\{\sup m(\Phi): \Phi \subset \mathcal{R},\ \sup \Phi \geq f\} \quad \text{and}$$

$$m_S^S(f) = \inf\{\sup m(\varphi_n): \{\varphi_n\} \subset \mathcal{R},\ \sup(\varphi_n) \geq f\} .$$

17.6. Let $m: \mathcal{R} \to F$ be an S-measure on X and $m = m_B + m_{PS}$ its Rickart decomposition. Then X is the disjoint union $X = X_B \cup X_{PS}$ of two subsets such that $m_B = X_B m$ and $m_{PS} = X_{PS} m$ (15A).

17.7. X is locally compact in the $\mathcal{R}$-topology if and only if $I_0(\mathcal{R}) = M^B(\mathcal{R})$.

17.8. If $n \in M_E^*(\mathcal{R})$ $(* = S, B)$ then n coincides with the measure induced by $\hat{n}$ on $\pi(X)$ in the obvious sense (15C). 17.1 shows how the integration theories of n and $\hat{n}_{\pi(X)}$ compare.

LITERATURE: Most of the results are due to Bauer [2].

III. MEASURABILITY

§1. Measurable functions and sets

The functions and maps that appear in the integration theory of an upper gauge $(\mathcal{R},M)$ all have a distinct local behaviour, which can be approximately described by saying that they "nearly" belong to $\mathcal{R}$, or behave "nearly as well" as do the functions in $\mathcal{R}$. The word "nearly" refers to the phenomenon that, upon removal of sets of arbitrarily small measure, they behave exactly like the functions of $\mathcal{R}$, in the sense that they do not oscillate more than the functions of $\mathcal{R}$. To compare the oscillation of a function f with values in an arbitrary space with that of functions in $\mathcal{R}$, it is natural to use the uniformity generated by $\mathcal{R}$ on the domain X, and to call a function f well-behaved with respect to $\mathcal{R}$ if it is uniformly continuous with respect to the $\mathcal{R}$-uniformity. A function on X having values in a uniform space will then be called $(\mathcal{R},M)$-measurable if, upon removal of sets of arbitrarily small measure, it becomes uniformly continuous.

This definition extends the one of Bourbaki, which is given only for Radon measures, to arbitrary weak upper gauges. It can be applied to set functions as well, and can replace the usual definition of Carathéodory-Halmos used most widely in probability theory. It has an advantage over the latter in that it is intuitive, easy to handle, and it places exactly the right restriction on functions with values in non-separable spaces.

18. $(\mathcal{R},M)$-Measurability

In this section, $(\mathcal{R},M)$ is a weak upper gauge on the set X.

A. Definition. If f is a map, M-a.e. defined on an integrable set A and with values in some uniform space, then f is said to be $(\mathcal{R},M)$-measurable (on A) if the family of integrable sets on which f is $\mathcal{R}$-uniformly continuous is dense in A.

A function defined almost everywhere on a subset B of X is said to be measurable on B if it is measurable on every integrable subset of B. A function defined a.e. on X is simply said to be $(\mathcal{R},M)$-measurable if it is measurable on X.

U(f) denotes the family of integrable sets on which f is uniformly continuous, and $\mathcal{L}_Y(\mathcal{R},M)$ the set of $(\mathcal{R},M)$-measurable maps with values in the uniform space Y.#

To say that f is uniformly continuous on the subset K of X means, of course, that the restriction $f|K$ is uniformly continuous in the uniformity induced on K. In view of the definition of dense families, we can restate the definition thus: f is measurable if for every integrable set A and every $\varepsilon > 0$ there is an integrable subset K of A on which f is uniformly continuous with $M(A - K) < \varepsilon$. Alternatively, f is measurable if for every integrable set A there is a sequence (K_n) of integrable subsets of A which converges almost everywhere and in M-mean to A such that f is uniformly continuous on every K_n (7.12). Finally, let F be any dense family. Clearly f is measurable if and only if the sets in F on which f is uniformly continuous are dense (8.20).

If $f: A \to Y$ is measurable and $\Phi: Y \to Z$ is a uniformly continuous map, then $\Phi \circ f: A \to Z$ is evidently also measurable.

18.1. Proposition. If $(\mathfrak{R},M)$ and $(\mathfrak{R},N)$ are equivalent weak upper gauges then they have the same measurable functions.

Proof. Suppose that f is $(\mathfrak{R},M)$-measurable and let $K \in \mathscr{C}_0(\mathfrak{R},M) = \mathscr{C}_0(\mathfrak{R},N)$ (8.18). There is a sequence of subsets K_n of K in $\mathscr{C}_0(\mathfrak{R},M)$, on which f is uniformly continuous and such that (K_n) converges to K, M-a.e. Hence (K_n) converges to K N-a.e., and thus (8.3) converges to K in N-mean. Since $\mathscr{C}_0(\mathfrak{R},N)$ is $(\mathfrak{R},N)$-dense (8.8), f is $(\mathfrak{R},N)$-measurable.#

18.2. Corollary. $(\mathfrak{R},M)$; $(\mathfrak{R},M^*)$; $(\mathfrak{R},\overline{M})$; and $(\mathfrak{R},M_p)$ all have the same measurable functions.#

18.3. Lemma. Let (f_n) be a sequence of measurable functions with values in uniform spaces Y_n, defined a.e. and measurable on the integrable set A.

The family $U((f_n)) = \cap\{U(f_n): n = 1, 2, \ldots\}$ of integrable sets K on which all the f_n are simultaneously uniformly continuous is dense in A.

Proof. Given an $\varepsilon > 0$, choose $K_1 \in U(f_1)$ such that $K_1 \subset A$ and $M(A - K_1) < \varepsilon 2^{-1}$. Then choose $K_2 \in U(f_2)$ such that $K_2 \subset K_1$ and $M(K_1 - K_2) < \varepsilon 2^{-2}$. Continue by induction and arrive at a decreasing sequence of sets $K_n \in U(f_n)$ satisfying $M(K_n - K_{n+1}) < \varepsilon 2^{-n-1}$. Evidently $K = \cap K_n$ belongs to $U((f_n))$ and satisfies $M(A - K) < \varepsilon$.#

18.4. Proposition. Let f, g be maps with values in the complete uniform spaces F, G, respectively, and defined a.e. and measurable on $A \in \mathcal{C}(\mathfrak{R}, M)$.

If $\Phi: F \times G \to H$ is a continuous map into a third uniform space H, then $\Phi(f, g)$ is measurable on A.

Proof. The family $U(f, g)$ of integrable sets on which both f and g are uniformly continuous is dense. For $K \in U(f, g)$, $f(K) \times g(K)$ is relatively compact in $F \times G$ (5.7), and so Φ is uniformly continuous on $f(K) \times g(K)$. Hence $\Phi(f, g)$ is uniformly continuous on all sets $K \in U(f, g)$.#

18.5. Corollary. The sum, difference, product, infimum, and supremum of two real-valued measurable functions are measurable.

If E is a Banach space and if $f, g \in \mathcal{L}_E = \mathcal{L}_E(\mathfrak{R}, M)$ and $h \in \mathcal{L} = \mathcal{L}(\mathfrak{R}, M)$, then $f + g$, $f - g$, and fh all belong to $\mathcal{L}_E(\mathfrak{R}, M)$; hence, $\mathcal{L}_E$ is a module over $\mathcal{L}$. Furthermore, $\|f\|$ is measurable.

Proof. The maps $(r,s) \to r+s$, $(r,s) \to r \vee s$, etc. are continuous from $\mathbb{R} \times \mathbb{R}$ to $\mathbb{R}$; and so are the maps $(\xi,\eta) \to \xi+\eta$ from $E \times E$ to E and $(r,\xi) \to r\xi$ from $\mathbb{R} \times E$ to E.#

The standard exhaustion argument 8.21, in conjunction with the fact that the dominated sets in $U((f_n)) \cap F$ are dense in A, yields the following result.

18.6. Lemma. Suppose $(\mathcal{R},M)$ is a weak upper gauge. If (f_n) is a sequence of functions defined and measurable on the integrable set A, then there is a countable[1)] family $F' \subset U((f_n))$ of mutually disjoint subsets of A such that $M(A - \bigcup F') = 0$.

Moreover, if a dense family F is given, the $K \in F'$ can be chosen to belong to F, to be dominated, and to have non-zero measure, if $M(A) > 0$.#

Here is a first connection between measurability and integrability.

18.7. Proposition. Suppose $(\mathcal{R},M)$ is a weak upper gauge, and f is a function on X having values in a Banach space E and vanishing outside an integrable set A.

If f is measurable on A and if it is majorized by an integrable function h, then f is integrable itself.

If $(\mathcal{R},M)$ is an upper gauge then f is integrable provided it is measurable on A and $M(f) < \infty$.

Proof. There is a sequence (K_n) of dominated integrable subsets of A in U(f) which converges to A a.e. The functions fK_n are the uniform

1) = countable or finite.

limits of functions φK_n, $\varphi \in \mathcal{R}_0 \otimes E$ (5.7) and are therefore integrable (8.5). From Lebesgue's theorems (8.3, 8.12), $f = \lim fK_n$, too, is integrable.#

The measurable sets are, of course, the measurable idempotents of $\mathcal{L}(\mathcal{R},M) = \mathcal{L}_{\mathbb{R}}(\mathcal{R},M)$.

<u>18.8. Proposition</u>. Let $(\mathcal{R},M)$ be, again, a weak upper gauge. Suppose A and B are two measurable sets and f is a function defined a.e. on $A \cup B$. If f is measurable on A and on B, then it is measurable on $A \cup B$.

<u>Proof</u>. $B \setminus A$ is measurable (18.5), and f is measurable on $B \setminus A$; hence we may assume that A and B are disjoint.

Let K be any integrable subset of $A \cup B$. Given any $\varepsilon > 0$, there is an integrable subset K' of K such that $M(K - K') < \varepsilon/2$ and such that both A and B are uniformly continuous on K' (18.3). There is a set U' in the $\mathcal{R}$-uniformity such that $(x,y) \in U'$ and $x,y \in K'$ imply $|A(x) - A(y)| < 1/2$ and $|B(x) - B(y)| < 1/2$. That is to say, x and y either both belong to A (or to B) or both don't. The sets $K'_A = AK'$ and $K'_B = BK'$ are integrable (18.7) and have union K'. There are integrable sets $K''_A \subset K'_A$ and $K''_B \subset K'_B$ on which f is uniformly continuous and such that $M(K' - K'') < \varepsilon/2$, where $K'' = K''_A \cup K''_B$. Since $M(K - K'') < \varepsilon$, the proof will be finished once it is shown that f is uniformly continuous on K''.

Given any element V of the uniformity on the range of f, find a set $U \subset U'$ belonging to the $\mathcal{R}$-uniformity and for which $x,y \in K''_A$ (or $x,y \in K''_B$) and $(x,y) \in U$ imply $(f(x),f(y)) \in V$. If $x,y \in K''$

and $(x,y) \in U$, then x and y lie either both in A or both in B; hence they lie either both in K''_A or both in K''_B, and so $(f(x),f(y)) \in V$ in either case: f is, indeed, uniformly continuous on K''.#

Definition. An $(\mathfrak{R},M)$-adequate cover by measurable sets is a family C of measurable sets such that every integrable set A can be covered a.e. by a countable subfamily $C' \subset C$. A is also said to be adequately covered by C or by C'.#

It will be established presently that the integrable sets are measurable, so that this definition is an extension of the one given in 8A.

18.9. Corollary. (Localization principle). If C is any $(\mathfrak{R},M)$-adequate cover, then a function f on X is $(\mathfrak{R},M)$-measurable if and only if it is measurable on every set in C.

Proof. The condition is clearly necessary. Suppose it is satisfied and let $A \in \mathfrak{C}(\mathfrak{R},M)$. There is a countable family (K_n) in C covering A almost everywhere. f is measurable on each of the sets $A \cap \bigcup\{K_n: 1 \leq n \leq k\}$ (18.8). Since these sets are dense in A, f is measurable on A.#

Example. Suppose X is locally compact in the $\mathfrak{R}$-uniformity (e.g., X is locally compact and $\mathfrak{R} = C_{oo}(X)$). If every point $x \in X$ has a neighborhood on which f is measurable, then f is measurable.

Indeed, if $C = \{U_x: x \in X\}$ is a family of such neighborhoods, one for each $x \in X$, then C is an $(\mathfrak{R},M)$-adequate cover: if K is a compact G_δ in X then K is contained in a finite union of sets in C; and every integrable set is adequately covered by a countable union of compact G_δ's.#

A <u>measurable step function</u> on X is any function (with values in a uniform space) taking only finitely many different values, each in a measurable set. Since the level sets of such a function form a (finite) adequate cover, the following is a consequence of 18.9.

<u>18.10. Corollary</u>. A measurable step function is measurable.

<u>B. Limit theorems</u>

The following theorem is in part justification for the definition of measurability, and is certainly the most significant result in this area.

<u>18.11. Theorem</u>. (Egoroff). Suppose $(\mathcal{R},M)$ is a weak upper gauge and F is an $(\mathcal{R},M)$-dense family, and let (f_n) be a <u>sequence</u> of functions, defined a.e. and measurable on the integrable set A and having values in a <u>metric</u> space (E,d). Then the following statements are equivalent:

(i) The sequence (f_n) converges a.e. on A.

(ii) (f_n) converges a.e. on every subset K of A in F.

(iii) For every $\varepsilon > 0$, there is a subset K of A in F, with $M(A - K) < \varepsilon$, on which all the f_n are uniformly continuous and on which they converge uniformly.

(iv) Those subsets of A in F on which all the f_n are uniformly continuous and uniformly convergent are dense in A.

Moreover, if one of these conditions is satisfied, then $f = \lim(f_n)$ is defined a.e. and measurable, on A.

<u>Proof</u>. The last statement follows from (iii), since the uniform limit of uniformly continuous functions is uniformly continuous. The implications (iii) $\Longleftrightarrow$ (iv) $\Rightarrow$ (i) $\Rightarrow$ (ii) are obvious, hence the proof will be complete once it has been shown that (ii) implies (iii):

Given an $\varepsilon > 0$, choose an integrable subset K' of A so that $M(A-K') < \varepsilon/4$ and all the f_n are uniformly continuous on K' (18.3). The sets

$$M^r_{n,m} = \{x \in K' : d(f_n(x), f_m(x)) \geq r^{-1}\} \qquad , \; r,m,n \in \mathbb{N},$$

are all integrable subsets of K'. Indeed, the functions

$$x \to d(f_n(x), f_m(x)) = F_{n,m}(x) \qquad , \; x \in K',$$

are uniformly continuous on K', and hence $K'F_{n,m}$ is integrable for all n and m (18.7); and the claim follows from 8.4. The sets

$$M^r_p = \cup\{M^r_{n,m} : m,n \geq p\} \qquad , \; r,p \in \mathbb{N},$$

are integrable (8.4), and decrease as p increases. Their intersection

$$\cap\{M^r_p : p = 1, 2, \ldots\},$$

is negligible since it is disjoint from the set C of points at which (f_n) converges. From the (weak) monotone convergence theorem (8.1), we deduce the existence of positive integers $p(r)$ for which

$$M(M^r_{p(r)}) < \varepsilon 2^{-r} \qquad , \text{ for each } r = 1, 2, \ldots \; .$$

The set $B = \cup\{M^r_{p(r)} : r = 2, 3, \ldots\}$ is an integrable subset of K' with $M(B) < \varepsilon/2$; and (f_n) converges uniformly on $K'' = C \cap (K' - B)$. If K'' is approximated to within $\varepsilon/4$ by an integrable subset $K \subset K''$ in F, then the (f_n) evidently converge uniformly on K, and $M(A-K) < \varepsilon$.#

Remark. Egoroff's theorem does not hold if the assumption of metrizability on E is dropped, or if (f_n) is replaced by an arbitrary net.

18.12. Corollary. (i) The pointwise limit (a.e.) of a sequence of measurable functions with values in a metric space is measurable.

(ii) The functions in $\mathcal{L}^p_E(\mathfrak{R},M)$ are measurable $(1 \le p < \infty$, E a Banach space).

(iii) If (f_n) is a sequence of numerical measurable functions, then $\inf(f_n)$, $\sup(f_n)$, $\lim\inf(f_n)$, and $\lim\sup(f_n)$ are all measurable. The functions in $\mathfrak{R}^S_\uparrow$ and in $\mathfrak{R}^S_\downarrow$ are measurable; and if $(\mathfrak{R},M)$ is B-continuous then so are the functions in $\mathfrak{R}^B_\uparrow$ and in $\mathfrak{R}^B_\downarrow$.

(iv) The $(\mathfrak{R},M)$-measurable sets form a tribe $T(\mathfrak{R},M)$ containing all locally integrable and therefore all locally negligible sets. $T(\mathfrak{R},M)$ contains all sets of the form $[f > r]$, $[f \ge r]$, $[f < r]$, and $[f \le r]$ for $f: X \to \overline{\mathbb{R}}$ measurable and $r \in \overline{\mathbb{R}}$. If $(\mathfrak{R},M)$ is B-continuous then $\mathcal{K}^B(\mathfrak{R})$ and $\mathcal{U}^B(\mathfrak{R})$ are in $T(\mathfrak{R},M)$.

(v) A set A is measurable if and only if AK is integrable for either all integrable sets K or all K in a dense family F.

(vi) If $f: X \to \overline{\mathbb{R}}$ is measurable then so is the function $1/f$, defined arbitrarily but constant on $[f = 0]$.

Proof. (i) is obvious, (ii) follows from 7.12 and 18.11, and all but the last statement of (iii) follow from 18.11. Suppose $(\mathfrak{R},M)$ is B-continuous and let $h \in \mathfrak{R}^B_\uparrow$. As $h = \sup\{h \wedge n : n \in \mathbb{N}\}$ it may be assumed that h is bounded. Since hK is integrable for every integrable Borel set K (8.17), h is measurable on such sets; and since these form an adequate cover (8.7), h is measurable (18.9).

(iv): It is obvious that $T(\mathfrak{R},M)$ forms a tribe, etc. For $f \in \mathcal{L}(\mathfrak{R},M)$ and $r \in \overline{\mathbb{R}}$, $[f > r] = [f - r > 0] = \lim\{(n(f-r) \wedge 1) : n = 1, 2, \ldots\}$ is measurable (18.5, 18.11). The remaining cases follow from consideration of $-f$ and complements.

(v): The condition is necessary (18.7). If it is satisfied then AK is integrable and so measurable for all $K \in \mathscr{C}_0(\mathscr{R},M) \cap F$; hence it is uniformly continuous on a dense family of integrable subsets $K' \subset K$. Since $AK|K' = A|K'$, A is uniformly continuous on a dense family, and so is measurable.

(vi): $1/f$ is clearly measurable on the measurable sets $A_0 = [f = 0]$ and $A_\infty = [|f| = \infty]$. Since the map $r \to 1/r$ from $\mathbb{R} \setminus \{0\}$ to $\mathbb{R}$ is continuous, $1/f$ is also measurable on $A_f = [f \in \mathbb{R},\ f \neq 0]$ (18.4). Since $A_0 \cup A_\infty \cup A_f = X$, f is measurable (18.9).#

C. Supplements

18.13. $T(\mathscr{R},M)$ is generally bigger than the tribe spanned by $\mathscr{C}(\mathscr{R},M)$ (6.12), even in cardinality.

18.14. Suppose for simplicity that $(\mathscr{R},M)$ is a finite upper gauge. A sequence (f_n) of $(\mathscr{R},M)$-measurable functions with values in a metric space is said to be <u>fundamental in measure</u> if, for every $\varepsilon > 0$,

$$\lim(M([d(f_n,f_m) > \varepsilon]) : m \geq n \to \infty) = 0.$$

It is said to <u>converge in measure</u> to f if, for every $\varepsilon > 0$,

$$\lim(M([d(f_n,f) > \varepsilon]) : n \to \infty) = 0.$$

A sequence fundamental in measure, (f_n), converges in measure, and the limit, f, is a measurable function. A subsequence of (f_n) converges to f a.e. An a.e. convergent sequence converges in measure.

18.15*. Let $(\mathscr{R},M)$ be an upper gauge and K an $(\mathscr{R},M)$-measurable set. A function f is then $(\mathscr{R},M)$-measurable on K if and only if its restriction to K is $(\mathscr{R}|K,M_K)$-measurable (15.5).

18.16. If $(\mathscr{R},M)$ and $(\mathscr{R},N)$ are two weak upper gauges that agree on $\mathscr{R}_+$ then they agree on $\mathscr{R}_+^\Sigma$ (6.16).

18.17. Let $(E_i)_{i\in I}$ be a family of uniform spaces, each with uniformity $\mathscr{U}_i$. The <u>product uniformity</u>, $\mathscr{U}$, on the cartesian product, $E = X\{E_i : i \in I\}$, is defined to be the coarsest uniformity such that all the natural projections from E to E_i, $i \in I$, are uniformly continuous.

(i) $\mathcal{U}$ has a basis of sets of the form $X\{V_i: i \in I\}$ where each V_i belongs to $\mathcal{U}_i$ and all but finitely many of the V_i equal $E_i \times E_i$.

(ii) Suppose I is countable and for each $i \in I$ let $f_i: X \to E_i$ be an $(\mathcal{R},M)$-measurable map. Then the product map

$$f = Xf_i: x \to (f_i(x))_{i \in I} \qquad , x \in X,$$

from X to E is measurable.

18.18. Suppose $(\mathcal{R},M)$ is a weak upper gauge and (f_n) is a sequence of $(\mathcal{R},M)$-measurable functions having values in a complete metric space. Then $\{x: \lim f_n(x) \text{ exists}\}$ is $(\mathcal{R},M)$-measurable.

18.19*. Let $(\mathcal{R},M)$ be a weak upper gauge and f a measurable Banach-valued function. If f has σ-finite carrier, in particular if f is integrable, then f is a.e. equal to an $\mathcal{R}$-Baire function $\tilde{f}$ (6.16). If f is dominated or a set, $\tilde{f}$ can be so chosen also.

18.20*. Suppose $(\mathcal{R},M)$ is a finite weak upper gauge (13A) and let $\mathcal{F}$ be any subfamily of $\mathcal{L}^1_E(\mathcal{R},M)$. Then $\mathcal{F}$ is said to be <u>uniformly integrable</u> if (i) $\lim\{M\|f\|[\|f\| > n]: n \to \infty\} = 0$ uniformly for all $f \in \mathcal{F}$. This happens if and only if

(iia) $\sup\{M(\|f\|): f \in \mathcal{F}\} < \infty$ and
(iib) $\lim\{M(Kf) : K \in \mathcal{C}(\mathcal{R},M),\ M(K) \to 0\} = 0$

uniformly for all $f \in \mathcal{F}$. Any finite subfamily of $\mathcal{L}^1_E$ is uniformly integrable. Condition (ii) says that $\mathcal{F}$ is bounded and that the upper S-norms $\{\|f\|M: f \in \mathcal{F}\}$ are uniformly absolutely continuous with respect to M, on $\mathcal{C}(\mathcal{R},M)$ (11B). We have the following generalization of Lebesgue's theorem.

<u>Theorem</u>. Let (f_k) be a sequence in $\mathcal{L}^1_E(\mathcal{R},M)$ that converges a.e. to a function f. Then (f_k) converges in mean to f if and only if $\{f_1, f_2, \ldots\}$ is uniformly integrable.

(Hint: for every $\varepsilon > 0$ there is a set $K \in \mathcal{C}(\mathcal{R},M)$ such that $M(Kf_k) < \varepsilon$ for all k and $\|f_k - f\| < \varepsilon$ uniformly on $X - K$.)

18.21. If $\mathcal{R}$ is an integration lattice then so is $\mathcal{R}^B_\uparrow - \mathcal{R}^B_\uparrow$. The Σ-closure (6.16) of the latter is denoted by $\mathcal{R}^\beta$ and consists of the $\mathcal{R}$-<u>Borel functions</u>. The sets in $\mathcal{R}^\beta$ are the $\mathcal{R}$-<u>Borel sets</u>, $\mathcal{C}(\mathcal{R}^\beta)$.

(i) $\mathcal{C}(\mathcal{R}^\beta)$ is the tribe generated by the $\mathcal{R}$-topology.

(ii) Let $(\mathcal{R},M)$ be a B-continuous weak upper gauge. Then (a) every Borel function is $(\mathcal{R},M)$-measurable and (b) an $(\mathcal{R},M)$-integrable function is equal M-a.e. to a Borel function.

LITERATURE: Proofs are adapted from [5, Ch. IV]. For 18.14 and for 18.20 see [1], [3], [11], [14], [19], [48], [49].

19. Integrability and measurability criteria

A. The integrability criterion

19.1. Proposition. Let $(\mathcal{R},M)$ be a weak upper gauge and let E be a Banach space. A function $f: X \to E$ is $(\mathcal{R},M)$-integrable if and only if it is measurable and majorized by an integrable function.

Proof. If f is integrable then it is measurable and majorized by the integrable function $\|f\|$ (18.12, 7.13).

Conversely, suppose f is measurable and majorized by the integrable function h. Then f is measurable on each of the integrable (8.4) sets $K_r = [h > r^{-1}]$, $r = 1, 2, \ldots$; and consequently fK_r is integrable (18.7). From Lebesgue's (weak) theorem (8.3), $f = \lim fK_r$ is integrable.#

This criterion can be somewhat sharpened if $(\mathcal{R},M)$ is an upper gauge:

Definition. An upper gauge $(\mathcal{R},M)$ is *regular* if every locally $(\mathcal{R},M)$-negligible set A has either $M(A) = 0$ or $M(A) = \infty$.

19.2. Proposition. If an upper gauge is essential or smooth then it is regular. Hence a strong upper gauge $(\mathcal{R},M)$ lies between two regular ones: $(\mathcal{R},\overline{M}) \leq (\mathcal{R},M) \leq (\mathcal{R},M^*)$. Also, if $1 \leq p < \infty$ then $(\mathcal{R},M_p)$ is regular if and only if $(\mathcal{R},M)$ is.

Proof. An essential upper gauge is regular by definition. Suppose $(\mathcal{R},M)$ is smooth, and let A be an $(\mathcal{R},\overline{M})$-negligible set. If $M(A) < \infty$, then A is majorized by an $(\mathcal{R},M)$-integrable function, and so vanishes off some σ-finite set B (8.7). Hence $M(A) = M(AB) \leq \overline{M}(A) = 0$. The last statement is obvious from 12.8.#

<u>19.3. Theorem</u>. (Integrability criterion). Let $(\mathcal{R},M)$ be a strong upper gauge on X, and let f be a Banach-valued function defined a.e. on X. Provided that either

(a) $(\mathcal{R},M)$ is regular or

(b) f vanishes outside a σ-finite set,

f is $(\mathcal{R},M)$-integrable iff it is measurable and has $M(f) < \infty$.

<u>Proof</u>. If f is integrable it is certainly measurable and has $M(f) < \infty$. Suppose these two conditions are satisfied. If f vanishes off a σ-finite set, it is $(\mathcal{R},M)$-integrable if and only if it is $(\mathcal{R},\overline{M})$-integrable. In view of the fact that $(\mathcal{R},\overline{M})$ is regular (19.2), only (a) need be considered.

For every natural number n, the sets $K_n = [\|f\| > n^{-1}]$ have finite upper measures and are measurable. If we can show that sets of this description are integrable, we are finished: from 18.7, the sequence $(K_n f)$ consists of integrable functions, and from Lebesgue's theorem 8.12, $f = \lim K_n f$ is integrable.

So let $K \in T(\mathcal{R},M)$ with $M(K) < \infty$. Choose an $(\mathcal{R},M)$-integrable subset $K' \subset K$ of maximal measure $M(K')$, and set $K'' = K - K'$. Evidently, K'' is measurable. Furthermore, $\overline{M}(K'') = 0$, since if $\overline{M}(K'') > 0$ then there is an $A \in \mathcal{C}(\mathcal{R},M)$ with $M(AK'')$, and then AK'' is a non-negligible integrable subset of K disjoint from K', in contradiction to the definition of K' (SUG). As $(\mathcal{R},M)$ is regular and $M(K'') < \infty$, $M(K'') = 0$, and K is $(\mathcal{R},M)$-integrable.#

<u>19.4. Corollary</u>. Let $(\mathcal{R},M)$ be a strong upper gauge, and let K be a measurable set with $\overline{M}(K) > 0$. Then K contains an $(\mathcal{R},M)$-integrable subset K' with $M(K') > 0$; and if $\overline{M}(K)$ is finite then K' can be chosen so that $\overline{M}(K - K') = 0$.

B. The measurability criterion

Henceforth $(\mathcal{R},M)$ is a fixed weak upper gauge on X.

Definition. A function f on X with values in a uniform space is said to be almost compact-valued if the family C(f) of integrable sets K such that f(K) is precompact is dense; and f is said to be almost separably-valued if the family S(f) of integrable sets K such that f(K) is separable is dense.#

In the spirit of this definition, a function f is measurable if it is almost uniformly continuous. From 5.7, $U(f) \subset C(f)$, and if the range of f is metrizable then

$$U(f) \subset C(f) \subset S(f).$$

(Note that C(f) is closed under finite unions and S(f) under countable unions.)

19.5. Theorem. (Measurability criterion). Let F be an $(\mathcal{R},M)$-dense family and let f be a map from X into some metric space (E,d). The following statements are all equivalent:

(i) f is measurable.

(ii) The family of sets $K \in F$ on which f is the uniform limit of measurable step functions is dense.

(iii) On every set in F, f is a.e. the pointwise limit of measurable step functions.

(iv) f is almost separably-valued and the pre-image $f^{-1}(U)$ is a measurable set, for every open set $U \subset E$.

(v) f is almost separably-valued and the pre-image $f^{-1}(B)$ is a measurable set, for every closed ball $B \subset E$.

(vi) f is almost separably-valued and the pre-image $f^{-1}(T)$ is a measurable set for every set T in the tribe on E generated by the open sets. (This tribe is also called the tribe of Borel sets on E).

Proof. (i) $\Rightarrow$ (ii): Given any integrable set A and any $\varepsilon > 0$, one can find a subset K of A in $F \cap U(f)$ such that $M(A - K) < \varepsilon$. Since $f(K)$ is totally bounded in E (5.7), there is, for arbitrary $\delta > 0$, a finite cover of it by δ-balls $B_1, \ldots, B_n$. Let $\xi_1, \ldots, \xi_n$ be the centers of these balls. For each i,

$$A_i = f^{-1}(B_i) \cap K = \{x \in K\colon d(f(x);\xi_i) < \delta\}$$

is a measurable set (18.12). Put $C_i = A_i \setminus \cup\{A_1, \ldots, A_{i-1}\}$, and let s denote the step function which has value ξ_i in C_i $(1 \leq i \leq n)$ and a fixed value ξ_0 outside K. Then s is clearly measurable (18.10), and $d(f(x), s(x)) \leq \delta$ on K.

(ii) $\Rightarrow$ (iii): Let $K \in F$. For each positive integer n, there is a $K_n \in F$ and a measurable step function s_n such that $M(K - K_n) < n^{-1}$ and $d(f(x), s_n(x)) < n^{-1}$ on K_n. The sequence $(s_n K_n)$ of measurable step functions converges almost everywhere to f on K; its limit is measurable on K, due to Egoroff's theorem (18.11).

(iii) $\Rightarrow$ (iv): If $K \in F$ and $s_n \to f$ on K except in the points of the negligible set N, then $f(K - N)$ lies in the closure of the countable set $\cup\{s_n(K)\colon n = 1, 2, \ldots\}$ and so is separable. This shows that $S(f)$ is dense. Let U be an open set of E. To show that $f^{-1}(U)$ is measurable, it suffices to prove that $f^{-1}(U) \cap K$ is integrable for each

K on which f is the pointwise limit of a sequence (s_n) as above (18.12(v)). This is evident from the equation

$$f^{-1}(U) \cap K = \bigcup_k \bigcap_{n>k} [s_n \in U] \cap K$$

since each of the sets $[s_n \in U] \cap K$ is integrable (18.7, 8.1).

(iv) $\iff$ (v) $\iff$ (vi): For $K \in S(f)$, $\{T \subset f(K): f^{-1}(T) \in T(\mathcal{R},M)\}$ is a tribe generated by the open sets or the closed balls $\subset f(K)$.

(vi) $\Rightarrow$ (i): It suffices to show that f is measurable on every set A in S(f) (18.9). Let $\{\xi_n\}$ be a countable dense subset of f(A), and for $n, p = 1, 2, \ldots$ set

$$A_n^p = \{x \in A: d(f(x), \xi_n) < p^{-1}\},$$

a measurable set. For fixed p define B_n^p inductively: $B_1^p = A_1^p$ and

$$B_{n+1}^p = A_{n+1}^p \setminus \bigcup\{A_k^p: 1 \leq k \leq n\}.$$

They form a measurable partition of A. Let s_m^p be the step function equal to ξ_i in B_i^p (if $B_i^p \neq \emptyset$; $m = 1, 2, \ldots$; $1 \leq i \leq m$) and to a fixed element ξ of E elsewhere. With increasing m, s_m^p tends to the measurable function s^p which is equal to ξ_i in B_i^p (if $B_i^p \neq \emptyset$; $i = 1, 2, \ldots$) and equal to ξ elsewhere. As p increases s^p converges to f pointwise on A. From Egoroff's theorem, f is measurable on A. This closes the circle of arguments and finishes the proof.#

Note that if $A \in C(f)$ then finitely many of the sets A_n^p will cover the precompact set f(A); s_n^p will be equal to s^p for sufficiently high n; and so s^p will be a measurable step function itself: The following result is obtained.

19.6. Corollary. A measurable function f with values in a metric space is, on every set in $C(f)$, the uniform limit of measurable step functions. In particular, if $f(X)$ is precompact then f can be approximated uniformly by measurable step functions.#

The statements (iii), (iv), and (vi) of the theorem show that only the topology of E and knowledge of the integrable and negligible sets is needed to determine which functions are measurable. This observation has the following consequence.

19.7. Corollary. Let E be a metrizable topological space, and let d_1, d_2 be metrics compatible with the topology of E.

A function $f: X \to E$ is $(\mathcal{R},M)$-measurable with respect to d_1 if and only if it is so with respect to d_2.

A function f with values in a metrizable space E is $(\mathcal{R},M)$-measurable if and only if it is $(\mathcal{L}^1(\mathcal{R},M),M)$-measurable.#

Remark. The last statement is, in general, false if E does not have a metrizable uniformity.

C. Scalar measurability

Let E be a Banach space with dual E'. Recall that a subspace $V \subset E'$ is norming if $\|\xi\| = \sup\{\langle \xi;v\rangle : v \in V, \|v\| \leq 1\}$ (4D).

19.8. Lemma. If V is a norming subspace of E' and E^c is a separable subspace of E, then V contains a countable $\mathbb{Q}$-vector subspace V^c which is norming on E^c, i.e., such that

$$\|\xi\| = \sup\{\langle \xi;v\rangle : v \in V^c, \|v\| \leq 1\} \qquad , \text{ for } \xi \in E^c.$$

Proof. Let $H = \{\xi_1, \xi_2, \ldots\}$ be a countable set in E spanning E^c. It may be assumed to be a vector space over the rationals $\mathbb{Q}$, and thus dense in E^c. There is a countable subset V^c of V such that

$$\|\xi_n\| = \sup\{\langle \xi_n; v\rangle : v \in V^c, \ \|v\| \leq 1\} \qquad , \text{ for all positive integers } n.$$

V^c may be assumed to be a $\mathbb{Q}$-vector space, since rational linear combinations can be adjoined if necessary. Evidently, V^c is norming on E^c. #

Definition. Let V be a subset of E', and $f: X \to E$ a function. Then f is said to be V-scalarly measurable if all the functions

$$\langle f;v\rangle : x \to \langle f(x);v\rangle \qquad , \text{ for } v \in V,$$

are measurable. It is said to be scalarly measurable if it is E'-scalarly measurable.

19.9. Proposition. If $f: X \to E$ is a map then the following statements are equivalent:

(i) f is $(\mathcal{R},M)$-measurable.

(ii) f is almost separably valued and scalarly measurable.

(iii) f is almost separably valued and there exists a norming subset $V \subset E'$ such that f is V-scalarly measurable.

Proof. The implications (i) $\Rightarrow$ (ii) $\Rightarrow$ (iii) are obvious. Suppose (iii) holds. It suffices to show that f is measurable on every set $K \in S(f)$ (18.9), and, merely, that f is measurable on K as a function into the separable space E^c spanned by $f(K)$.

Let V^c be a countable subset of V consisting of vectors of norm ≤ 1 and norming on E^c; and let B be a closed ball of radius

$r \geq 0$ around $\xi \in E^c$. The set

$$f^{-1}(B) \cap K = \bigcap\{x \in K: \langle f(x) - \xi; v\rangle \leq r,\ v \in V^c\}$$

is evidently measurable (18.12), and since B was arbitrary, f is measurable on K (19.5(v)).#

The following generalization is proved analogously.

<u>19.10. Proposition</u>. Let $f: X \to Y$ be a map into a metrizable uniform space Y. Then f is $(\mathcal{R},M)$-measurable if and only if f is almost separably valued and $\varphi \circ f$ is measurable for all uniformly continuous functions $\varphi: Y \to \mathbb{R}$.

D. Supplements

19.11. If $(\mathcal{R},M)$ is a regular upper integral then M is countably additive on $\mathcal{L}_+(\mathcal{R},M)$.

19.12. If $(\mathcal{R},M)$ is a regular strong upper gauge and $f: X \to E$ is $(\mathcal{R},M)$-measurable, then f is p-integrable if and only if $M_p(f) < \infty$ $(1 \leq p < \infty)$.

19.13. Let $(Y,\mathcal{Y})$ be a uniform space. The sets in the Σ-closure (6.16) of the family of uniformly continuous real-valued functions are the <u>Baire sets</u> of $(Y,\mathcal{Y})$, and their collection is denoted by $\mathcal{B}^S(\mathcal{Y})$. It is a σ-algebra

(i) If $\mathcal{Y}$ is generated by an integration lattice $\mathcal{R}$ then $\mathcal{B}^S(\mathcal{Y}) = \mathfrak{C}(\mathcal{R}^\Sigma)$.

(ii) If $\mathcal{Y}$ is generated by a clan $\mathfrak{C}$ then $\mathcal{B}^S(\mathcal{Y})$ is the σ-algebra spanned by $\mathfrak{C}$.

(iii) If $\mathcal{Y}$ is the uniformity of a metric d, then $\mathcal{B}^S(\mathcal{Y})$ is the tribe spanned by the open balls.

(iv) Let $(X,\mathcal{U})$ be another uniform space and $f: X \to Y$ a map. Then f is said to be <u>Baire-measurable</u> if $f^{-1}(K) \in \mathcal{B}^S(\mathcal{U})$ for all $K \in \mathcal{B}^S(\mathcal{Y})$. This is the case iff f is $\mathcal{B}^S(\mathcal{U})$-$\mathcal{B}^S(\mathcal{Y})$-uniformly continuous. A Baire function (6.16) is Baire measurable. A Baire measurable map from $(X,\mathcal{R})$ to some metric space is measurable for every weak upper gauge on $(X,\mathcal{R})$ for which it is almost separably valued.

19.14. Let $(Y,\mathcal{T})$ be a topological space. The Borel sets of $\mathcal{T}$ are the sets in the tribe $\mathfrak{A}^B(\mathcal{T})$ spanned by $\mathcal{T}$. A map $f\colon (X,\mathcal{S}) \to (Y,\mathcal{T})$ from a second topological space $(X,\mathcal{S})$ is called Borel-measurable if $f^{-1}(K) \in \mathfrak{A}^B(\mathcal{S})$ for all $K \in \mathfrak{A}^B(\mathcal{T})$.

(i) If $\mathcal{T}$ underlies a uniformity $\mathcal{Y}$ then $\mathfrak{A}^S(\mathcal{Y}) \subset \mathfrak{A}^B(\mathcal{T})$.

(ii) A Borel function (18.21) is Borel-measurable. A Borel-measurable map is measurable for every B-continuous weak upper gauge for which it is almost compact valued.

19.15. Let $(\mathcal{R},M)$ be a smooth upper integral. A set $A \subset X$ is then $(\mathcal{R},M)$-measurable if and only if

$$(*) \qquad M(A \cap K) + M(K \setminus A) = M(K)$$

for all $K \subset X$, or for all K in an $(\mathcal{R},M)$-dense family.

Note. In almost all textbooks, the starting point is an S-measure $m > 0$ on a clan. The outer measure m^S is constructed as in 9.11, and the measurability of sets A is defined by equation (*). It is shown that the measurable sets so defined form a tribe T, and subsequently a function f with values in a space Y equipped with a clan $\mathcal{C}'$ is defined to be measurable if $f^{-1}(K') \in T$ for all $K' \in \mathcal{C}'$. If Y is a Banach space, $\mathcal{C}'$ is taken to be the clan generated by the open balls. More Banach-valued functions are measurable in this sense than in ours, since almost separably-valuedness is not required. However, this generality is of no avail and has to be removed by additional hypotheses in each application. The integrability criterion, for instance, does not hold if measurability is understood in this wide sense. It should also be noted that the definition of measurability described above is restricted to smooth upper integrals--and so misses the M_p-norms and the upper gauges $\|m^*\|$ (9C) and $\|U\|^*$ (11C). As there are uniform spaces whose uniformity does not derive from a clan on them, our definition is applicable also to functions with a wider class of permissible ranges.

20. The spaces $\mathcal{L}_E^\infty$ and L_E^∞

In this section, $(\mathfrak{R},M)$ is a fixed upper gauge, and E is a Banach algebra with multiplication $(\xi,\eta) \to \xi\eta$. ($\mathbb{R}$ and $\mathbb{C}$ are examples; every Banach space E becomes a Banach algebra under the trivial multiplication $\xi\eta = 0$ for all $\xi,\eta \in E$.)

A. Definitions. $\mathcal{L}_E^\infty(\mathfrak{R},M)$ denotes the vector space of all essentially bounded functions, defined a.e. and measurable on X, and with values in E; and $L_E^\infty(\mathfrak{R},M)$ denotes the set of their classes modulo locally negligible functions. If $E = \mathbb{R}$, the suffix is suppressed as usual. For $f \in \mathcal{L}_E^\infty$,

$$M_\infty(f) = \inf\{r \in \mathbb{R} : \overline{M}([\|f\| > r]) = 0\}$$

is the essential sup-norm of f. This definition is extended unambiguously to classes $f^\bullet \in L_E^\infty$ by $M_\infty(f^\bullet) = M_\infty(f)$. $M_\infty(f)$ is, by definition, the smallest number r such that

$$f(x) \in [-r,r] \qquad \text{, for locally almost all } x \in X.\#$$

As trivial consequences of this definition we state the inequalities of the mean:

(20.1) $\quad -M_\infty(f)g \mathrel{\dot{\le}} fg \mathrel{\dot{\le}} M_\infty(f)g \qquad$, for $f \in \mathcal{L}^\infty$, $g \in \mathcal{L}_+^\infty$, and

$$M_p(fg) \le M_\infty(f)M_p(g) \qquad \text{, for } f \in \mathcal{L}_E^\infty,\ g \in \mathcal{L}_E^p,\ 1 \le p < \infty.$$

The following result is evident from the definitions and (18.1, 18.2).

20.2. Proposition. If $(\mathcal{R},M)$ and $(\mathcal{R},N)$ are equivalent upper gauges then

$$\mathcal{L}_E^\infty(\mathcal{R},M) = \mathcal{L}_E^\infty(\mathcal{R},N), \quad \text{and} \quad M_\infty = N_\infty .$$

In particular, $(M_p)_\infty = (M^*)_\infty = \overline{M}_\infty$, for $1 \leq p < \infty$.

20.3. Proposition. M_∞ is a complete seminorm on $\mathcal{L}_E^\infty$; $\mathcal{L}^\infty$ is a Riesz space; L_E^∞ is a Banach algebra; and L^∞ is a Banach lattice satisfying $M_\infty(f^p) = (M_\infty(f))^p$, for $1 \leq p < \infty$, $f \in L_+^\infty$.

Proof. M_∞ is clearly positively homogeneous. Let $f,g \in \mathcal{L}_E^\infty$ and $r,s \in \mathbb{R}$. If the measurable sets $[\|f\| > r]$ and $[\|g\| > s]$ are both $(\mathcal{R},\overline{M})$-negligible then so are the subsets $[\|f+g\| > r+s]$ and $[\|fg\| > rs]$ of their union; hence $M_\infty(f+g) \leq M_\infty(f) + M_\infty(g)$ and $M_\infty(fg) \leq M_\infty(f) \cdot M_\infty(g)$.

Let $f \in \mathcal{L}_+^\infty$ and $0 \leq a < M_\infty(f)$. The set $[f > a]$ $(= [f^p > a^p])$ is not locally negligible, and so $M_\infty(f^p) > a^p$. This is true for $0 < p < \infty$ and yields $M_\infty(f^p) \geq (M_\infty(f))^p = (M_\infty((f^p)^{1/p}))^p \geq M_\infty(f^p)$, as desired.

Everything is now clear except the completeness of $(\mathcal{L}_E^\infty, M_\infty)$. For that, let (f_n) be a Cauchy sequence in $\mathcal{L}_E^\infty$. For each n there is an integer $k(n)$ such that $M_\infty(f_r - f_s) < n^{-1}$, for all $r,s \geq k(n)$, i.e., there is a locally negligible set N_n^{rs} such that $\|f_r - f_s\| \leq n^{-1}$, except on N_n^{rs}. The set $N = \cup\{N_n^{rs}: k(n) \leq r,s;\ n = 1, 2, \ldots\}$ is also locally negligible. The sequence (f_n) converges uniformly on $X - N$ to a function f', whose extension to a function f which is zero on N is measurable and bounded and satisfies $M_\infty(f - f_n) \to 0$. This proves that $(\mathcal{L}_E^\infty, M_\infty)$ is complete. Since the set $\mathcal{N}_E^\infty$ of bounded locally negligible functions is an ideal of $\mathcal{L}_E^\infty$ with quotient L_E^∞, L_E^∞ is a Banach algebra and L^∞ is a Banach lattice.#

B. Properties of L^∞

Several facts about $L^\infty(\mathcal{R},M)$ will be collected, mainly for use in the last chapter.

20.4. Lemma. A function f in $\mathcal{L}^\infty$ is invertible in $\mathcal{L}^\infty$ if and only if $[|f| < \varepsilon]$ is locally negligible for some $\varepsilon > 0$.

Proof. If $fg = 1$ loc. a.e. for some g in $\mathcal{L}^\infty$, then $[|f| < (2M_\infty(g))^{-1}]$ is locally negligible. On the other hand, if the condition is satisfied with $\varepsilon > 0$, then the function f^{-1}, defined to equal zero on $[|f| < \varepsilon]$ and as usual elsewhere, is an inverse for f in $\mathcal{L}^\infty$. This shows, in particular, that $|f^\bullet| - M_\infty(f^\bullet)$ is never invertible in L^∞, for any $f^\bullet \in L^\infty$.#

20.5. Proposition. The ideals (= solid subspaces) of the Riesz space L^∞ and the ideals of the algebra L^∞ coincide.

Proof. Suppose I is a lattice ideal of L^∞, and let $f \in L^\infty$ and $g \in I$. Then $|fg|$ is majorized by $M_\infty(f)|g| \in I$, and so $fg \in I$; hence I is an algebra ideal. Conversely, assume that I is an algebra ideal of L^∞, and let $g \in L^\infty$ with $|g| \leq f \in I$. Then $g/f = g\cdot(1/f)$ is measurable (18.5, 18.12) and is smaller than 1, hence $g = f\cdot(g/f)$ is in I, and so I is a lattice ideal.#

20.6. Proposition. A proper ideal I in L^∞ is contained in a maximal proper ideal J which is automatically closed.

Proof. Indeed, let $f \in L^\infty$ with $a = M_\infty(f^\bullet - 1^\bullet) < 1$. Then $0 < 1 - a \leq |f^\bullet|$ and thus $f^\bullet$ is invertible, and not in I. Hence I does not meet the open unit ball around $1^\bullet$ in L^∞. The closure of any proper ideal is therefore itself a proper ideal. A maximal proper ideal containing I (it exists, by Zorn's lemma) is therefore closed.#

20.7. Lemma. If J is a maximal proper ideal, then the quotient L^∞/J is algebraically and order isomorphic to the reals.

Proof. Denote by $\bar{f}$ the class of $f \in L^\infty$, in the quotient field $Q = L^\infty/J$. Clearly, $\bar{1}$ is the identity of Q. As $f - M_\infty(f)$ is not invertible in L^∞ (20.4), $\bar{f} - M_\infty(f)\bar{1}$ is not invertible in Q, and so is zero. Hence $\bar{f} = M_\infty(f)\bar{1}$ in Q, and $Q = \mathbb{R}\bar{1} \cong \mathbb{R}$.#

A character of L^∞ is a non-trivial, linear, and multiplicative map from L^∞ to $\mathbb{R}$.

20.8. Corollary. A character of L^∞ is a norm-decreasing Riesz-space homomorphism, and every proper ideal of L^∞ lies in the kernel of some character.

20.9. Proposition. Suppose $(\mathcal{R},M)$ is a B-continuous and essential upper gauge, and let $\underline{X}$ denote the support of $(\mathcal{R},M)$ (16C).

The points $x \in \underline{X}$ are exactly those points $x \in X$ for which the character $\varphi \to \varphi(x)$ of $\mathcal{R}$ can be extended to a character of L^∞.

Proof. If $x \notin \underline{X}$ then there is a set $U \in \mathcal{U}^B(\mathcal{R})$ containing x and of measure $M(U) = 0$. There is a $\varphi \in \mathcal{R}_+$ such that $\varphi(x) > 0$ and $\varphi \leq U$. Since $\varphi^\bullet$ is negligible, $t(\varphi^\bullet) = 0$ for all characters t of L^∞, and so $t(\varphi^\bullet) \neq \varphi(x)$.

Conversely, if $x \in X$, let J denote the set of $f \in L^\infty$ which, in some neighborhood $U \in \mathcal{U}^B$ of x, are majorized by a function in $\mathcal{R}$ that vanishes in x. Clearly J is an ideal and is proper: If $1U \leq \varphi$ for some $U \in \mathcal{U}^B$ and some $\varphi \in \mathcal{R}$ such that $\varphi(x) = 0$, then $U \cap [\varphi < 1/2]$ is negligible and in $\mathcal{U}^B$, and so does not intersect $\underline{X}$. Any character t of L^∞ annihilating J satisfies $t(\varphi - \varphi(x)) = 0$, and so $t(\varphi) = \varphi(x)$, for $\varphi \in \mathcal{R}$.#

C. Supplements

20.10. Let f be a positive and $(\mathcal{R},M)$-measurable function. The function $\theta \to \log M_{1/\theta}(f)$ is convex and continuous from $[0,1]$ to $\overline{\mathbb{R}}$.

20.11. A function $f: X \to E$ is $(\mathcal{R},M_\infty)$-integrable if and only if it is, loc. a.e., an $\mathcal{R}$-uniformly continuous function vanishing at infinity.

20.12. The set of all characters of L^∞ is a locally compact space, $\hat{X}$, in the topology of pointwise convergence. If L^∞ is a complete Riesz space (in which case $(\mathcal{R},M)$ is called localizable), then $\hat{X}$ is hyperstonean: the closure of an open set is open. $\hat{X}$ is called the <u>Stonean space of</u> L^∞. For $f \in L^\infty$, the map $\hat{f}: t \to t(f)$ on $\hat{X}$ is the <u>Gelfand transform</u> of f. It is continuous. If $m: \mathcal{R} \to E$ is a measure majorized by $(\mathcal{R},M)$ then $\hat{m}(\hat{f}) \to \int f dm$ $(f \in L^1 \cap L^\infty)$ is the <u>Radon measure associated with</u> m in the way of Stone-Kakutani. If $M = m^*$, then the spaces $L^p_E(\mathcal{R},m^*)$ and $L^p_E(X,m^B)$ are isomorphic (E a Banach space, $1 \leq p \leq \infty$).

20.13. The completeness of $\mathcal{L}^\infty_E$ follows also from 7.8, 18.11, and the fact that M_∞ is an upper S-norm.

§2. Relations between spaces of measurable functions

Henceforth, $(\mathcal{R},M)$ is a fixed strong upper gauge on the set X. Furthermore, E, F, G are three Banach spaces with norms $\| \ \|$, and there is a bilinear map $E \times F$ to G, denoted by juxtaposition, of norm not exceeding one:

$$\|\xi\eta\| \le \|\xi\| \cdot \|\eta\| \qquad , \text{ for } \xi \in E, \ \eta \in F.$$

21. Relations between the L^p $(1 \le p \le \infty)$

A. Norm-relations

Let $1 \le r,p,q \le \infty$ be three numbers satisfying $1/r = 1/p + 1/q$. The product fg of a function f in $\mathcal{L}_E^p$ with a function g in $\mathcal{L}_F^q$ is measurable, and vanishes off a σ-finite set if either p or q is finite. From (18.4), the inequality of the mean, and the integrability criterion (19.3), fg is in $\mathcal{L}_G^r$, and

21.1. $$M_r(fg) \le M_p(f) \cdot M_q(g).$$

Therefore, every element g of L_F^q gives rise, by multiplication on the right, to a bounded linear map $R_g : f \to fg$ from L_E^p to L_G^r of operator norm $\|R_g\| \le M_q(g)$. In other words, there is a natural linear map

$$R: L_F^q \to L(L_E^p, L_G^r)$$

from L_F^q to the space of bounded linear maps from L_E^p to L_G^r. Also, this map has norm not exceeding one. We want to show that in a large number of instances R is an isometry, so that L_F^q can be identified with a closed linear subspace of $L(L_E^p, L_G^r)$.

This will clearly not be so if $E \times F \to G$ is the zero map; we need a condition on the bilinear map $E \times F \to G$:

21.2. $\|\xi\| = \sup\{\|\xi \times \eta\|: \eta \in F, \|\eta\| \leq 1\}$ for all $\xi \in E$.

Here are some instances where this is satisfied.

(1) $F = E'$, G is the coefficient field $\mathbb{R}$ or $\mathbb{C}$, and $E \times E' \to G$ is the duality $\langle ; \rangle$.

(2) $E = F'$, and $F' \times F \to \mathbb{R}, \mathbb{C}$ is the duality.

(3) $E = L(F,G)$, and $E \times F \to G$ is the application of $\xi \in E$ to $\eta \in F$.

The claim that R is an isometric embedding will evidently follow from the following theorem.

21.3. Theorem. If $f: X \to E$ is a measurable map and condition 21.2 holds, then

$$\overline{M}_p(f) = \sup\{M_r(fg): g \in \mathcal{L}_F^q,\ M_q(g) \leq 1\},$$

for $1 \leq r,p,q \leq \infty$ and $1/p + 1/q = 1/r$.

Proof. We shall assume that $\overline{M}_p(f) < \infty$ and leave the slight alterations necessary when $\overline{M}_p(f) = \infty$ to the reader.

Notice that both sides of the equation depend on f positive-homogeneously and continuously in $\overline{M}_p$-mean. As the measurable step functions are dense in $\mathcal{L}_E^p(\mathcal{R},\overline{M})$ (8.4), it may be assumed that f is a step function of norm $\overline{M}_p(f) = 1$. From 21.1, it need only be shown that the left hand side of the equation is no larger than the right hand side.

Suppose that $r = \infty$, so that $p = q = \infty$. Given an $\varepsilon > 0$, there is a non-negligible integrable set A on which f is constant: $f = \xi \in E$ on A say, with $\|\xi\| > 1 - \varepsilon$. There is an $\eta \in F_1$ with $\|\xi\eta\| > 1 - \varepsilon$. If $g = A\eta \in \mathcal{L}_F^\infty$ (18.10), then $M_\infty(g) \leq 1$ and $M_\infty(fg) > 1 - \varepsilon$.

When r is finite, define $N = M_r$. Then $M_p = N_{p/r}$, $M_q = N_{q/r}$, and $1/(p/r) + 1/(q/r) = 1$: it may be assumed that $r = 1$.

The subcase $p = \infty$ (and so $q = 1$) is proved as above: Given an $f \in L_E^\infty$ and an $\varepsilon > 0$, choose A, ξ, η as above and define $g = (M(A))^{-1}A\eta$. Then $M(g) = 1$ and $M(fg) = (M(A))^{-1}M(A\|\xi\|) \geq 1 - \varepsilon$.

Finally, assume that $r = 1$ and that both p and q are finite. Given a step function $f = \sum A_i\xi_i$ in $\mathcal{L}_E^p$, of M_p-norm one, and given an $\varepsilon > 0$, choose sets $B_i \in \mathscr{C}(\mathcal{R},M)$ such that $B_i \subset A_i$ and $\overline{M}(A_i - B_i) = 0$ (19.4); then $\overline{M}_p(f - \sum B_i\xi_i) = 0$. Next, choose vectors $\eta_i \in F$, of norm one and such that $\|\xi_i\eta_i\| \geq \|\xi_i\|(1 - \varepsilon)$, and put

$$g = \sum B_i\|\xi_i\|^{p/q}\eta_i.$$

Clearly, $M_q(g) = 1$. Since $p = p/q + 1$,

$$\|fg\| = \sum B_i\|\xi_i\|^{p/q}\|\xi_i\eta_i\| \geq (1 - \varepsilon) \sum B_i\|\xi_i\|^p,$$

and so $M(fg) \geq (1 - \varepsilon)M(\sum B_i\|\xi_i\|^p) = (1 - \varepsilon)\overline{M}(f^p) = (1 - \varepsilon)$. The proof is finished.#

<u>21.4. Corollary</u>. Let $(\mathcal{R},M)$ be an upper integral for the positive measure m, let $E = F'$, and let $G = \mathbb{R}$ or $\mathbb{C}$ be the coefficient field. Then

$$\overline{M}_{p'}(f) = \sup\{\int fg\,dm: g \in \mathcal{L}_F^p,\ M_p(f) \leq 1,\ fg \geq 0\}$$

for $f: X \to F'$ measurable and $1/p + 1/p' = 1$, $1 \leq p \leq \infty$. In particular, the duality

$$(f,g) \to \int fg\,dm \qquad , (f \in \mathcal{L}_{F'}^{p'},\ g \in \mathcal{L}_F^p)$$

identifies $L_{F'}^{p'}$ isometrically with a closed subspace of the dual $(L_F^p)'$ of L_F^p.

Proof. Given a measurable function $f\colon X \to F'$ and an $r < \overline{M}_{p'}(f)$, choose a $g \in \mathcal{L}_F^p$ with $M_p(g) = 1$ and $M(fg) > r$. The function $h = |fg|/fg$ is measurable and of modulus one, if suitably defined (18.12), and so $g' = hg$ is in $\mathcal{L}_F^p$ and $M_p(g') = 1$. fg' is positive and satisfies

$$\int fg'dm = M(fg') = M(fg) > r.\#$$

Numbers $p,p' \in [1,\infty]$ that satisfy $1/p + 1/p' = 1$ are called conjugate. It will be shown that $L^{p'}$ is the dual of L^p, for $1 < p < \infty$, following an investigation of the order relations between the L^p.

B. Order relations

21.5. Proposition. Let $(\mathcal{R},M)$ be a strong upper gauge, and let $1 \leq p,q,r \leq \infty$ be such that $1/r = 1/p + 1/q$.

Right multiplication $(f \to fg)$ by an element $g \in L^q$ is a positive map from L^p to L^r if and only if g is positive. In this instance, it is a normal Riesz space isomorphism from L^p into L^r.

If $S \subset L^p$ and $T \subset L^q$ are increasingly directed sets with suprema $s_o \in L^p$ and $t_o \in L^q$, respectively, then

$$ST = \{st\colon s \in S,\ t \in T\} \subset L^r$$

is also increasingly directed and has supremum $s_o t_o$ in L^r.

The simple proof is left to the reader.

21.6. Lemma. If $(\mathcal{R},M)$ is a strong upper gauge, then the order dual of the Riesz space $L^1(\mathcal{R},M)$ coincides with the dual of the Banach space $L^1(\mathcal{R},M)$.

Proof. A continuous linear functional t of the Banach space L^1 can be identified with an S-continuous measure on $\mathcal{L}^1$ that vanishes on the negligible functions. Since $\mathcal{L}^1$ is full, t has finite variation (8.12, 6.5).

Conversely, suppose t belongs to the order dual of L^1. It may be split into its positive and negative parts, and it suffices to show that these are norm-continuous; therefore t may be assumed to be positive. If t is not norm-continuous, then there is a sequence (φ_n) of positive integrable functions of norm one such that $t(\varphi_n) \geq 2^n$. The function $\varphi = \sum 2^{-n}\varphi_n$ is integrable and has $t(\varphi) = \infty$, a contradiction.#

21.7. Theorem. If $(\mathcal{R},M)$ is an upper integral for the positive measure m, and if $F = \mathbb{R}$ or $F = \mathbb{C}$, then $L_F^{p'}(\mathcal{R},M)$ is the dual of the Banach space $L_F^p(\mathcal{R},M)$ under the duality

$$(f,g) \to \int fg\,dm \qquad , \; f \in L_F^p, \; g \in L_F^{p'},$$

provided that $1 < p < \infty$ and $1/p + 1/p' = 1$.

L_F^∞ is the dual of L_F^1 if and only if $L_{\mathbb{R}}^\infty$ is a complete Riesz space.

Proof. We shall assume that $F = \mathbb{R}$ and leave the simple extension to the case $F = \mathbb{C}$ to the reader. For $1 \leq p < \infty$, let $t\colon L^p \to \mathbb{R}$ be a continuous linear functional. From the lemma, t may be assumed to be positive. Identifying it with an S-continuous measure on $\mathcal{R}^S$ that vanishes on the M-negligible sets, we see that it is absolutely continuous with respect to m (6.7), and so $t = \sup\{\varphi m\colon \varphi \in \mathcal{R}_+^S, \varphi m \leq t\}$ (6.19).

Let g denote the supremum, in $L^{p'}$, of the classes $\{\varphi^{\bullet}: \varphi \in \mathcal{R}_+^S, \varphi m \leq t\} = \Phi$. Since $L^{p'}$ is complete for $1 < p < \infty$ (8.13), and by assumption for $p = 1$, g exists. Indeed, for $\varphi \in \Phi$,

$$M_{p'}(\varphi) = \sup\{\int \varphi\psi\, dm: \psi \in \mathcal{R}_+, M_p(\psi) \leq 1\} \leq \|t\|$$

so that Φ is bounded. Now,

$$t(\psi) = \sup\{\int \varphi\psi\, dm: \varphi \in \Phi\} = \int g\psi dm,$$

by continuity in M_p-mean. Hence $t(f) = \int fgdm$ for all $f \in L^p$, as claimed.

Since the dual of any Riesz space is complete (3.4), the condition that L^{∞} be complete is also necessary.#

C. Localizable upper gauges

An upper gauge $(\mathcal{R},M)$ is said to be localizable if $L^{\infty}(\mathcal{R},M)$ is a complete Riesz space.

21.8. Proposition. A strictly localizable upper gauge $(\mathcal{R},M)$ is localizable.

In particular, a strong upper gauge $(\mathcal{R},M)$ is localizable if it is B-continuous, or σ-finite, or if $\mathcal{R}$ is first-uncountably generated (14A).

Proof. Let P be an $(\mathcal{R},M)$-adequate partition and let S be an increasingly directed family in L^{∞}, bounded from above. For $K \in P$, $SK^{\bullet}$ has a supremum $f_K^{\bullet}$ in L^1 (21.5). For each $K \in P$ select an $f_K \in f_K^{\bullet}$ with $f_K = f_K \cdot K$, and define f equal to f_K on each $K \in P$ and zero outside $\cup P$. Since P is an adequate cover and f is measurable on each K in P, f is measurable (18.9). Clearly, $f^{\bullet}$ is the supremum of S in L^{∞}.#

D. Supplements

21.9. A linear form t on a Riesz space R is said to be <u>order continuous</u> if whenever $\Phi \subset R_+$ is decreasingly directed with infimum zero then $\lim t(\Phi) = 0$. 21.7 identifies L^1 with the set of order continuous linear forms on L^∞.

21.10. If $(\mathcal{R},m)$ is any positive S-measure, then the dual of $L^1(\mathcal{R},m^S)$ is isomorphic with $L^\infty(\hat{\mathcal{R}},\hat{m}^B)$.

21.11. Let $1 < p < \infty$ and $1/p + 1/p' = 1$. If $g \in \mathcal{L}_F^{p'}(\mathcal{R},M)$, where $(\mathcal{R},M)$ is an upper integral for the positive measure m, then

$$\varphi \to \int \varphi g dm \qquad , \varphi \in \mathcal{R},$$

is a (norm) M_p-compact F-valued measure on $\mathcal{R}$ (Adapt the proof of 10.9).

LITERATURE: [14], [19]. Cf also [25].

22. The theorem of Radon-Nikodym

A. Measures defined by densities

In this subsection, $(\mathcal{R},M)$ is a fixed strong upper gauge on X.

Definition. A Banach-valued function $g\colon X \to F$ is said to be locally $(\mathcal{R},M)$-integrable if $g\varphi$ is $(\mathcal{R},M)$-integrable for every bounded and $\mathcal{R}$-dominated function $\varphi \in \mathcal{R}_0$.

Using repeatedly linearity and Lebesgue's theorem, the following characterization is easily verified.

22.1. Lemma. The function $g\colon X \to F$ is locally $(\mathcal{R},M)$-integrable if and only if either of the following conditions is satisfied.

(i) gf is integrable for each $\mathcal{R}$-dominated, bounded, and integrable function $f\colon X \to \mathbb{R}$.

(ii) gK is integrable for all K in some dense family of dominated integrable sets, or for all K in an adequate cover consisting of dominated sets.

It follows that a locally integrable function g is measurable (18.9), and that $\|g\|$ is also locally integrable.

Let m be a positive scalar measure majorized by M. By

$$n(\varphi) = (gm)(\varphi) = \int \varphi g dm \qquad (\varphi \in \mathcal{R}_0)$$

a new, F-valued, measure $n = gm$ is defined on $\mathcal{R}_0$. Since

$$\|n(\varphi)\| \leq \int |\varphi|\,\|g\|dm = \sup\{\int |\varphi|(\|g\| \wedge k)dm\colon k = 1, 2, \ldots\}$$

n has finite variation, $|n| \leq \|g\| \cdot m$, and is absolutely continuous with respect to m (2.20). Hence gm is *-continuous if m is (* = S or B; (3.10)). Evidently, gm depends only on the class $g^{\cdot}$ of g modulo locally negligible functions.

<u>Definition</u>. gm is called the measure <u>with density</u> g <u>and</u> <u>base</u> m.

Conversely, let $n: \mathcal{R} \to F$ be a measure which is absolutely continuous with respect to m. Then n is said to have a <u>locally integrable derivative</u> $g = dn/dm$ with respect to n, provided there is a $g \in \mathcal{L}_F^{\ell}(\mathcal{R},M)$ with $n = gm$. Here $\mathcal{L}_F^{\ell}(\mathcal{R},M)$ denotes the vector space of all E-valued locally integrable functions, and $L_F^{\ell}(\mathcal{R},M)$ denotes their classes modulo locally negligible functions. As usual, the suffix is suppressed if $F = \mathbb{R}$.#

<u>22.2. Lemma</u>. gm has variation $\|g\|m$.

<u>Proof</u>. Let $A \in U(g)$ be a dominated set. Given any $\varepsilon > 0$, approximate g uniformly on A to within ε by an integrable step function $\sum \xi_i B_i$ ($\xi_i \in F$, $B \in \mathfrak{C}_0(\mathcal{R},M)$, the sum finite: 19.6), and put $A_i = AB_i$. Then

$$\int Ad|n| \geq \sum \|\int A_i dn\| = \sum \|\int A_i g dm\| \geq (1 - \varepsilon) \sum \|\int \xi_i A_i dm\|$$

$$= (1 - \varepsilon) \sum \int \|\xi_i\| A_i dm \geq (1 - \varepsilon)^2 \int \|g\| A dm \geq (1 - \varepsilon)^2 \int Ad|n|.$$

Hence $\int Ad|n| = \int A\|g\|dm$ for all dominated sets $A \in U(g)$. Linearity and continuity yield $\int fd|n| = \int f\|g\|dm$ for all bounded, dominated, integrable functions; in particular, for all $f \in \mathcal{R}_0$. Finally $|n|(\varphi) = \int \varphi\|g\|dm$ for $\varphi \in \mathcal{R}_0$.#

22.3. Proposition. Let $(\mathcal{R},M)$ be an upper integral for the positive measure m on $\mathcal{R}$. The map $g \to gm$ from $L^{\ell}_F(\mathcal{R},M)$ to $(m)_F$ is linear and one-to-one and satisfies

$$|gm| = \|g\|m.$$

It is a normal Riesz space isomorphism of the Riesz space $L^{\ell}(\mathcal{R},M)$ into (m).

Proof. The first statement is obvious. The lemma shows that $g \to gm$ is a Riesz space homomorphism from L^{ℓ} into (m) (2D). To see that it is normal, let $S \subset L^{\ell}$ be an increasingly directed family with supremum $s_0 \in L^{\ell}$. Then s_0m majorizes the set $Sm = \{sm: s \in S\}$, and

$$(s_0m)(\varphi) = \sup\{(sm)(\varphi): s \in S\}$$

for all $\varphi \in \mathcal{R}_{o+}$. Indeed, if $\varphi \in \mathcal{R}_{o+}$ satisfies $\sup\{(sm)(\varphi): s \in S\} < (s_0m)(\varphi)$, then there is a supremum $f^{\cdot}$ in L^1 of the classes $\{(s\varphi)^{\cdot}: s \in S\}$; and $f < s_0\varphi - \varepsilon$ on some set $K \in \mathscr{C}_0(\mathcal{R},M)$ with $Kdm = M(K) > 0$ for some $\varepsilon > 0$ (8.6). Then $s_0 - \varepsilon K\varphi^{-1}$ is an upper bound for S and is genuinely smaller than s_0, which is a contradiction.#

B. The integration theory of gm

We shall now assume that $(\mathcal{R},M)$ is a strong and essential upper gauge majorizing the positive measure m. Furthermore, E, F, G are three Banach spaces, and there is a bilinear map from $E \times F$ to G, of norm not exceeding one and denoted by juxtaposition $(\|\xi\eta\| \leq \|\xi\|\cdot\|\eta\|$ for $\xi \in F$ and $\eta \in F)$. For the integration of $n = gm$ $(g \in \mathcal{L}^{\ell}_F(\mathcal{R},M))$ we use the upper gauge $(\mathcal{R},N)$ defined by

$$N(f) = (\|g\|M)(f) = M(\|g\|f) \qquad , \ f: X \to \overline{\mathbb{R}_+}.$$

22.4. Theorem. (i) $(\mathcal{R},N)$ is an essential upper gauge majorizing n in variation; it is B-continuous or an upper integral if $(\mathcal{R},M)$ is.

(ii) A function $f: X \to E$ is $(\mathcal{R},N)$-integrable (N-negligible) if and only if $\|g\|f$ is $(\mathcal{R},M)$-integrable, and in this instance

$$\int f dn = \int f g dm \in G.$$

(iii) A function f on X with values in a uniform space is $(\mathcal{R},N)$-measurable if and only if it is $(\mathcal{R},M)$-measurable on $[g \neq 0]$.

Proof. N is evidently an upper S-norm and is finite on $\mathcal{R}_{o+}$. Let (φ_n) be a sequence in $\mathcal{R}_0 \otimes E$. If $(\varphi_n\|g\|)$ converges M-almost everywhere (or is M-Cauchy) then (φ_n) clearly converges N-almost everywhere (or is N-Cauchy), and vice versa. This shows that $(\mathcal{R},N)$ is an upper gauge, and that the equation $\int f dn = \int f g dm$ holds. If $(\mathcal{R},M)$ is an upper integral, N is additive on $\mathcal{R}_{o+}$. From 15.2 $(\mathcal{R},N)$ is B-continuous if $(\mathcal{R},M)$ is.

There is only (iii) left to prove. Since the $\mathcal{R}$-dominated $(\mathcal{R},M)$-integrable (-negligible) and the $(\mathcal{R},N)$-integrable (-negligible) subsets of $[g \neq 0]$ coincide, the family U(f) is $(\mathcal{R},M)$-dense in $[g \neq 0]$ if and only if it is $(\mathcal{R},N)$-dense in $[g \neq 0]$.#

C. The existence of scalar derivatives

It will now be investigated when a scalar measure n absolutely continuous with respect to m has a derivative dn/dm.

22.5. Lemma. $L^{\ell}(\mathcal{R},M)$ is a complete Riesz space if and only if $L^{\infty}(\mathcal{R},M)$ is, that is if and only if $(\mathcal{R},M)$ is localizable.

Proof. Suppose that $(\mathcal{R},M)$ is localizable and let $S \subset L^{\ell}_{+}$ be increasingly directed and bounded from above. For $k = 1, 2, \ldots$ the sets $S \wedge k = \{s \wedge k: s \in S\} \subset L^{\infty}_{+}$ each have a supremum s_k, and $\sup\{s_k: k = 1, 2, \ldots\} = \lim(s_k)$ is clearly the supremum of S. The converse is obvious.#

22.6. Theorem. (Radon-Nikodym). Suppose $(\mathcal{R},M)$ is an upper integral for the positive measure m.

Then every scalar measure n absolutely continuous with respect to m has a locally $(\mathcal{R},M)$-integrable derivative $g = dn/dm$ if and only if $(\mathcal{R},M)$ is localizable.

Proof. If every measure $n \in (m)$ has a derivative $dn/dm \in L^{\ell}$, then L^{ℓ} is Riesz space isomorphic with (m) and thus is complete (3.4, 2.18).

Conversely, suppose $(\mathcal{R},M)$ is localizable and let $n \in (m)$. It may be assumed that n is positive. For every integer $k > 0$, the measure $n \wedge km$ $(\leq km)$ is continuous in M-mean and so can be extended to L^1. The linear functional $n \wedge km$ on L^1 is of the form $\varphi \to \int \varphi f_k dm$ for some $f_k \in L^{\infty}$ (21.7), and so

$$n(\varphi) = \sup\{(n \wedge km)(\varphi): k = 1, 2, \ldots\} = \sup \int \varphi f_k dm = \int (\sup f_k)\varphi dm$$

for all $\varphi \in \mathcal{R}_{o+}$. This shows at once that $\sup\{f_k: k = 1, 2, \ldots\}$ is locally integrable and is a derivative for n.

If n is complex-valued, then $dn^r/dm + idn^i/dm$ is clearly a derivative dn/dm (3.14).#

If $(\mathcal{R},M)$ is not localizable, there still exists a derivative for a large number of measures absolutely continuous with respect to m.

Definition. A measure n of finite variation on $\mathcal{R}$ is bounded (Cf. 13.12) if

$$\|n\| := \sup\{|n|(\varphi): \varphi \in \mathcal{R},\ 0 \leq \varphi \leq 1\}$$

is finite. The measure n is σ-bounded if there is an increasing sequence (φ_k) in $\mathcal{R}$ with $0 \leq \varphi_k \leq 1$ such that

(1) $$|n|(\varphi) = \sup\{|n|(\varphi\varphi_k): k = 1, 2, \ldots\}\ , \text{ for all } \varphi \text{ in } \mathcal{R}_+.$$

An S-measure n is bounded or σ-bounded iff $(\mathcal{R}, \overline{n}^S)$ is (13.12). A bounded measure is σ-bounded.#

22.7. Proposition. Let $(\mathcal{R},M)$ be an upper integral for the positive measure m. Every σ-bounded scalar measure $n \ll m$ has a locally $(\mathcal{R},M)$-integrable derivative with respect to m.

Proof. We may assume that n is positive. Let (φ_k) be as in (1). If $\varphi_k n$ has a derivative g_k for all k, the proof is finished; for then $n(\varphi) = \sup n(\varphi_k\varphi) = \sup \int \varphi f_k dm$, which shows at once that $\sup\{f_k: k = 1, 2, \ldots\}$ is locally integrable and is a derivative for n. It may be assumed therefore that n is bounded. If so, let F denote the increasingly directed set of classes $f^\cdot$ in $L^1_+(\mathcal{R},M)$ satisfying $fm \leq n$. Then $M(f^\cdot) = \sup\{\int f\varphi dm: 0 \leq \varphi \leq 1,\ \varphi \in \mathcal{R}\} \leq \|n\|$, so that F is bounded from above. Any representative in the supremum of F in L^1 is evidently a derivative for n with respect to m.#

D. Supplements

22.8*. $(\|g\|m)^{*-} = \|g\|\widetilde{m}^{*}$ for $m \in M_+^*(\mathcal{R})$.

22.9*. Chain rule. If $g = dn/dm \in \mathcal{L}^{\ell}(\mathcal{R},M)$ and $h = dr/dn \in \mathcal{L}^{\ell}(\mathcal{R}, |g|M)$, then $dr/dm = gh \in \mathcal{L}^{\ell}(\mathcal{R}, |gh|M)$ and $|gh|M = |g|(|h|M)$.

22.10. A locally $(\mathcal{R},M_q)$-integrable function is locally $(\mathcal{R},M_p)$-integrable, for $1 \leq p \leq q \leq \infty$. (See 12.10)

22.11. $\cup\{\mathcal{L}_E^p : 1 \leq p \leq \infty\} \subset \mathcal{L}_E^{\ell}$.

22.12. Suppose m and n are (purely additive) scalar measures of finite variation on $\mathcal{R}$, n is absolutely continuous with respect to m, and $|n|$ is finite. Then there exists a sequence (φ_n) in $\mathcal{R}_0$ such that

$$n(\varphi) = \lim(m(\varphi\varphi_n)) \qquad \text{for all } \varphi \in \mathcal{R}_0.$$

(Hint: approximate $d\hat{n}/d\hat{m}$ in $\hat{m}^B$-mean.)

22.13. Let $(\mathcal{R},M)$ be an upper integral for the positive measure m and let $g \in \mathcal{L}_F^{\ell}(\mathcal{R},M)$. The measure

$$\varphi \to \int \varphi g dm \qquad , \varphi \in \mathcal{R},$$

is norm-compact. (Use 10.9.)

LITERATURE: [11], [19].

23. The Riesz-Thorin convexity theorem

The main result of this number is a celebrated tool in harmonic analysis, and is a model for a large part of approximation theory. It will not be used in the sequel, though.

A. The three lines theorem

This is a corollary to the maximum principle for strips in the plane. The latter reads as follows.

23.1. Theorem. Let $S = \{z \in \mathbb{C} : x_0 \leq x = \mathrm{Re}(z) \leq x_1\}$ be a vertical strip and let f be a bounded and continuous complex function on S, analytic in the interior of S. For $x_0 \leq x \leq x_1$ put

$$M_f(x) = \sup\{|f(z)| : \mathrm{Re}(z) = x, \ -\infty < \mathrm{Im}(z) < \infty\}.$$

Then

$$|f(z)| \leq \max\{M_f(x_0), M_f(x_1)\} \qquad , \text{ for } z \in S.$$

Proof. Set $A = \max\{|x_0|, |x_1|\}$, and for $n \in \mathbb{N}$ put $f_n(z) = f(z)\cdot\exp(n^{-1}(z^2 - A^2))$. As $|y| \to \infty$, $\mathrm{Re}(z^2) = (x^2 - A^2) - y^2 \to -\infty$ and so $f_n(z) \to 0$ uniformly in x. For sufficiently large η, $|f_n(z)|$ becomes arbitrarily small on $|y| = \eta$. From the maximum principle for compact sets, $|f_n|$ attains its maximum in a point of the vertical boundary, $x = x_0$ or $x = x_1$, of the rectangle $x_0 \leq x \leq x_1$, $-\eta \leq y \leq \eta$, and so the conclusion of the theorem holds for f_n. As $n \to \infty$, the f_n converge to f, and, since $\mathrm{Re}(z^2) < 0$, $|f_n(z)| = |f(z)|\exp(n^{-1}\mathrm{Re}(z^2))$ tends increasingly to $|f(z)|$. Hence

$$|f(z)| = \sup|f_n(z)| \leq \sup(M_{f_n}(x_0) \vee M_{f_n}(x_1)) = \sup M_{f_n}(x_0) \vee \sup M_{f_n}(x_1) = M_f(x_0) \vee M_f(x_1)$$

for $z \in S$, as desired.#

23.2. Corollary. (Three lines theorem). The function $x \to \log M_f(x)$ is convex for $x_0 \leq x \leq x_1$.

Proof. Put $x_0 \leq a \leq c \leq x_1$, $0 \leq \Theta \leq 1$, and $b = \Theta a + (1-\Theta)c$. For any real number t, the function $F(z) = e^{tz}f(z)$ is also analytic and bounded in S. By choosing

$$t = (\log M_f(a) - \log M_f(c))/c - a,$$

we get $M_F(a) = e^{ta}M_f(a) = e^{tc}M_f(c) = M_F(c)$. (If $f \equiv 0$ then $\log M_f(x) \equiv 0$, and there is nothing to prove. If $f \not\equiv 0$ then $M_f(x) > 0$ for all x as the zeroes of f do not cluster.) From the theorem,

$$e^{tb}M_f(b) = M_F(b) \leq M_F(a) = M_F(c).$$

Taking logarithms, using $c - b = \Theta(c - a)$, and substituting for t,

$$\log M_f(b) \leq -tb + tc + \log M_f(c) = \Theta t(a-c) + \log M_f(c) = \Theta \log M_f(a) + (1-\Theta)\log M_f(c).$$

B. The Riesz-Thorin theorem

This theorem applies to the following situation. $(\mathcal{R},M)$ and $(\mathcal{S},N)$ are two essential upper integrals; E,F are two complex Banach spaces; and T is a complex-linear map from $\mathcal{R} \otimes E$ to $\mathcal{L}_F(\mathcal{S},N)$. The problem answered by the Riesz-Thorin theorem is this: For which p,q can T be extended to a continuous linear map from $L_E^p(\mathcal{R},M)$ to $L_E^q(\mathcal{R},M)$?

For this to be possible, T must map M-negligible functions to zero, and the operator norm

(1) $$\|T\|_{p,q} = \sup\{N_q(T\varphi): \varphi \in \mathcal{R} \otimes E,\ M_p(\varphi) \leq 1\}$$

must be finite. Except if $p = \infty$, these conditions are also sufficient, as $\mathcal{R} \otimes E$ is dense in $\mathcal{L}_E^p(\mathcal{R},M)$.

<u>23.3. Theorem</u>. (Riesz-Thorin). The function $G: (a,b) \to \log \|T\|_{1/a,1/b}$ is a convex function of (a,b) in the unit rectangle $0 \le a, b \le 1$. In other words, if $0 \le \theta \le 1$, $1/p = \theta/p_0 + (1-\theta)/p_1$, and $1/q = \theta/q_0 + (1-\theta)/q_1$, then $\|T\|_{p,q} \le \|T\|^{\theta}_{p_0 q_0} \|T\|^{1-\theta}_{p_1 q_1}$.

<u>Proof</u>. If $\|T\|_{p,q} = 0$ for some pair p,q, then $T = 0$, and $G(a,b) = \log \|T\|_{1/a,1/b}$ is identically equal to $-\infty$. This case may therefore be excluded.

It is clearly sufficient to show that the set C of points (a,b) where $G(a,b) < \infty$ is convex and that the restriction of G to C is convex.

Next, if $\|T\|_{p,q} < \infty$ for some pair p,q then T can be extended to all p-integrable step functions. It may therefore be assumed that $\mathfrak{R}$ contains all integrable step functions. Let m,n denote the restrictions of M,N to $\mathfrak{R},\mathfrak{S}$ respectively.

From 21.4 and (1)

$$(2) \qquad \|T\|_{1/a,1/b} = \sup\{\,|\int \langle Tf;g\rangle dn| : M_{1/a}(f) \le 1,\ N_{1/1-b}(g) \le 1\}$$

where the supremum is taken over all step functions $f \in \mathfrak{R} \otimes E$ and all $(\mathfrak{S},N)$-integrable step functions g with values in F'. Let (a_0,b_0) and (a_1,b_1) be in C, and put

$$a(\zeta) = (1-\zeta)a_0 + \zeta a_1 \quad \text{and} \quad b(\zeta) = (1-\zeta)b_0 + \zeta b_1$$

where $\zeta = \xi + i\eta$ lies in the strip $0 \le \xi \le 1$. The point $(a,b) = (a(t),b(t))$ lies on the line joining (a_0,b_0) with (a_1,b_1), for $0 < t < 1$. Note that $a \ne 0$ and $b \ne 1$.

It has to be shown that $G(a,b) \leq (1-t)G(a_0,b_0) + tG(a_1,b_1)$. For this, it is sufficient to show that, for all step functions f,g as above, the integral $I = \int \langle Tf;g\rangle dn$ obeys

(3) $$\log|I| \leq (1-t)G(a_0,b_0) + tG(a_1,b_1).$$

Choose f and g as follows:

$$f = \sum r_i^a \xi_i A_i \quad \text{with} \quad \sum r_i M(A_i) \leq 1,$$

where the $\xi_i \in E$ have norm one, the A_i are disjoint in $\mathscr{C}(\mathscr{R},M)$, and where $r_i > 0$. The second equation expresses the fact that $M_{1/a}(f) \leq 1$. Similarly,

$$g = \sum s_j^{1-b} \eta_j B_j \quad \text{with} \quad \sum s_j N(B_i) \leq 1,$$

where the B_j are disjoint in $\mathscr{C}(\mathscr{S},N)$, the $\eta_j \in F'$ are of norm one, and where $s_j > 0$. The desired inequality (3) will follow from the three lines theorem applied to the analytic function $G(\zeta)$ defined next. Put

$$f_\zeta = \sum r_i^{a(\zeta)} \xi_i A_i, \quad g_\zeta = \sum s_j^{1-b(\zeta)} \eta_j B_j, \quad \text{and}$$

$$G(\zeta) = \int \langle Tf_\zeta; g_\zeta\rangle dn = \sum_{i,j} r_i^{a(\zeta)} s_j^{1-b(\zeta)} \int \langle TA_i\xi_i; \eta_j B_j\rangle dn .$$

Clearly, $G(\zeta)$ is analytic and bounded in the strip $0 \leq \xi = \mathrm{Re}(\zeta) \leq 1$. From Hölders inequality and the definition of $\|T\|_{1/\alpha,1/\beta}$,

(4) $$|G(\zeta)| \leq \|T\|_{1/\alpha,1/\beta}\, M_{1/\alpha}(f_\zeta)\, N_{1/1-\beta}(g_\zeta),$$

for arbitrary (α,β) in the unit rectangle. The two norms on the right can be calculated:

$$M_{1/\alpha}(f_\zeta) = \begin{cases} M\left(\sum\left(r_i^{(1-\xi)a_0+\xi a_1} A_i\right)^{1/\alpha}\right)^{\alpha} & \text{for } \alpha \neq 0 \\ \sup\left\{r_i^{(1-\xi)a_0+\xi a_1} : i\right\} & \text{for } \alpha = 0 \end{cases}$$

(5)

$$N_{1/1-\beta}(g_\zeta) = \begin{cases} N\left(\sum\left(s_j^{1-(1-\xi)b_0-\xi b_1} B_j\right)^{1-1/\beta}\right)^{1-\beta} & \text{for } \beta \neq 1 \\ \sup\left\{s_j^{1-(1-\xi)b_0-\xi b_1} : j\right\} & \text{for } \beta = 1 \end{cases}$$

(4) and (5) reduce at $\xi = 0$, $\alpha = a_0$, $\beta = b_0$ to

$$M_{1/a_0}(f_{i\eta}) \leq 1 , \quad N_{1/1-b_0}(g_{i\eta}) \leq 1 , \quad \text{and} \quad |G(i\eta)| \leq \|T\|_{1/a_0, 1/b_0} ,$$

while at $\xi = 1$, $\alpha = a_1$, $\beta = b_1$ they become

$$M_{1/a_1}(f_{1+i\eta}) \leq 1 , \quad N_{1/1-b_1}(g_{1+i\eta}) \leq 1 , \quad \text{and} \quad |G(1+i\eta)| \leq \|T\|_{1/a_1, 1/b_1} .$$

By using the three lines theorem on $G(\zeta)$, the desired inequality (3) is obtained

$$\log|I| = \log|G(t)| \leq \log M_G(t) \leq (1-t)\log M_G(0) + t \log M_G(1)$$

$$\leq (1-t)G(a_0,b_0) + tG(a_1,b_1) .\#$$

<u>23.4. Corollary</u>. Let $T: \mathcal{R} \otimes E \to \mathcal{L}_F(\mathcal{S},N)$ be a linear map such that $N_\infty(Tf) \leq M_\infty(f)$ and $N_1(Tf) \leq M_1(f)$ for $f \in \mathcal{R} \otimes E$. Then T has a unique continuous extension to a map from $L_E^p(\mathcal{R},M)$ to $L_F^p(\mathcal{S},N)$, for all p in $[1,\infty)$.

LITERATURE: [14].

24. L. Schwartz's tight measures

A. Definition. Let $\mathcal{R}$ be again a dominated integration lattice on X. Unlike 4B, each $\mathcal{R}[K]$ will now be equipped with the uniformity of uniform convergence on compact sets, for $K \in \mathcal{K}(\mathcal{R})$. The inductive limit of these uniformities, on $\mathcal{R} = \cup\{\mathcal{R}[K] : K \in \mathcal{K}(\mathcal{R})\}$, is called the uniformity of dominated uniform convergence on compacta.#

Compactness is understood, of course, in the sense of the $\mathcal{R}$-uniformity on X. If X is locally compact and $\mathcal{R} = C_{oo}(X)$, then this new uniformity coincides with the uniformity of dominated uniform convergence.[1)]

Definition. (i) A Banach-valued measure on $\mathcal{R}$ is tight if its variation is continuous on the unit ball

$$\mathcal{R}_1 = \{\varphi \in \mathcal{R} : |\varphi| \leq 1\}$$

in the uniformity of dominated uniform convergence on compacta. $M^T(\mathcal{R};E)$ denotes the set of tight E-valued measures of finite variation.

(ii) An upper gauge $(\mathcal{R},M)$ is tight if the compact sets are $(\mathcal{R},M)$-dense.#

A Radon measure is tight. The proof, 4.11, that Radon measures are B-continuous applies literally to tight measures and shows that a tight measure is B-continuous.

B. Example: Point measures. If $x \in X$ and $\xi \in E$, then

$$\varphi \to \xi\varphi(x) := \xi\delta_x(\varphi) \qquad , \varphi \in \mathcal{R}$$

defines the point measure $\xi\delta_x$. It has variation $\|\xi\|\delta_x$ and is clearly

1) "compact" shall include "closed".

tight, the closure of $\{x\}$ being a compact set. The measure $\delta_x: \varphi \to \varphi(x)$ is called the Dirac measure in $x \in X$. Any measure in the band $M^A(\mathcal{R},E)$ spanned by the E-valued point measures is called atomic or discrete. It is easy to check that every atomic measure is of the form

$$m = \sum \xi_i \delta_{x_i}$$

with $\xi_i \in E$, $x_i \in X$ and $\sum\{\|\xi_i\|: x_i \in K\} < \infty$, for any $K \in \mathcal{K}(\mathcal{R})$. Its variation is

$$|m| = \sum \|\xi_i\| \delta_{x_i}$$

and is tight. A measure m whose variation $|m|$ is disjoint from $M^A(\mathcal{R})$ is called diffuse or non-atomic. The set of diffuse E-valued measures is denoted by $M^D(\mathcal{R};E)$. $M^D(\mathcal{R}) = M^D(\mathcal{R};\mathbb{R})$ is a band, and

$$M^D \perp M^A \subset M^T \subset M^B \subset M^S .$$

C. Characterization of tightness

24.1. Theorem. Suppose $\mathcal{R}$ is a dominated integration lattice and m is a positive B-measure on $\mathcal{R}$. The following are equivalent:

(i) m is tight.

(ii) $(\mathcal{R}, m^B)$ is tight.

(iii) There is a tight upper gauge majorizing m.

(iv) $\pi(X)$ is $\hat{m}^B$-measurable in $S(\mathcal{R})$.

(v) $S(\mathcal{R}) - \pi(X)$ is locally $\hat{m}^B$-negligible.

Proof. (i) $\Rightarrow$ (ii): Let A be in the $(\mathcal{R},m^B)$-dense family $\mathcal{K}^B(\mathcal{R})$ (9.4), and choose a $\varphi \in \mathcal{R}$ such that $A \leq \varphi \leq 1$. Every compact (and closed) subset K of X belongs to $\mathcal{K}^B(\mathcal{R})$. Indeed, $\pi(K)$ is compact in $S(\mathcal{R})$ (5D), and so is the infimum of the functions in $C_{oo}(S(\mathcal{R}))$ majorizing it (Urysohn's lemma). Since $\pi(K)$ is also the infimum of the functions in $\hat{\mathcal{R}}$ majorizing it (5.6), $K = \pi^{-1}(\pi(K))$ belongs to $\mathcal{K}(\mathcal{R})$.

Given any $\varepsilon > 0$, we can therefore find a $\psi_K \in \mathcal{R}$ such that $K \leq \psi_K$ and $m(\psi_K) \leq \varepsilon + m^B(K)$ (9.3). For every compact set K, define $\varphi_K = \varphi \wedge \psi_K$. The (φ_K) converge to φ uniformly on compact sets, and lie in $\mathcal{R}_1 \cap \mathcal{R}[[\varphi \neq 0]]$; and so

$$m^B(A) \leq m(\varphi) = \lim m(\varphi_K) \leq \varepsilon + \sup m^B(A \cap K),$$

where K ranges over the compact (closed) subsets of X. Since the sets $A \cap K$ are compact and closed, too, this shows that the compact subsets of A are dense in A, and hence are dense in every integrable set.

(ii) $\Rightarrow$ (iii) is clear.

(iii) $\Rightarrow$ (i): Let $(\mathcal{R}, M)$ be a tight upper gauge majorizing m. It has to be shown that m is continuous on $\mathcal{R}_1[K]$ in the topology of uniform convergence on compacta, for all $K \in \mathcal{K}(\mathcal{R})$ (4.1). Since the integrable sets are cofinal in $\mathcal{K}(\mathcal{R})$, it may be assumed that K is integrable. Suppose, then, that the net (φ_i) in $\mathcal{R}_1[K]$ converges to zero uniformly on compact sets. Given any $\varepsilon > 0$, choose a compact set $L \subset K$ for which $M(K - L) < \varepsilon$. For all indices i such that $|\varphi_i| < \varepsilon/M(L)$ on L,

$$|m(\varphi_i)| \leq M(\varphi_i) \leq M(\varphi_i L) + M(\varphi_i (K - L)) \leq \varepsilon + \varepsilon = 2\varepsilon,$$

and so $m(\varphi_i) \to 0$.

(ii) $\Rightarrow$ (v): Let a $K \in \mathcal{K}^B(\hat{\mathcal{R}})$ and an $\varepsilon > 0$ be given. Then $\pi^{-1}(K)$ is in $\mathcal{K}^B(\mathcal{R})$, and there is a compact set $L \subset \pi^{-1}(K)$ such that

$m^B(\pi^{-1}(K) - L) < \varepsilon$. Clearly, $\hat{m}^B(K - \pi(L)) < \varepsilon$. Since L is compact it is $(\hat{\mathcal{R}}, \hat{m}^B)$-integrable, and thus $\hat{m}^B(K \setminus \pi(X)) = 0$. Therefore $\pi(X)$ has locally $\hat{m}^B$-negligible complement in $S(\mathcal{R})$.

(v) $\Rightarrow$ (iv) is clear.

(iv) $\Rightarrow$ (ii): Let an $A \in \mathcal{K}^B(\mathcal{R})$ and $\varepsilon > 0$ be given. Since $\pi(A) = \overline{\pi(A)} \cap \pi(X)$ is $\hat{m}^B$-integrable, there is a compact subset K of $\pi(A)$ such that $\hat{m}^B(\pi(A) - K) < \varepsilon$. Hence $\pi^{-1}(K) \subset A$ and $m^B(A - \pi^{-1}(K)) < \varepsilon$. Since the $\mathcal{R}$-open sets are exactly the pre-images, under π, of the open sets in $S(\mathcal{R})$ (5.11), $\pi^{-1}(K)$ is compact in X. The compact sets are therefore $(\mathcal{R}, m^B)$-dense.#

<u>24.2. Corollary</u>. $M^T(\mathcal{R})$ is a band in $M^B(\mathcal{R})$.

<u>Proof</u>. To show that M^T is an ideal, let $m, n \in M^T_+$ be given. Since $(m+n)^{\wedge B-} = \hat{\bar{m}}^B + \hat{\bar{n}}^B$ (13.10), $S(\mathcal{R}) - \pi(X)$ is locally $(\hat{m}+\hat{n})^B$-negligible, and so $m+n$ is tight.

If $S(\mathcal{R}) - \pi(X)$ is locally $\hat{m}^B$-negligible for all m in an increasingly directed subset G of M^T_+, then it is locally $\hat{n}^B$-negligible, where $n = \sup G$ (13.10). Hence $n \in M^T$, and M^T is a band.#

<u>24.3. Example</u>. Suppose X is a completely regular topological space and let $\mathcal{K}_c$ denote the family of compact subsets of X. Bourbaki [5, Ch. IX] calls <u>pre-measure</u> a map μ which associates with every $K \in \mathcal{K}_c$ a positive Radon measure μ_K on K in such a way that, for $K \subset L$ in $\mathcal{K}_c$, μ_K is the measure induced on K by μ_L (15C). A pre-measure μ is called locally bounded if every point $x \in X$ has a neighborhood U such that the upper S-norm (encombrement)

$$M = \sup\{\bar{\mu}_K^B : K \in \mathcal{K}_c\}$$

is finite on U. A locally bounded pre-measure is also called a measure.

Let μ be a measure in this sense and let M be defined as above. If $\mathcal{R}$ denotes the set of all bounded continuous functions φ on X such that $M(|\varphi|) < \infty$, then $\mathcal{R}$ is evidently an integration lattice on X. Since μ is locally bounded and X completely regular, the $\mathcal{R}$-topology coincides with the original topology on X.

Let m denote the unique positive measure on $\mathcal{R}$ which coincides with M on $\mathcal{R}_+$ (3.2). It is the supremum of the tight measures m_K, viewed as measures on $\mathcal{R}$ in the obvious way, and is therefore a tight measure itself. The standard essential upper integral $\overline{m}^B$ associated with it is easily seen to coincide with M (13.10).

If, in particular, $\sup\{\mu_K(K): K \in \mathcal{K}\} < \infty$, then μ can be identified with a tight measure on the space $C_b(X)$ of all continuous bounded functions on X.

D. Radonian spaces

While it is often difficult to decide whether a given measure is tight, it is possible to exhibit a large class of spaces on which every B-measure is tight.

For simplicity's sake all integration lattices $(X,\mathcal{R})$ are henceforth assumed to be dominated and not to vanish identically in any one point of X. As before, "compact" stands for "compact and closed."

Definition. An integration lattice $(X,\mathcal{R})$ is called radonian if every positive B-measure on $\mathcal{R}$ is tight, i.e., if $M^T(\mathcal{R}) = M^B(\mathcal{R})$. A completely regular topological space X (not necessarily Hausdorff) is radonian if $(X,C_b(X))$ is.

Remark. Since the tight measures form a band (24.2), an integration lattice $(X,\mathcal{R})$ is radonian if all bounded B-measures are tight. Indeed, every B-measure $m > 0$ is the supremum of the bounded B-measures φm, $\varphi \in \mathcal{R}_+$.

Note also that $(X,\mathcal{R})$ is radonian if and only if every strong and B-continuous upper gauge $(\mathcal{R},M)$ on $\mathcal{R}$ is tight (see 15.8, 9.1).#

Example. If X is locally compact, both $(X,C_{oo}(X))$ and $(X,C_b(X))$ are radonian integration lattices, and X is radonian. Indeed, the B-measures on $C_b(X)$ coincide with the bounded B-measures on $C_{oo}(X)$ via extension, and $C_b(X)$ and $C_{oo}(X)$ both define the same topology and hence the same compact sets.

This example has an analogue for arbitrary integration lattices.

24.4. Proposition. An integration lattice $(X,\mathcal{R})$ is radonian if and only if X is radonian in the $\mathcal{R}$-topology.

Proof. Suppose $(X,\mathcal{R})$ is radonian, and let $m \in M_+^B(C_b(X))$. From 5.11, $C_{b+}(X) \subset \mathcal{R}_\uparrow^B$, and there is a unique $m_0 \in M_+^B(\mathcal{R})$ whose extension coincides with m on $C_b(X)$, namely $m_0 = m|\mathcal{R}$. Clearly $m_0^B = m^B$ and $\mathfrak{F}(\mathcal{R},m_0^B) = \mathfrak{F}(C_b(\mathcal{R}),m^B)$. Since the compact sets for $\mathcal{R}$ and for $C_b(\mathcal{R})$ coincide and are dense for $(\mathcal{R},m_0^B)$, they are dense for $(C_b(\mathcal{R}),m^B)$. Hence m is tight; and so X is radonian.

Conversely, assume that X is radonian and let $m \in M_+^B(\mathcal{R})$. By the remark above we may assume that m is bounded. The standard integral extension m' of m is then B-continuous on $C_b(X) \subset \mathcal{R}_\uparrow^B$ (9A(1)), hence tight. As above we see that $(\mathcal{R},M)$ is tight. Hence $(X,\mathcal{R})$ is radonian.#

The argument shows that being radonian depends only on the $\mathcal{R}$-topology and does not change if $\mathcal{R}$ is replaced by another integration lattice $\mathcal{S}$, provided that $\mathcal{R}^B_\uparrow = \mathcal{S}^B_\uparrow$.

Let Y be a subset of a completely regular space X. We say that Y is a <u>radonian subset</u> if it is radonian in the relative topology.

If the topology of X comes from the integration lattice $\mathcal{R}$ (it always comes from $C_b(X)$), then the topology of Y comes from $\mathcal{R}|Y$, and so by 24.4 Y is radonian if and only if $(Y,\mathcal{R}|Y)$ is a radonian integration lattice. Any compact subset of X is radonian.

Recall that a subset $Y \subset X$ is <u>universally B-measurable</u> if it is $(\mathcal{R},m^B)$-measurable for all $m \in M^B(\mathcal{R})$.

<u>24.5. Theorem</u>. Let $(X,\mathcal{R})$ be an integration lattice. Then

(i) The radonian subsets of X are universally B-measurable and form a σ-ring.

(ii) If $(X,\mathcal{R})$ is radonian then the radonian subsets of X coincide with the universally B-measurable subsets and hence form a tribe.

<u>Proof</u>. (i) Suppose $Y \subset X$ is radonian and let $m \in M^B_+(X,\mathcal{R})$. Dealing first with the case that m is bounded, we find a least $(\mathcal{R},m^B)$-integrable majorant $\tilde{Y}$ of Y (14.1). It exists since $(\mathcal{R},m^B)$ is smooth and $m^B(Y) < \infty$. For any $\varphi \in \mathcal{R}|Y$ we define

$$(Ym)(\varphi) = \int \tilde{Y}\varphi' dm$$

where φ' is any element of $\mathcal{R}$ with $\varphi'|Y = \varphi$. $(\mathcal{R}|Y,Ym)$ is a B-measure (15C) and hence is tight.

Let $\varepsilon > 0$ be given. There is a compact set $K \subset Y$ such that $(Ym)^B(Y-K) < \varepsilon$. Then $\tilde{Y}-K$ is a least $(\mathfrak{R},m^B)$-integrable majorant for $Y-K$ (14.3), and from 15.7

$$m^B(\tilde{Y}-K) = \int \tilde{Y}dm - \int Kdm = \int Yd(Ym) - \int Kd(Ym) = (Ym)^B(Y-K) < \varepsilon.$$

Since $K \subset Y \subset \tilde{Y}$, Y is $(\mathfrak{R},m^B)$-measurable.

If m is not bounded, we find an $(\mathfrak{R},m^B)$-adequate partition $P \subset \mathcal{K}^B(\mathfrak{R})$ (16.4, 16.2, 9.4). Then Y is $(\mathfrak{R},(Km)^B)$-measurable for all K in P, and from 18.9 is $(\mathfrak{R},m^B)$-measurable. Hence Y is universally measurable.

Next, let (Y_n) be a sequence of radonian subsets of X and set $Y = \bigcup Y_n$. Let a positive bounded B-measure m on $\mathfrak{R}|Y$ and an $\varepsilon > 0$ be given. As the Y_n are all $(\mathfrak{R}|Y,m^B)$-integrable (19.2-3), there exist compact subsets $K_n \subset Y_n$ having $m^B(Y_n - K_n) < \varepsilon 2^{-n}$. Hence $m^B(Y - \bigcup K_n) < \varepsilon$, and there is an $N \in \mathbb{N}$ such that $m^B(Y - \bigcup\{K_n : 1 \le n \le N\}) < \varepsilon$ (8.9). The compact sets are thus $(\mathfrak{R}|Y,m^B)$-dense: Y is radonian.

To finish the proof of (i), it is left to be shown that the difference of two radonian sets is radonian. This will follow from the statement that <u>a universally B-measurable set in a radonian space is radonian</u>. For if Y_1, Y_2 are radonian, clearly $Y_1 \setminus Y_2$ is universally B-measurable in the radonian set $Y_1 \cup Y_2$ and hence is radonian. Also, the underlined statement implies (ii), so that it is all that is left to be proved.

Suppose, then, that $(X,\mathfrak{R})$ is radonian and $Y \subset X$ universally B-measurable. Let m be a bounded positive B-measure on $\mathfrak{R}|Y$. Then $m': \varphi \to m(\varphi|Y)$ is a B-measure on $\mathfrak{R}$, and is tight. Y, being universally B-measurable, is $(\mathfrak{R},m'^B)$-integrable (19.3) and can be exhausted from within by compact sets up to an $(\mathfrak{R},m'^B)$-negligible set, say $A \subset Y$. For any

$\varepsilon > 0$ there is an $h \in \mathcal{R}_{\uparrow}^{B}$ with $h \geq A$ and $m'^{B}(h) < \varepsilon$. Evidently $h|Y \geq A$ and

$$m^{B}(h|Y) = \sup\{m(\varphi|Y): h \geq \varphi \in \mathcal{R}\} = m'^{B}(h) < \varepsilon;$$

and so A is m^{B}-negligible: The compact sets are $(\mathcal{R}|Y, m^{B})$-dense in Y. Hence $(Y, \mathcal{R}|Y)$ is radonian.#

24.6. Example. A complete metric Hausdorff space X and its homeomorphs are radonian.

Proof. Visualize X homeomorphically embedded in its Stone-Čech compactification X_{β}, which is the spectrum of $C_b(X)$. Then X is a G_{δ} in X_{β} and hence radonian. Indeed, for $n = 1, 2, \ldots$ let O_n be the set of all $t \in X_{\beta}$ that have a neighborhood whose intersection with X has diameter less than n^{-1}. It is easily seen that O_n is open in X_{β} and $X = \bigcap O_n$.#

24.7. Example. A polish space is a separable Hausdorff space whose topology admits a complete metric. From the above, a polish space X is radonian. Moreover, since the uniformly continuous and bounded functions $UC_b(X)$ are the closure of a countably generated integration lattice (generated, for instance, by the functions $x \to d(x, x_n)$, where d is a bounded metric for the topology and $\{x_n\}$ a countable dense subset), every S-measure on $UC_b(X)$ is B-continuous (3.16) and hence tight. Since evidently $C_b(X) \subset UC_b(X)_{\uparrow}^{S}$, we find that:

> If X is a polish space then every S-measure on $C_b(X)$ is tight.#

24.8. Example. A separable metric space Y is called a Souslin space (or an analytic space) if there is a polish space X and a continuous map f from X onto Y. If f is also one-to-one, Y is called a Lusin space.

> A Souslin or Lusin space Y is radonian; moreover, every S-measure on $C_b(Y)$ is tight.

Proof. Suppose X is a polish space, Y is separable and metrizable, and $f: X \to Y$ is a continuous surjection. Put $\mathcal{S} = UC_b(Y)$, and let $\mathcal{R}$ denote the integration lattice on X generated by $UC_b(X)$ and $UC_b(Y)\circ f$. Then both $\mathcal{S}$ and $\mathcal{R}$ are (the uniform closures of) countably generated integration lattices. Let $\tilde{X}$ and $\tilde{Y}$ denote the spectra of $\mathcal{R}$ and $\mathcal{S}$, respectively. They are compact metrizable spaces containing X and Y homeomorphically, and f has a unique continuous extension $\tilde{f}$ from $\tilde{X}$ to $\tilde{Y}$, which maps X onto Y. We have to show that every S-measure m on $\mathcal{S}$ is tight.

Let $\hat{m}: \hat{\mathcal{S}} \to \mathbb{R}$ be the Gelfand-Bauer transform of m and $\hat{m}^B = \hat{m}^S$ the associated upper integral. For any set $A \subset \tilde{X}$ put

$$M(A) = \hat{m}^B(\tilde{f}(A)).$$

Clearly

(1) M is subadditive, increasing, and if $A_n \uparrow A$ then $M(A_n) \uparrow M(A)$.

(2) If K_n is a decreasing sequence of compact sets with intersection K, then $M(K_n) \downarrow M(K)$.

To see this, note that $\tilde{f}(K_n) \downarrow \tilde{f}(K)$. Indeed, since obviously $\tilde{f}(K) \subset \cap \tilde{f}(K_n)$, only the reverse inequality needs to be shown. Now, if $y \in \cap \tilde{f}(K_n)$, then $\tilde{f}^{-1}(y) \cap K_n \neq \emptyset$ for all n. From compactness, there is a point $x \in \cap \tilde{f}^{-1}(y) \cap K_n = \tilde{f}^{-1}(y) \cap K$, and so $y \in f(K)$.

Since X is complete in some metric space, it is a G_δ in $\tilde{X}$ (24.6). There is a decreasing sequence (O_n) of open sets with intersection X. Since $\tilde{X}$ is compact and metrizable, each O_n is a K_σ, and there is an increasing sequence $(K_{n,k})$ of compacta such that $O_n = \cup \{K_{n,k}: k = 1, 2, \ldots\}$.

Let $r < M(X) = \hat{m}^B(Y) = \|m\|$. Since $X \cap K_{1,k} \uparrow X$, there is a $k(1)$ such that $M(X \cap K_{1,k(1)}) > r$. Since

$$X \cap K_{1,k(1)} \cap K_{2,k} \uparrow X \cap K_{1,k(1)},$$

there is a $k(2)$ such that

$$M(X \cap K_{1,k(1)} \cap K_{2,k(2)}) > r.$$

Continuing by induction we find a sequence $k(n)$, $n = 1, 2, \ldots,$ such that

$$M(X \cap K_{1,k(1)} \cap \ldots \cap K_{n,k(n)}) > r$$

for all n. Clearly $K = \cap \{ K_{n,k(n)}: n = 1, 2, \ldots\}$ is a compact subset of X, and from (2), $M(K) \geq r$. Hence $\tilde{f}(K) = f(K)$ is a compact subset of Y having $\hat{m}^B(f(K)) \geq r$. This shows that Y is $\hat{m}^B$-integrable (9.4) and hence that m is tight (24.1).#

The last two examples appear usually in the following form. A set X together with a σ-algebra $\mathcal{U}$ of subsets is called a <u>standard Borel</u> space (an <u>analytic Borel</u> space) if there is a Lusinian (Souslinian) topology τ on X which spans $\mathcal{U}$. In view of the fact that the S-measures on $\mathcal{U}$ coincide with the S-measures on $C_b(X,\tau)$ via extension, we may restate the last two examples thus: If m is any positive S-measure on $\mathcal{U}$, then for all $A \subset \mathcal{U}$

$$m(A) = \sup\{m(K): \ K \subset A, \ K \text{ compact}\}.$$

In other words, the compact sets are dense. It is exactly this property (and the fact that $\mathcal{U}$ is countably generated) that makes standard and analytic Borel spaces so useful (Cf. 36.5).

E. Supplements

Tight measures have been studies extensively by L. Schwartz [5,39] in the context of example 24.3.

If $(\mathcal{R},M)$ is tight, then $(\mathcal{R},M)$-measurability can be defined for any map with values in a locally compact (topological) space Y. Any two uniformities which give the topology on Y give rise to the same measurable functions; in fact, a function $f: X \to Y$ is measurable with respect to any compatible uniformity (there are!) if the family of $\mathcal{R}$-compact sets on which f is continuous, Cont(f), is $(\mathcal{R},M)$-dense. Bourbaki [5] therefore defines a map from the locally compact space X to another one, Y, to be measurable with respect to the positive Radon measure m on X, if Cont(f) is $(C_{00}(X),m^B)$-dense.

24.9. A dominated subset K of X is $\mathcal{R}$-compact if and only if $\mathcal{R} \supset \Phi \downarrow^B 0$ on K implies that Φ converges uniformly to zero on K.

24.10. If $\mathcal{C}$ is a finitely generated clan then every positive measure on $\mathcal{S}(\mathcal{C})$ is tight.

24.11. If $(\mathcal{R},M)$ is tight so are $(\mathcal{R},M^*)$ and $(\mathcal{R},\overline{M})$.

24.12. The uniformity of <u>dominated pointwise convergence</u> on X is defined to be the inductive limit of the uniformities of pointwise convergence on each of the sets $K \in \mathcal{K}(\mathcal{R})$. Its topology is simply the topology of pointwise convergence. A measure on $\mathcal{R}$ is discrete if and only if it is continuous in this topology.

24.13. A positive S-measure m on $\mathcal{R}$ is said to have the <u>Darboux-property</u> if every non-negligible $(\mathcal{R},m^S)$-integrable set A can be written as the disjoint union of two non-negligible integrable sets B and C. If this is so, one can even choose $\int B dm = \int C dm$. The positive S-measures with the Darboux-property are exactly the positive diffuse S-measures.

24.14. $I_0(\mathcal{R})$ carries a natural uniformity, the one generated by the functions $n \to n(\varphi)$ $(n \in I_0(\mathcal{R});\ \varphi \in \mathcal{R})$. It is called the <u>vague uniformity</u>, and its gauge has a basis of pseudometrics of the form

$$d_\varphi(n,m) = |n(\varphi) - m(\varphi)| \qquad (\varphi \in \mathcal{R};\ n,m \in I_0(\mathcal{R})).$$

Its underlying topology is the <u>vague topology</u> on $I_0(\mathcal{R})$.

(i) The map $m \to \overline{m}^B(f)$ from $M_+^B(\mathcal{R})$ to the reals is vaguely lower semicontinuous, for all $f \in \mathcal{R}_\uparrow^B$. It is vaguely upper semicontinuous for $f \in \mathcal{R}_{\downarrow+}^B$.

(ii) Suppose $\mathcal{R}$ contains the constants and let $f \geq 0$ be a bounded function and $m \in M_+^B(\mathcal{R})$. If f is continuous $\overline{m}^B$-a.e. then the map $n \to \overline{n}^B(f)$ from $M_+^B(\mathcal{R})$ to the reals is vaguely continuous at m.

24.15. Suppose $\mathcal{R}$ contains the constants and separates the points of X, and let H be a family in $M^T(\mathcal{R})$. H is said to be equi-tight, or to satisfy the condition of Prokhorov, if

(a) $\sup\{|m|(1): m \in H\} < \infty$ and

(b) For every $\varepsilon > 0$ there exists a compact subset K of X such that $|\overline{m}|^B(X-K) < \varepsilon$ for all $m \in H$.

Theorem. (Prokhorov). If $H \subset M^T(\mathcal{R})$ is equi-tight then so is its vague closure $\overline{H}$ in $M^T(\mathcal{R})$; and $\overline{H}$ is vaguely compact. Conversely, if X is locally compact or polish in the $\mathcal{R}$-topology, and if H is vaguely compact then it is equi-tight. (Cf. [5], Ch. IX)

24.16. If m is a bounded positive Radon measure on a locally compact space X then every bounded continuous function on X is $\overline{m}^B$-integrable and $\int dm$ is a B-continuous and tight measure on $C_b(X)$. Conversely, every tight measure on $C_b(X)$ arises in this way.

24.17. If $(\mathcal{R},M)$ and $(\mathcal{R},N)$ are equivalent weak upper gauges and $(\mathcal{R},M)$ is tight, then so is $(\mathcal{R},N)$.

24.18. An integration lattice $(X,\mathcal{R})$ is radonian if and only if the radonian subsets of X form an $(\mathcal{R},m^B)$-adequate cover, or are $(\mathcal{R},m^B)$-dense, for every $m \in M_+^B(\mathcal{R})$.

24.19. (Capacities). An (outer) capacity on a topological Hausdorff space X is a function M from the power set of X to $\overline{\mathbb{R}}_+$ satisfying (1) and (2) in 24.8. A subset A of X is said to be M-capacitable if $M(A) = \sup\{M(K): K \subset A, K \text{ compact}\}$. The proof of 24.8 shows that (i) A $K_{\sigma\delta}$ in a compact space X is capacitable for all capacities; and can be modified to show that (ii) A continuous image Y of a $K_{\sigma\delta}$ X in a compact space $\tilde{X}$ (a $\mathcal{K}$-analytic set) which is contained in a σ-compact space $\tilde{Y}$ is M-capacitable for all capacities M on $\tilde{Y}$. (iii) The $\mathcal{K}$-analytic subsets of a Hausdorff space are closed under countably unions and intersections. A continuous image of a $\mathcal{K}$-analytic set is $\mathcal{K}$-analytic.

The statements (ii) and (iii) are Choquet's capacitability theorem. For more information see [9] and [33].

LITERATURE: [5], Ch. IX, [39], [43], [44], [47].

IV. ELEMENTARY OPERATIONS ON MEASURE SPACES

There are various ways to construct new measures and upper gauges from given ones and from other data. Measures defined by densities (22A) and measures induced on a subset (15C) are two examples. In this chapter, integrals of measure fields and also products, images, and projective limits of measures will be studied.

To avoid repetitions throughout this chapter, E, F, G will be three fixed Banach spaces each of whose norms will be written as $\| \ \|$. There will also be given a bilinear map from $E \times F$ to G, of norm not exceeding one and denoted by juxtaposition.

§1. Fields of measures and of upper gauges

25. Integrable fields

A. Definitions. Let X be a set, and let $(Y,\mathcal{S})$ be an integration lattice.

(i) A field on X of measures on $(Y,\mathcal{S})$ is a map

$$\lambda: X \to I(\mathcal{S};E) \ ; \quad \lambda: x \to \lambda_x \ ,$$

which associates with every point $x \in X$ an (E-valued) measure λ_x on $\mathcal{S}$. The field λ is said to be of finite variation (*-continuous, compact, etc.) if all the λ_x are.

(ii) Let $(\mathcal{R},M)$ be a weak upper gauge on X. The field λ is said to be $(\mathcal{R},M)$-integrable if

$$\langle \lambda;\varphi \rangle : x \to \langle \lambda_x;\varphi \rangle = \lambda_x(\varphi)$$

is an $(\mathcal{R},M)$-integrable function for all $\varphi \in \mathcal{S}$. The field λ is said to be of integrable variation if

$$\langle |\lambda|;\varphi \rangle : x \to \langle |\lambda_x|;\varphi \rangle = |\lambda_x|(\varphi)$$

is in $\mathcal{L}^1(\mathcal{R},M)$ for all $\varphi \in \mathcal{S}$.

(iii) Suppose $U: \mathfrak{R} \otimes E \to G$ is a linear map majorized by $(\mathfrak{R},M)$, and λ is an $(\mathfrak{R},M)$-integrable E-valued field of measures on $\mathcal{S}$. <u>The integral</u>, $\int \lambda dU$, is defined by

$$(\int \lambda dU)(\varphi) = \int \langle \lambda;\varphi\rangle dU \in G \qquad , \text{ for } \varphi \in \mathcal{S}.$$

(iv) Suppose $m: \mathfrak{R} \to F$ is a measure, majorized in variation by $(\mathfrak{R},M)$, and λ is an $(\mathfrak{R},M)$-integrable E-valued field of measures. For every $\varphi \in \mathcal{S}$,

$$(\int \lambda dm)(\varphi) = \int \langle \lambda;\varphi\rangle dm$$

exists (10.11), and the G-valued measure $\int \lambda dm$ on $\mathcal{S}$ so defined is called <u>the integral of</u> λ <u>over</u> m.

(v) Let D be another Banach space and $\Lambda: x \to \Lambda_x$ a field of linear maps $\Lambda_x: \mathcal{S} \otimes D \to E$. Λ is said to be of finite semi-variation, compact, etc. if all the Λ_x are. We denote the function $x \to \Lambda_x(\varphi) = \langle \Lambda_x;\varphi\rangle$ by $\langle \Lambda;\varphi\rangle$, for $\varphi \in \mathcal{S} \otimes D$. The field Λ is said to be integrable for the weak upper gauge $(\mathfrak{R},M)$ if each of the functions $\langle \Lambda;\varphi\rangle$ is, $\varphi \in \mathcal{S} \otimes D$. If Λ is $(\mathfrak{R},M)$-integrable and $U: \mathfrak{R} \otimes E \to G$ is a linear map[1)] majorized by M, then

$$(\int \Lambda dU)(\varphi) = \int \langle \Lambda;\varphi\rangle dU$$

exists for all $\varphi \in \mathcal{S} \otimes D$ and defines a linear map $\int \Lambda dU: \mathcal{S} \otimes D \to G$. It is called <u>the integral of</u> Λ <u>over</u> U. Finally, Λ is said to have <u>integrable semivariation</u> if the function $\langle \|\Lambda\|;\varphi\rangle: x \to \langle \|\Lambda_x\|;\varphi\rangle$ belongs to $\mathcal{L}^1(\mathfrak{R},M)$ for all $\varphi \in \mathcal{S}_+$.#

---

1) E.g., defined by a measure $m: \mathfrak{R} \to F$ of finite variation as in (iv).

In order to compare the integration theories of λ_x, $\int \lambda dm$, m and of Λ_x, $\int \Lambda dU$, U respectively, suitable upper gauges majorizing the λ_x and Λ_x will be needed:

<u>Definition</u>. <u>A field</u> on X <u>of weak upper gauges</u> on $(Y,\mathcal{B})$ is a map

$$L: x \to L_x \qquad , x \in X,$$

which associates with each point $x \in X$ a weak upper gauge $(\mathcal{B},L_x)$ on Y. The field L is said to be <u>integrable</u> for the weak upper gauge $(\mathcal{R},M)$ on X if

$$\langle L;\varphi\rangle : x \to \langle L_x;\varphi\rangle = L_x(\varphi)$$

is $(\mathcal{R},M)$-integrable for all φ in $\mathcal{B}$. The field L is said to <u>majorize</u> the field λ of measures if

$$\|\langle \lambda_x;\varphi\rangle\| \leq \langle L_x;\varphi\rangle \qquad , \text{ for all } \varphi \text{ in } \mathcal{B}_+.$$

Analogously, L is said to majorize the field Λ of linear maps from $\mathcal{B} \otimes D$ to E if

$$\|\langle \Lambda;\varphi\rangle\| \leq \langle L;\varphi\rangle \text{ everywhere} \qquad , \text{ for all } \varphi \in \mathcal{B} \otimes D. \#$$

In order that the field L of upper gauges majorize the field λ of measures, λ_x has to be at least S-continuous for each x in X. Conversely, if λ_x is $*$-continuous and of finite variation for each x in X ($*$ = S or B), then

$$\lambda^*: x \to |\lambda_x|^*$$

is a field of upper integrals majorizing λ. If $|\lambda|$ is $(\mathcal{R},M)$-integrable, so is λ^*. The analogous statement applies if Λ is a (weakly *-continuous and weakly compact) field of linear maps $\Lambda_x: \mathcal{S} \otimes D \to E$; if Λ has integrable semivariation then $\|\Lambda\|^*$ is an integrable field of weak upper gauges majorizing Λ (11.8)

<u>25.1. Lemma</u>. Let M be any upper S-norm on X which is finite on $\mathcal{R}_+$, and let $L: x \to L_x$ be a field on X of upper S-norms on Y such that

$$\langle L;\varphi\rangle: x \to \langle L_x;\varphi\rangle$$

is $(\mathcal{R},M)$-integrable for all φ in $\mathcal{S}_+$. Then

$$N(f) = (\int L dM)(f) = M\langle L;f\rangle \qquad f: Y \to \overline{\mathbb{R}_+}$$

defines on Y an upper S-norm, $N = \int L dM$, which is finite on $\mathcal{S}_+$. Furthermore, if $g: Y \to E$ is an $(\mathcal{S},N)$-integrable function, then g is $(\mathcal{S},L_x)$-integrable for M-almost all $x \in X$, and

$$\langle L;g\rangle: x \to \langle L_x;\|g\|\rangle$$

is $(\mathcal{R},M)$-integrable.

<u>Proof</u>. The first contention is obvious upon inspection. Let $g \in \mathcal{L}_E(\mathcal{S},N)$. There is a sequence (φ_k) in $\mathcal{S} \otimes E$ such that $g = \sum \varphi_k$ N-almost everywhere and in N-mean, and $\sum N(\varphi_k) < \infty$ (7.12). Let A denote the set of points $y \in Y$ such that $g(y) \neq \sum \varphi_k$. Since

$$M(\sum\langle L;\varphi_k\rangle) \le \sum M(\langle L;\varphi_k\rangle) < \infty \quad \text{and} \quad M(\langle L;A\rangle) = 0,$$

the set B of points $x \in X$ such that either $\sum \langle L_x;\varphi_k\rangle = \infty$ or $\langle L_x;A\rangle \neq 0$ is M-negligible; g is evidently $(\mathcal{S},L_x)$-integrable for all points $x \in X - B$.

To show that $\langle L;g\rangle$ ($= \langle L;\|g\|\rangle$) is $(\mathcal{R},M)$-integrable, we may assume that g is positive and real-valued, and that $\varphi_k \in \mathcal{S}_+$ for all positive integers k. The result then follows from

$$M(\langle L;g\rangle - \langle L;\sum^{n}\varphi_k\rangle) \leq M(\langle L;g - \sum^{n}\varphi_k\rangle) \xrightarrow[n\to\infty]{} 0$$

and the fact that $\langle L;\sum^{n}\varphi_k\rangle$ is $(\mathcal{R},M)$-integrable for all n.#

25.2. Lemma. If $(\mathcal{R},M)$ and the $(\mathcal{S},L_x)$ are all weak upper gauges (upper gauges, strong upper gauges, upper integrals), then so is $(\mathcal{S},N)$.

Proof. The case of upper integrals is evident. In the case of weak upper gauges, let (φ_n) be an increasing sequence in $\mathcal{S}_+$ whose pointwise supremum g is majorized by an integrable function h. The sequence $f_n = \langle L;\varphi_n\rangle$ of integrable functions has supremum $\langle L;g\rangle$ and is majorized a.e. by the integrable function $\langle L;h\rangle$. From lemma 25.1 and the monotone convergence theorem, $\langle L;g - \varphi_n\rangle$ converges to zero a.e., and from Lebesgue's theorem

$$N(g - \varphi_n) = M(\langle L;g - \varphi_n\rangle) \to 0.$$

Hence $(\mathcal{S},N)$ is a weak upper gauge. The same argument shows that $(\mathcal{S},N)$ is an upper gauge provided $(\mathcal{R},M)$ and the $(\mathcal{S},L_x)$ are: one replaces the assumption $h \in \mathcal{L}^1(\mathcal{S},N)$ by $N(h) < \infty$ and finds $M(\langle L;h\rangle) < \infty$.

Assume, then, that $(\mathcal{R},M)$ and the $(\mathcal{S},L_x)$ are all strong upper gauges, and let $0 \leq f \leq g$ in $\mathcal{L}^1(\mathcal{S},N)$ be such that $N(f) = N(g)$. Then $0 \leq \langle L;f\rangle \leq \langle L;g\rangle$ in $\mathcal{L}^1(\mathcal{R},M)$ and $M(\langle L;f\rangle) = M(\langle L;g\rangle)$, so that $\langle L;f\rangle = \langle L;g\rangle$ M-a.e. Hence $f = g$ L_x-a.e. for M-almost all $x \in X$, that is $f = g$ N-a.e.#

For the following important result, $(\mathcal{R},M)$ is a weak upper gauge on X, L is an $(\mathcal{R},M)$-integrable field of weak upper gauges, and $N = \int LdM$.

Secondly, $\lambda: X \to M_E^S(\mathcal{S})$ is an $(\mathcal{R},M)$-integrable field majorized by L, $m: \mathcal{R} \to F$ is a measure majorized in variation by M, and $n = \int \lambda dm$. Clearly n is majorized by N.

To treat the most general situation, we consider at the same time an $(\mathcal{R},M)$-integrable field Λ of linear maps from $\mathcal{S} \otimes D$ to E majorized by L, a linear map U from $\mathcal{R} \otimes E$ to G majorized by M, and $V = \int \Lambda dU$. Clearly V is majorized by N.

25.3. Theorem. (Fubini).

(i) Let f be a Banach-valued function. Then f is N-negligible if and only if it is L_x-negligible for M-almost all $x \in X$.

(ii) Let f be a Banach-valued $(\mathcal{S},N)$-integrable function. Then f is $(\mathcal{S},L_x)$-integrable for M-almost all x in X, and $\langle L;f\rangle$ is $(\mathcal{R},M)$-integrable.

(iii) Let f be a numerical $(\mathcal{S},N)$-integrable function. Then

$$\int fd\lambda: x \to \int fd\lambda_x \in E$$

exists for almost all x, is $(\mathcal{R},M)$-integrable and

$$\int fdn = \int (\int fd\lambda)dm \in G.$$

(iv) In the special case where $E = F = G$ equals $\mathbb{R}$ or $\mathbb{C}$, and f is a Banach-valued $(\mathcal{S},N)$-integrable function, $\int fd\lambda$ exists almost everywhere, is $(\mathcal{R},M)$-integrable, and satisfies

$$\int fdn = \int (\int fd\lambda)dm.$$

(v) If $f \in \mathcal{L}^1_D(\mathcal{S},N)$ then $\int f d\Lambda$ exists a.e., is $(\mathcal{R},M)$-integrable, and

$$\int f dV = \int (\int f d\Lambda) dU .$$

Proof. Only (iii), (iv), and (v) have not yet been proved. If a sequence (φ_k) in $\mathcal{S}$ is selected as in lemma 25.1, then $\int f d\lambda = \Sigma \int \varphi_k d\lambda$ M-almost everywhere, and

$$M(\| \int f d\lambda - \sum^n \langle\lambda;\varphi_k\rangle \|) \leq M(\langle L; f - \sum^n \varphi_k\rangle) \to 0 \quad \text{as} \quad n \to \infty$$

$\int f d\lambda$ is thus $(\mathcal{R},M)$-integrable, and equal to $\Sigma \langle\lambda;\varphi_k\rangle$ M-almost everywhere and in M-mean. Hence

$$\int (\int f d\lambda) dm = \int \Sigma \langle\lambda;\varphi_k\rangle dm = \Sigma \int \langle\lambda;\varphi_k\rangle dm = \Sigma\, n(\varphi_k) = \int f dn.$$

In case (iv), if f is E-valued, select (φ_k) in $\mathcal{S} \otimes E$ instead, and then use the same argument. For (v), select (φ_k) in $\mathcal{S} \otimes D$ instead.#

25.4. Example. Suppose that m and λ are of finite variation, and that $M = m^S$ and $L = \lambda^S$. Then $N = \int \lambda^S dm^S$ is an upper integral majorizing $n = \int \lambda dm$, and so n has finite variation $|n| \leq \int |\lambda| d|m|$. If both m and λ are positive then n^S, the biggest upper integral for n (9.1), exceeds N. An $(\mathcal{S},n^S)$-integrable (-negligible) function f is then $(\mathcal{S},N)$-integrable (-negligible). The conclusions of the theorem hold for such functions and constitute the classical theorem of Fubini.

B. Tame fields

For the sake of simplicity it is now assumed that $(\mathcal{R},M)$ is a strong upper gauge on X and that L is an integrable field of strong upper gauges $(\mathcal{S},L_x)$ on Y.

We let $N = \int L dM$ and ask how measurability for $(\mathcal{R},M)$, $(\mathcal{S},N)$, and the $(\mathcal{S},L_x)$ compare. A new notion is required.

Definition. Let K be an $(\mathcal{R},M)$-integrable set. We say that L is tame on K if there is a sequence (B_n) of $(\mathcal{S},N)$-integrable subsets of Y such that $\langle L_x; Y - \cup B_n\rangle = 0$ for almost all x in K; we say that L is tame on the measurable set $A \subset X$ if the family $T(L)$ of integrable sets on which L is tame is dense in A; and, finally, we say that L is tame if it is tame on X.

Remarks. (1) The two definitions are readily seen to coincide if A is integrable.

(2) If $1 = X$ belongs to $\mathcal{S}$ (or to $\mathcal{S}^S$), then every integrable field L is tame.

(3) The supremum in $L^\infty(\mathcal{R},M)$ of any family of classes of sets on which L is tame is also a class of sets on which L is tame. In particular, if $(\mathcal{R},M)$ is localizable then there is a maximal class of measurable sets on which L is tame. Any set in this class will be called a maximal set on which L is tame.#

25.5. Theorem. Suppose L is tame on $A \in T(\mathcal{R},M)$, and let f be any $(\mathcal{S},N)$-measurable function on Y with values in a uniform space S.

Then f is $(\mathcal{S},L_x)$-measurable for locally almost all $x \in A$. Furthermore, if S is $\overline{\mathbb{R}}$ or a Banach space, then $\langle L;f\rangle$ is $(\mathcal{R},M)$-measurable on A.

Proof. Let K be any integrable subset of A on which L is tame, and let (B_n) be a sequence in $\mathcal{F}(\mathcal{S},N)$ such that $L_x(Y - \cup B_n) = 0$ for almost all $x \in K$. The negligible set of points $x \in K$ where $L_x(Y - \cup B_n) > 0$ will be denoted by K'. The B_n may be chosen to be mutually disjoint.

For every natural number n, there is a sequence (B_n^k) of mutually disjoint, integrable subsets of B_n on which f is uniformly continuous and such that $N(B_n - \cup B_n^k) = 0$. Let K_n^k be the negligible set of points $x \in X$ for which B_n^k is not $(\mathcal{S}, L_x)$-integrable, and put $K'' = \cup\{K_n^k: k,n = 1,2,\ldots\}$. Lastly, let K''' denote the set, also negligible, of points $x \in K$ where $L_x(B_n - \cup\{B_n^k: k = 1,2,\ldots\}) > 0$. For $x \in K - (K' \cup K'' \cup K''')$, $\{B_n^k\}$ is clearly an $(\mathcal{S}, L_x)$-adequate partition, and so f is $(\mathcal{S}, L_x)$-measurable for these x (18.9). The first statement is proved by letting K range over the dense family of all integrable subsets of A on which L is tame.

Now, suppose that S is a Banach space. The fB_n^k are $(\mathcal{S}, N)$-integrable (18.7), and so the functions $K\langle L; fB_n^k\rangle$ are $(\mathcal{R}, M)$-integrable. The functions $K\langle L; \sum\{fB_n^k: k,n \leq N\}\rangle$ are measurable, as is their supremum $K\langle L; \sum B_n^k f\rangle \doteq K\langle L; f\rangle$. This proves the second statement.#

<u>25.6. Corollary</u>. Suppose $(\mathcal{R}, M)$ and all the $(\mathcal{S}, L_x)$ are regular. If L is tame then $N = \int L dM$ is regular.

<u>Proof</u>. Let $A \in T(\mathcal{S}, N)$ with $N(A) < \infty$ and $\overline{N}(A) = 0$. For K and (B_n) as above,

$$M(K\langle\overline{L}; A\rangle) = M(K\langle L; \cup B_n A\rangle) \leq N(A \cup B_n) = 0,$$

and hence $\langle\overline{L}; A\rangle = 0$ locally a.e. As $\langle L; A\rangle < \infty$ a.e., it follows that $\langle L; A\rangle = 0$ locally a.e. Finally, $N(A) = 0$ follows from $N(A) = M(\langle L; A\rangle) <$ and the measurability of $\langle L; A\rangle$.#

C. Supplements

25.7. In the situation of 25.3(v), assume U is (weakly) compact. Then so is $\int \Lambda dU$.

25.8. Suppose $\mathcal{S}_*$ is dominated, m is a finite, positive *-measure on $\mathcal{R}$, and $\lambda: X \to M_+^*(\mathcal{S})$ is an $(\mathcal{R}, m^*)$-integrable field. For every $1 \leq p, q < \infty$, $\int \lambda_p^* dm_q^* = N_{pq}$ is an upper gauge on $\mathcal{S}$.

If λ is an integrable field on X of finite variation such that $\lambda_x \ll \rho$ for all $x \in X$ and some $\rho \in M^S(\mathcal{S})$, then $\int \lambda dm \ll \rho$ for all $m \in M_+^S(\mathcal{R})$. If λ, λ' are two integrable fields on X of positive measures on $(Y, \mathcal{S})$ such that $\lambda_x \ll \lambda'_x$ for all $x \in X$, then $\int \lambda dm \ll \int \lambda' dm$ for all $m \in M_+^S(\mathcal{R})$.

LITERATURE: Proofs are adapted from [5], Ch. V.

26. Adequate fields

It will be proved that B-continuity is preserved under the formation of integrals of fields, provided the latter are measurable.

A. The vague uniformity

The spaces $M_E^*(\mathcal{S})$ of measures on $\mathcal{S}$ and $U(\mathcal{S})$ of upper gauges on $\mathcal{S}$ both carry a natural uniformity, namely the one generated by the functions

$$m \to m(\varphi) \quad \text{from} \quad M_E^*(\mathcal{S}) \quad \text{to} \quad E, \quad \text{or}$$

$$M \to M(\varphi) \quad \text{from} \quad U(\mathcal{S}) \quad \text{to} \quad \overline{\mathbb{R}_+}, \quad \text{respectively,}$$

for all φ in $\mathcal{S}_+$. Its gauge has a base of pseudometrics of the form

$$d_\varphi(m,n) = \|m(\varphi) - n(\varphi)\| \qquad , m,n \in M_E^*(\mathcal{S}), \quad \text{or}$$

$$d_\varphi(M,N) = |M(\varphi) - N(\varphi)| \qquad , M,N \in U(\mathcal{S}), \quad \text{respectively,}$$

where φ ranges over $\mathcal{S}_+$.

<u>Definition</u>. This uniformity is called the <u>vague uniformity</u> on $M_E^*(\mathcal{S})$ or on $U(\mathcal{S})$, respectively.

If $(\mathfrak{R},M)$ is a weak upper gauge on X, a field $\lambda\colon X \to M_E^*(\mathcal{S})$ of measures or a field $L\colon X \to U(\mathcal{S})$ of upper gauges is said to be $(\mathfrak{R},M)$-measurable if it is measurable with respect to the vague uniformity on its range.#

A field $L\colon X \to U(\mathcal{S})$ of upper gauges is then $(\mathfrak{R},M)$-measurable if there is a dense family $U(L)$ of integrable sets on each of which $\langle L;\varphi\rangle$ is uniformly continuous for all $\varphi \in \mathcal{S}$. For this to be true it is <u>not</u> sufficient that $\langle L;\varphi\rangle$ be measurable for all $\varphi \in \mathcal{S}$, unless $\mathcal{S}$ is countably generated. A similar remark applies to fields of measures.

B. <u>Definition</u>. A field of measures or of upper gauges is said to be $(\mathcal{R},M)$-<u>adequate</u> if it is both $(\mathcal{R},M)$-integrable and $(\mathcal{R},M)$-measurable.

An integrable field of upper gauges, $L: X \to U(\mathcal{B})$, is said to be $(\mathcal{R},M)$-B-<u>adequate</u> if there is a dense family $U^B(L)$ such that

$$K\langle L;\varphi\rangle \in K \cdot \mathcal{R}_{\uparrow}^{B}$$

for all $K \in U^B(L)$ and all $\varphi \in \mathcal{B}$.#

Note that a measurable field of integrable variation is integrable, and thus adequate (19.3).

An $(\mathcal{R},M)$-adequate field of upper gauges is clearly $(\mathcal{R},M)$-B-adequate.

An $(\mathcal{R},M)$-integrable field is $(\mathcal{L}^1,M)$-uniformly continuous, and so $(\mathcal{L}^1,M)$-adequate; however, B-continuity is lost in the transition from $(\mathcal{R},M)$ to $(\mathcal{L}^1,M)$; and this is exactly what the notion of adequacy is designed to preserve.

<u>26.1. Lemma</u>. If λ is an $(\mathcal{R},M)$-measurable field of B-measures with integrable variation $|\lambda|$, then λ^B is an $(\mathcal{R},M)$-B-adequate field of upper integrals.

<u>Proof</u>. Since $|\lambda|$ is integrable, $\langle|\lambda|;\varphi\rangle$ is finite almost everywhere for $\varphi \in \mathcal{B}_+$. On any set $K \in U(\lambda)$,

$$\langle\lambda^B;\varphi\rangle = \langle|\lambda|;\varphi\rangle K = \sup\{K \textstyle\sum \|\lambda(\varphi_i)\|: \varphi_i \in \mathcal{R}_+, \sum \varphi_i = \varphi, \text{ the sum finite}\},$$

and this is the product of K with an element of $\mathcal{R}_{\uparrow}^{B}$.#

<u>26.2. Corollary</u>. λ_p^B is an $(\mathcal{R},M)$-B-adequate field of upper gauges $(1 \le p < \infty)$.

<u>26.3. Proposition</u>. Let $(\mathcal{R},M)$ be a B-continuous upper gauge and let $L: X \to U(\mathcal{S})$ be an integrable field of B-continuous upper gauges. Then

(i) $(\mathcal{S}, \int LdM)$ is B-continuous, provided either that L is uniformly continuous or that $\langle L;\varphi\rangle \in \mathcal{R}_{\uparrow}^{B}$ for all φ in $\mathcal{S}_{+}$; and

(ii) $(\mathcal{S}, \int Ld\overline{M})$ is B-continuous, provided that L is either $(\mathcal{R},M)$-B-adequate or $(\mathcal{R},M)$-adequate.

<u>Proof</u>. (i): Let $\mathcal{S}_{+} \supset \Phi \uparrow^{B} h \in \mathcal{S}_{\uparrow}^{B}$. Then

$$\mathcal{R}_{\uparrow}^{B} \supset \langle L;\Phi\rangle \uparrow^{B} \langle L;h\rangle \in \mathcal{R}_{\uparrow}^{B} ;$$

and so

$$M(\langle L;\Phi\rangle) \uparrow M(\langle L;h\rangle) \qquad (8.14).$$

(ii): If $K \in U^{B}(L)$, and if Φ and h are chosen as above, then

$$(K\langle L;\Phi\rangle) \uparrow^{B} K\langle L;h\rangle \quad \text{and} \quad M(K\langle L;\Phi\rangle) \uparrow M(K\langle L;h\rangle)$$

(15.2). The desired equality is then proved by taking the supremum over all K in the dense family $U^{B}(L)$ (13.3).#

<u>26.4. Corollary</u>. Let $m \in M_{F}^{B}(\mathcal{R})$, and let $\lambda: X \to M_{E}^{B}(\mathcal{S})$ be an $(\mathcal{R},\overline{m}^{B})$-adequate field of integrable variation. Then $\int \lambda dm$ is B-continuous.

<u>Proof</u>. $\int \lambda dm$ is majorized by $\int \lambda^{B} d\overline{m}^{B}$.#

<u>26.5. Example</u>. Let $m \in M_{F}^{B}(\mathcal{R})$, and let $\lambda: X \to M_{E}^{B}(\mathcal{S})$ be an $(\mathcal{R},\overline{m}^{B})$-adequate field of integrable variation. Then λ^{B} is $(\mathcal{R},M)$-B-adequate, by lemma 26.1, and so $(\mathcal{S}, \int \lambda^{B} d\overline{m}^{B})$ is B-continuous. $\int \lambda^{B} d\overline{m}^{B}$ is an upper integral for $\underline{n} = \int |\lambda| d|m|$ and hence is smaller than $\underline{n}^{B}$ (9.1)

An $(\mathcal{S},\underline{n}^B)$-integrable function is then $(\mathcal{S}, \int \lambda^B d\bar{m}^B)$-integrable, and, from Fubini's theorem, is $(\mathcal{S},\lambda_x^B)$-integrable for locally almost all x in X, etc. This is the classical theorem of Fubini for B-continuous measures.

C. Supplement

26.6. An $(\mathcal{R},M)$-adequate map is $(\mathcal{R},M^*)$-adequate and $(\mathcal{R},\bar{M})$-adequate.

26.7. Suppose $\mathcal{R}$ and $\mathcal{S}$ contain the constants and λ is an equi-tight (24.8) and $(\mathcal{R},m^B)$-adequate field of positive measures on $\mathcal{S}$ $(m \in M_+^B(\mathcal{R}))$. Then $\int \lambda dm$ is tight.

LITERATURE: [5], Ch. VI.

27. Products of elementary integrals

In this section, $(X_1,\mathcal{R}_1)$ and $(X_2,\mathcal{R}_2)$ are two dominated integration lattices, and $m_1\colon \mathcal{R}_1 \to F$, $m_2\colon \mathcal{R}_2 \to E$ are two measures of finite variation.

A. Definition. The product of the integration lattices,

$$(Y,\mathcal{S}) = (X_1 \times X_2,\ \mathcal{R}_1 \times \mathcal{R}_2) = (X_1,\mathcal{R}_1) \times (X_2,\mathcal{R}_2),$$

is defined in the natural way: Its underlying space, $X_1 \times X_2$, is the cartesian product of X_1 with X_2; its integration lattice, $\mathcal{R}_1 \times \mathcal{R}_2$, is the one spanned by $\mathcal{R}_1 \otimes \mathcal{R}_2$, the set of functions of the form

$$\sum_k \varphi_1^k \otimes \varphi_2^k\colon (x_1,x_2) \to \sum_k \varphi_1^k(x_1)\varphi_2^k(x_2) \qquad , \ (x_1,x_2) \in Y,$$

where $\varphi_1^k \in \mathcal{R}_1$, $\varphi_2^k \in \mathcal{R}_2$, and where the sum is finite.#

In order to get a feel for the size of $\mathcal{R}_1 \times \mathcal{R}_2$, consider $\overline{\mathcal{R}}_1 \otimes \overline{\mathcal{R}}_2$, $\overline{\mathcal{R}}_i$ being the closure of $\mathcal{R}_i$ under dominated uniform convergence. $\overline{\mathcal{R}}_i$ is a dominated integration lattice and an algebra, of bounded functions, and therefore

$$\mathcal{A} = \overline{\mathcal{R}}_1 \otimes \overline{\mathcal{R}}_2$$

is an algebra of functions on $X_1 \times X_2$ (5.6). $\mathcal{A}$, too, is dominated. For if $\varphi = \sum \varphi_1^k \otimes \varphi_2^k \in \mathcal{A}$, then there are functions $\psi_i \in \mathcal{R}_i$ such that all the φ_i^k vanish outside $L_i = [\psi_i \geq 1]$ for $i = 1, 2$, and so φ vanishes outside $L_1 \times L_2 = [\psi_1 \otimes \psi_2 \geq 1]$. Note here that the $\mathcal{R}_1 \otimes \mathcal{R}_2$-dominated and the $\mathcal{A}$-dominated subsets of $X_1 \times X_2$ coincide, and that they are exactly the subsets of the products of $\mathcal{R}_i$-dominated sets $(i = 1, 2)$. Each of the functions φ_i^k is the uniform limit of functions in $\mathcal{R}_i$ vanishing outside L_i (5.7), and thus $\mathcal{A}$ is contained in the closure of $\mathcal{R}_1 \otimes \mathcal{R}_2$ under dominated uniform convergence. Since the closure of $\mathcal{A}$ under dominated uniform convergence is an integration lattice (5.6), $\mathcal{R}_1 \times \mathcal{R}_2$ is contained in the closure of $\mathcal{R}_1 \otimes \mathcal{R}_2$ under dominated uniform convergence. The following result has now been established.

<u>27.1. Lemma</u>. $\mathcal{R}_1 \otimes \mathcal{R}_2$ is dense in $\mathcal{R}_1 \times \mathcal{R}_2$ in the topology of dominated uniform convergence.#

<u>Example</u>. If $\mathcal{C}_i$ is a clan on X_i and $\mathcal{R}_i = \mathcal{S}(\mathcal{C}_i)$, then $\mathcal{R}_1 \times \mathcal{R}_2$ consists of the step functions $\mathcal{S}(\mathcal{C})$ over the clan $\mathcal{C} = \mathcal{C}_1 \times \mathcal{C}_2$ generated by the collection of products

$$P = \{K_1 \times K_2 \colon K_i \in \mathcal{C}_i;\ i = 1, 2\}$$

$\mathcal{C}$ consists of the finite unions of elements of P and is commonly called the <u>product</u> of the clans $\mathcal{C}_i$, $i = 1, 2$. Moreover, $\mathcal{S}(\mathcal{C}) = \mathcal{R}_1 \times \mathcal{R}_2 = \mathcal{R}_1 \otimes \mathcal{R}_2$.#

Suppose $m_1 \colon \mathcal{R}_1 \to E$ and $m_2 \colon \mathcal{R}_2 \to F$ are any two measures of finite variation. Their product is defined initially on $\mathcal{R}_1 \otimes \mathcal{R}_2$ by

$$(m_1 \times m_2)(\Sigma\, \varphi_1^k \otimes \varphi_2^k) = \Sigma\, m_2(\varphi_2^k) m_1(\varphi_1^k) \in G.$$

If $\varphi(x_1, x_2) = \Sigma\, \varphi_1^k(x_1)\varphi_2^k(x_2)$ (in $\mathcal{R}_1 \otimes \mathcal{R}_2$), this formula can be read as follows

$$(1) \qquad (m_1 \times m_2)(\varphi) = \int \left(\int \varphi(x_1, x_2)\, dm_2(x_2) \right) dm_1(x_1).$$

That is, $m_1 \times m_2$ is the integral of the field on X_1 of measures λ_{x_1} $(x_1 \in X_1)$ on $(X_2, \mathcal{R}_2)$ defined by

$$\langle \lambda_{x_1}; \varphi \rangle = \int \varphi(x_1, x_2)\, dm_2(x_2) \qquad (\varphi \in \mathcal{R}_1 \otimes \mathcal{R}_2).$$

Note that $\langle\lambda_{x_1};\varphi\rangle = \int \varphi(x_1,x_2)dm_2(x_2)$ belongs to $\mathcal{R}_1 \otimes E$, and so the iterated integral is well-defined, even if m_1 is not S-continuous, since m_1 has finite variation (17.2). It satisfies

$$\|\langle\lambda_{x_1};\varphi\rangle\| \le \int |\varphi(x_1,x_2)|d|m_2|(x_2).$$

The functions on both sides of this equation belong to $\overline{\mathcal{R}}_1$, and so

$$\|(m_1 \times m_2)(\varphi)\| \le \int (\int |\varphi(x_1,x_2)|d|m_2|(x_2))d|m_1|(x_1).$$

This shows that $m_1 \times m_2$ is continuous on $\mathcal{R}_1 \otimes \mathcal{R}_2$ in the topology of dominated uniform convergence, and so has a unique extension to $\mathcal{R}_1 \times \mathcal{R}_2$ (27.1, 4.7). This extension is also denoted by $m_1 \times m_2$, and is called <u>the product</u> of m_1 and m_2.

<u>Definition</u>. $(X_1 \times X_2, \mathcal{R}_1 \times \mathcal{R}_2, m_1 \times m_2) = (X_1,\mathcal{R}_1,m_1) \times (X_2,\mathcal{R}_2,m_2)$ is called the <u>product of the elementary integrals</u> $(X_i,\mathcal{R}_i,m_i)$.#

The following result is implicit in the above.

<u>27.2. Lemma</u>. The field $\lambda: X_1 \to I_0(\mathcal{S})$ given by

$$\langle\lambda_x;\varphi\rangle = \int \varphi(x,x_2)dm_2(x_2) \quad , \; x \in X_1, \; \varphi \in \mathcal{S} = \mathcal{R}_1 \times \mathcal{R}_2,$$

is uniformly continuous and has uniformly continuous and finite variation $|\lambda|$ given by

$$\langle|\lambda_x|;\varphi\rangle = \int \varphi(x,x_2)d|m_2|(x_2) \quad , \; x,\varphi \text{ as above.}$$

Moreover, $m_1 \times m_2 = \int \lambda dm_1$ has finite variation, and

$$|m_1 \times m_2| \le |m_1| \times |m_2| = \int |\lambda| d|m_1|.$$

B. Integration theory of the product

27.3. Theorem. If both m_1 and m_2 are $*$-continuous ($*$ = S or B) or tight, then so is $m_1 \times m_2$.

Proof. $*$-continuity: Define a field L on X_1 of upper integrals on $(Y,\mathcal{S})$ by

$$\langle L;f\rangle(x_1) = m_2^*(f_{x_1}) \qquad , \text{ where } f_{x_1}(x_2) = f(x_1,x_2), \quad f \in \overline{\mathbb{R}}_+^Y \;.$$

Since

$$\langle L;\varphi\rangle = |m_2|(\varphi_{x_1}) = \int \varphi(x_1,x_2)d|m_2|(x_2) = |\lambda|(\varphi) \in \overline{\mathcal{R}}_1$$

for $\varphi \in \mathcal{S}_+$, L is uniformly continuous. Also,

$$\int L dm_1^* = m_1^* \times m_2^*$$

coincides with $|m_1| \times |m_2|$ on $\mathcal{S}_+$, hence majorizes $m_1 \times m_2$, and is $*$-continuous, and so $m_1 \times m_2$ is $*$-continuous.

Tightness: Since the tight measures form a band (24.2) and since $|m_1 \times m_2| \leq |m_1| \times |m_2|$, it may be assumed that both m_1 and m_2 are positive. Let us write $n = m_1 \times m_2$.

Let $L_i \in \mathcal{K}^B(\mathcal{R}_i)$ and let K_i be any compact subset of L_i. Both $L = L_1 \times L_2$ and the compact set $K = K_1 \times K_2$ are in $\mathcal{K}^B(\mathcal{S})$ and therefore are $(\mathcal{S},n^B)$-integrable. From Fubini's theorem applied to the field λ,

$$n^B(L-K) = \int \left(\int (L-K)d\lambda\right)dm_1 = \int \left(L_1 \int L_2 dm_2 - K_1 \int K_2 dm_2\right)dm_1$$

$$= \int L_1 dm_1 \int L_2 dm_2 - \int K_1 dm_1 \int K_2 dm_2 .$$

This difference can be made arbitrarily small if the K_i are chosen sufficiently large. The compact sets are thus dense in $\mathcal{K}^B(\mathcal{S})$. As the latter family is dense, n is tight (24.1).#

<u>27.4. Corollary</u>. (classical Fubini theorem). Assume that the m_i are $*$-continuous, $* = S$ or B, and let f be any $(|m_1| \times |m_2|)^*$-integrable or $m_1^* \times m_2^*$-integrable function on $X_1 \times X_2$. Then $f_{x_1}: x_2 \to f(x_1,x_2)$ is m_2^*-integrable for almost all $x_1 \in X_1$, $\int f(x_1,x_2)dm(x_2)$ is m_1^*-integrable, and

$$\int fd(m_1 \times m_2) = \int (\int f(x_1,x_2)dm_2(x_2))dm_1(x_1).$$

Furthermore, if $(\mathcal{R}_2, m_2^*)$ is σ-finite and f is either an $(|m_1| \times |m_2|)^*$-measurable or an $m_1^* \times m_2^*$-measurable function, then f_{x_1} is m_2^*-measurable for locally almost all $x_1 \in X_1$.#

From 9.1, the upper integral $(|m_1| \times |m_2|)^*$ is bigger or equal to $m_1^* \times m_2^*$.

<u>27.5. Lemma</u>. For $f_i: X_i \to \overline{\mathbb{R}}_+$ $(i = 1, 2)$,

$$(|m_1| \times |m_2|)^*(f_1 \otimes f_2) = m_1^*(f_1)m_2^*(f_2) = (m_1^* \times m_2^*)(f_1 \otimes f_2).$$

<u>Proof</u>. If $m_i^*(f_i) < a_i$ then there is an $h_i \in \mathcal{R}_{i\uparrow}^*$ with $h_i \geq f_i$ and $m_i^*(h_i) < a_i$. Clearly

$$f_1 \otimes f_2 \leq h_1 \otimes h_2 \in (\mathcal{R}_1 \times \mathcal{R}_2)_\uparrow^* .$$

If $\mathcal{R}_i \supset \Phi_i \uparrow^* h_i$ then $\mathcal{R}_1 \times \mathcal{R}_2 \supset \Phi_1 \otimes \Phi_2 \uparrow^* h_1 \otimes h_2$ and $(|m_1| \times |m_2|)^*(h_1 \otimes h_2) = \sup |m_1|(\Phi_1)|m_2|(\Phi_2) = m_1^*(h_1)m_2^*(h_2) < a_1 a_2$. Hence

$$(|m_1| \times |m_2|)^*(f_1 \otimes f_2) \leq (m_1^* \times m_2^*)(f_1 \otimes f_2).\#$$

<u>27.6. Proposition</u>. Let $(X_i, \mathcal{R}_i, m_i)$ be two $*$-continuous elementary integrals of finite variation and put $(X, \mathcal{R}, m) = (X_1, \mathcal{R}_1, m_1) \times (X_2, \mathcal{R}_2, m_2)$ and $\underline{m} = |m_1| \times |m_2|$.

(i) If C_i is an $(\mathcal{R}_i, m_i^*)$-adequate cover $(i = 1, 2)$, then

$$C = C_1 \times C_2 = \{K_1 \times K_2 : K_i \in C_i;\ i = 1, 2\}$$

is an $(\mathcal{R}, \underline{m}^*)$-adequate cover.

(ii) If $f_i : X_i \to \mathbb{R}$ is an $(\mathcal{R}_i, m_i^*)$-measurable function $(i = 1, 2)$ then $f = f_1 \otimes f_2 : X \to \mathbb{R}$ is $(\mathcal{R}, \underline{m}^*)$-measurable. If f_i is $(\mathcal{R}, m_i^*)$-integrable then f is $(\mathcal{R}, \underline{m}^*)$-integrable and

$$\int f d(m_1 \times m_2) = \int f_1 dm_1 \times \int f_2 dm_2 .$$

<u>Proof</u>. (i): Let $K_i \in C_i$ $(i = 1, 2)$ and put $K = K_1 \times K_2$. There is a sequence (φ_n^i) in $\mathcal{R}_{i+}$ with $m_i^*(|K_i - \varphi_n^i|) \underset{n\to\infty}{\to} 0$. From the lemma,

$$\underline{m}^*(|K - \varphi_n^1 \otimes \varphi_n^2|) \le \underline{m}^*(|K - K_1 \otimes \varphi_n^2|) + \underline{m}^*(|K_1 \otimes \varphi_n^2 - \varphi_n^1 \otimes \varphi_n^2|)$$

$$= m_1^*(K_1) m_2^*(|K_2 - \varphi_n^2|) + m_1^*(|K_1 - \varphi_n^1|) m_2^*(|\varphi_n^2|) \underset{n\to\infty}{\to} 0,$$

and so K is integrable: C consists of $(\mathcal{R}, \underline{m}^*)$-integrable sets. In the beginning of the section we have seen that every $\mathcal{R}$-dominated set is contained in the product of two dominated $\mathcal{R}_i$-Baire sets: $\mathcal{C}_0 = \mathcal{C}(\mathcal{R}_1, m_1^*) \times \mathcal{C}(\mathcal{R}_1, m_2^*)$ is an $(\mathcal{R}, \underline{m}^*)$-adequate cover. It is sufficient to show that every set $L = L_1 \times L_2 \in \mathcal{C}_0$ can be $(\mathcal{R}, \underline{m}^*)$-adequately covered by C.

If $\{K_i^n : n = 1, 2, \ldots\} \subset C_i$ covers L_i $(\mathcal{R}_i, m_i^*)$-adequately, then

$$\{K_1^n \times K_2^m : n, m = 1, 2, \ldots\} \subset C$$

covers L $(\mathcal{R}, \underline{m}^*)$-adequately. Indeed, the points of L that do not belong to the union of this family are contained in the product of two m_i^*-negligible sets and by 27.5 form an $\underline{m}^*$-negligible set.

(ii): $U(f_1)\times U(f_2)$ consists of $(\mathcal{R},\underline{m}^*)$-integrable sets on which $f_1 \otimes f_2$ is uniformly continuous and is an adequate cover. Hence (18.9) $f_1 \otimes f_2$ is $(\mathcal{R},\underline{m}^*)$-measurable. The second statement follows as in (i), upon replacing K_i by f_i.#

27.7. Application. Let $(X,\mathcal{R},m)$ be a positive elementary *-integral and $f\colon X \to \overline{\mathbb{R}}_+$ an $(\mathcal{R},m^*)$-measurable function. If $\varphi\colon \mathbb{R}_+ \to \mathbb{R}_+$ is an increasing function with continuous derivative and satisfying $\varphi(0) = 0$, then

$$\int \varphi\circ f dm = \int_{\mathbb{R}_+} \varphi'(s)\cdot m^*([f \geq s])d\lambda(s)$$

(in the sense that if either integral exists then so does the other and equality obtains; λ denotes Lebesgue measure on $\mathbb{R}_+$).

Proof. The function

$$(x,s) \to \varphi'(s)[0,f(x)](s) = \varphi'(s)\{(x,s)\colon f(x) - s \geq 0\}$$

is $(m\times\lambda)^*$-measurable from $X\times\mathbb{R}_+$ to $\mathbb{R}_+$. This follows from 18.12(iv) and the fact that $(x,s) \to f(x) - s$ is measurable (27.6). Hence

$$m^*(\varphi\circ f) = m^*(\int_0^f \varphi'(s)d\lambda(s)) = m^*(x \to \int [0,f(x)](s)\varphi'(s)d\lambda(s))$$

$$= (m\times\lambda)^*((x,s) \to \varphi'(s)[f(x) \geq s]) = \lambda^*(s \to \varphi'(s)m^*([f \geq s])).\#$$

C. Supplements

27.8. The product of finitely many (scalar-valued) elementary integrals $(X_i,\mathcal{R}_i,m_i)$ $(i = 1,\dots,n)$ is defined by induction. $\mathcal{R}_1 \otimes \cdots \otimes \mathcal{R}_n$ is dense in $\mathcal{R}_1\times\cdots\times\mathcal{R}_n$ in the topology of dominated uniform convergence. If $1 \leq k < n$ then

$$(X_1,\mathcal{R}_1,m_1)\times\cdots\times(X_n,\mathcal{R}_n,m_n) =$$
$$= ((X_1,\mathcal{R}_1,m_1)\times\cdots\times(X_k,\mathcal{R}_k,m_k))\times((X_{k+1},\mathcal{R}_{k+1},m_{k+1})\times\cdots\times(X_n,\mathcal{R}_n,m_n)).$$

LITERATURE: Standard textbooks, e.g., [1], [3], [19], [21], [35], [36].

§2. Images of measures

We shall consider two dominated integration lattices $(X,\mathcal{R})$ and $(Y,\mathcal{S})$, and a map $p: X \to Y$. These, as well as the three Banach spaces E, F, and G, will be fixed in the following sections. Moreover, $(\mathcal{R},M)$ is a fixed weak upper gauge.

28. Adequate and adapted maps

A. Adequate and tame maps

Definition. The map $p: X \to Y$ is said to be $(\mathcal{R},M)$-adequate for the strong upper gauge $(\mathcal{R},M)$ on X, if it is measurable and if $\varphi\circ p$ is $(\mathcal{R},M)$-integrable for every $\varphi \in \mathcal{S}$.#

With p there is associated the field δ_p of point measures on $(Y,\mathcal{S})$ given by

$$\langle \delta_{p(x)};\varphi\rangle = \varphi\circ p(x) \qquad , \ x\in X,\ \varphi\in\mathcal{S},$$

and δ_p is evidently adequate if and only if p is.

The natural field of upper integrals majorizing δ_p is δ'_p, defined by

$$\langle \delta'_p;f\rangle = f\circ p \qquad , \text{ for } f: Y \to \overline{\mathbb{R}}_+ .$$

Note that it differs from the standard fields δ^*_p, and that it, too, is $(\mathcal{R},M)$-adequate. Its integral, $N = \int \delta'_p dM$, is given by

$$\left(\int \delta'_p dM\right)(f) = M(f\circ p) \qquad , \text{ for } f: Y \to \overline{\mathbb{R}}_+ .$$

If $(\mathcal{R},M)$ is B-continuous and essential, then $(\mathcal{S},N)$ is also B-continuous (26.3).

Definition. We write $N = pM$ and call it the image of M under p.

Let us find the sets on which δ'_p is tame. First, consider any $(\mathcal{S},N)$-integrable set B. The set $p^{-1}(B) = B\circ p = \int Bd(\delta_p)$ is $(\mathcal{R},M)$-integrable, by Fubini's theorem; and, for $x \in p^{-1}(B)$, $\langle\delta'_{p(x)};Y - B\rangle = 0$; hence δ'_p is tame on $p^{-1}(B)$ for $B \in \mathcal{C}(\mathcal{S},N)$.

Conversely, if δ'_p is tame on $K \in \mathcal{C}(\mathcal{R},M)$, then there is a sequence (B_n) in $\mathcal{C}(\mathcal{S},N)$ such that

$$\langle\delta'_p;Y - \cup B_n\rangle = p^{-1}(Y - \cup B_n) = 0,$$

almost everywhere on K. We have the following result.

<u>28.1. Lemma</u>. δ'_p is tame exactly on those measurable sets which are $(\mathcal{R},M)$-adequately covered by

$$p^{-1}(\mathcal{C}(\mathcal{S},N)) = \{p^{-1}(B): B \in \mathcal{C}(\mathcal{S},N)\}.$$

In particular, δ'_p is tame if and only if $p^{-1}(\mathcal{C}(\mathcal{S},N))$ is an adequate cover. If $1 \in \mathcal{S}^S_\uparrow$ then δ'_p is tame.#

It will henceforth be assumed that $p: X \to Y$ is $(\mathcal{R},M)$-adequate and <u>tame</u> (in the sense that δ'_p is tame).

<u>28.2. Lemma</u>. If C is an $(\mathcal{S},N)$-adequate cover, then $p^{-1}(C) = \{p^{-1}(K): K \in C\}$ is an $(\mathcal{R},M)$-adequate cover.

<u>Proof</u>. Let $A \in \mathcal{C}(\mathcal{R},M)$. There is a countable family $\{B_n\} \subset \mathcal{C}(\mathcal{S},N)$ such that $A \subset \cup p^{-1}(B_n)$ almost everywhere. Each B_n may be itself replaced by a countable family in C.#

<u>28.3. Corollary</u>. If $(\mathcal{S},pM)$ is strictly localizable, so is $(\mathcal{R},M)$.#

Now, let $m: \mathcal{R} \to F$ be a measure majorized by M. The integral $n = \int \delta_p dm$ is given by

$$n(\varphi) = \int \langle \delta_p ; \varphi \rangle dm = \int \varphi \circ p dm \qquad , \text{ for } \varphi \in \mathcal{S}.$$

Definition. $\int \delta_p dm$ is called the image of m under p, and is denoted by pm or by $p(m)$.#

The following results are obtained from 25.3 and 26.3.

28.4. Lemma. If m has finite variation then so does pm, and $|pm| \leq p|m|$.

If $|m|$ is majorized by M then $|pm|$ is majorized by pM; and if $(\mathcal{R}, M)$ is, in addition, B-continuous then so is pm.#

28.5. Theorem. Let $f: Y \to E$ be a map.

(i) If f is $(\mathcal{S}, pM)$-measurable then $f \circ p$ is $(\mathcal{R}, M)$-measurable.

(ii) If f is $(\mathcal{S}, pM)$-integrable then $f \circ p$ is $(\mathcal{R}, M)$-integrable, and

$$\int f d(pm) = \int f \circ p dm.$$

Proof. (ii) follows immediately from Fubini's theorem applied to the fields δ_p, δ'_p.

(i): The sets $B \in \mathcal{C}(\mathcal{S}, N)$ on which f is uniformly continuous form an $(\mathcal{S}, N)$-adequate cover, and their preimages, $p^{-1}(B)$, form an $(\mathcal{R}, M)$-adequate cover. Each of the $p^{-1}(B)$ is, a.e., a countable union of sets on which p is uniformly continuous, i.e., on which $f \circ p$ is uniformly continuous; $f \circ p$ is thus uniformly continuous on the members of an $(\mathcal{R}, M)$-adequate cover, and so is measurable (18.9).#

Remark. Suppose $M = m^*$ for some $m \in M^*(\mathfrak{R})$. Then $pM = pm^* \leq p(m^*)$ (9.1), and the conclusions (i) and (ii) hold for $(\mathfrak{R}, (pm)^*)$-integrable and -measurable functions, respectively.

28.6. Corollary. If $(\mathfrak{R},M)$ is strong and localizable so is $(\mathcal{S},pM)$.

Proof. Let F be an increasingly directed and bounded subset of $L^\infty_+(\mathcal{S},pM)$. Then $F\circ p \subset L^\infty(\mathfrak{R},M)$ has a supremum $g^\bullet$. Furthermore, if $B \in \mathcal{C}(\mathcal{S},pM)$ then $FB^\bullet$ has a supremum $f^\bullet_B$ in $L^1(\mathcal{S},pM)$, and this is the limit in pM-mean of a sequence $(f^\bullet_n B^\bullet)$ with $f^\bullet_n \in F$ (8.13). Hence $f_n\circ p\cdot B\circ p$ converges in M-mean to $f_B\circ p$. This shows that $g = f_B\circ p$ a.e. on $B\circ p$, and so g is constant on locally almost all fibres of p; that is, $g \in g^\bullet$ may be assumed to be of the form $g = f\circ p$, where f is some real-valued function on Y. Since $f = f_B$ almost everywhere on B, f is measurable. Clearly, $f^\bullet$ is the supremum of F in $L^\infty(\mathcal{S},pM)$.#

28.7. Corollary. The map $f \to f\circ p$ is a linear isometry of $L^q_E(\mathcal{S},pM)$ into $L^q_E(\mathfrak{R},M)$, for $1 \leq q \leq \infty$. Its image is exactly the set of classes in $L^q_E(\mathfrak{R},M)$ that contain a function constant on the fibres of p. It is normal on $L^q(\mathcal{S},pM)$.#

B. Adapted maps

Definition. Let $(\mathfrak{R},M)$ be a weak upper gauge on X. The map $p: X \to Y$ is $(\mathfrak{R},M)$-adapted if it is adequate and the family $A(p)$ of $(\mathfrak{R},M)$-integrable sets K such that $p(K)$ is $(\mathcal{S},pM)$-integrable is dense.

28.8. Lemma. If p is adapted then $p(A(p))$ is $(\mathcal{S},pM)$-dense. Moreover, if F is a dense subfamily of $A(p)$ then $p(F)$ is also dense.

Proof. Let $B \in \mathscr{C}(\mathcal{S},pM)$. Then $p^{-1}(B) \in \mathscr{C}(\mathcal{R},M)$, hence $p^{-1}(B) \subset A(p)$, and there is a sequence (K_n) in F of subsets of $p^{-1}(B)$ with $M(p^{-1}(B) - \cup K_n) = 0$. Clearly, $(pM)(B - \cup p(K_n)) = M(p^{-1}(B) - \cup p^{-1}p(K_n)) = 0$.#

28.9. Corollary. An adapted map is tame.

28.10. Theorem. Suppose $(\mathcal{R},M)$ is a strong upper gauge and $p: X \to Y$ is $(\mathcal{R},M)$-adapted.

(i) A map $f: Y \to E$ is $(\mathcal{S},pM)$-integrable if and only if $f \circ p$ is $(\mathcal{R},M)$-integrable.

(ii) A map f on Y with values in a metrizable space S is $(\mathcal{S},pM)$-measurable if and only if $f \circ p$ is $(\mathcal{R},M)$-measurable.

Proof. The "only if" parts are known from 28.5.

(i): Assume first that f is real-valued and positive, and that $f \circ p$ is $(\mathcal{R},M)$-integrable. As $(pM)(f) = M(f \circ p) < \infty$, we may subtract from f an $(\mathcal{S},pM)$-integrable function $f' \leq f$ of maximal measure $(pM)(f')$. It has to be shown that $f'' = f - f'$ is negligible: If this were not so but rather $M(f'' \circ p) > 0$, then there would be an integrable set $K \subset X$ of non-zero measure and an $a > 0$ such that $f'' \geq aK$. K may be chosen such that $p(K)$ is integrable. Clearly $f' + ap(K) \leq f$ is $(\mathcal{S},pM)$-integrable, also, and has bigger measure than f' (SUG): a contradiction. Hence $(pM)(f - f') = 0$. The case for Banach-valued functions f follows from (ii): If $f \circ p$ is $(\mathcal{R},M)$-integrable then f is $(\mathcal{S},N)$-measurable and majorized by the $(\mathcal{S},N)$-integrable function $\|f\|$.

(ii): Suppose $f \circ p: X \to S$ is measurable. It will be assumed initially that f is a subset $B \subset Y$. For $K \in A(p)$,

$$(B \circ p)(p^{-1}p(K)) = (Bp(K)) \circ p$$

is $(\mathcal{R},M)$-integrable, and hence $Bp(K)$ is $(\mathcal{S},pM)$-integrable. From lemma 28.8, B is measurable.

Now, let $f: Y \to S$ be any function such that $f \circ p$ is measurable. The sets $K \in A(p)$ on which $f \circ p$ is uniformly continuous are dense, and their images are $(\mathcal{S},pM)$-dense (28.8). If K is one of them, then the range of $f \circ p$ on K, and so of f on $p(K)$, is contained in a separable subspace of S (5.7). Furthermore, if U is an open subset of S, then $(f \circ p)^{-1}(U) = f^{-1}(U) \circ p$ is measurable (19.5), and therefore $f^{-1}(U)$ is measurable. From (19.5), f is measurable.#

C. Images of tight measures

28.11. Proposition. Suppose the strong upper gauge $(\mathcal{R},M)$ is B-continuous and tight, and $p: X \to Y$ is $(\mathcal{R},M)$-adequate.

Then p is adapted and tame; $(\mathcal{S},p\bar{M})$ is B-continuous and tight; and a map f from Y to some uniform space Y is $(\mathcal{S},p\bar{M})$-measurable if and only if $f \circ p$ is $(\mathcal{R},M)$-measurable.

Proof. $(\mathcal{S},p\bar{M})$ is B-continuous. The sets $p(K) \subset Y$ such that $K \subset X$ is compact and p is uniformly continuous on K are compact, and so $(\mathcal{S},p\bar{M})$-integrable; and they are dense (28.8). Hence p is adapted and tame, and $(\mathcal{S},p\bar{M})$ is tight. The "only if" part of the last statement is contained in 28.5. Suppose then that $f \circ p$ is measurable. The compact sets $K \subset X$ on which both $f \circ p$ and p are uniformly continuous are dense, and so their images, $p(K)$, also are dense (28.8). As $p(K)$ has the quotient topology of $p: K \to p(K)$, f is continuous on $p(K)$; hence f is uniformly continuous on $p(K)$.#

28.12. Corollary. The image (under an $(\mathcal{R},\bar{m}^B)$-adequate map) of a tight measure m is tight.#

28.13. Example. Suppose m is a positive Radon measure on the locally compact space X, and p is a map from X to another locally compact space Y; put $\mathcal{R} = C_{oo}(X)$ and $\mathcal{S} = C_{oo}(Y)$. Bourbaki [5] calls the map p m-proper if it is $(\mathcal{R},m^B)$-measurable (19D) and if $\varphi\circ p \in \mathcal{L}^1(\mathcal{R},m^B)$ for all $\varphi \in \mathcal{S}$; that is if it is $(\mathcal{R},m^B)$-adequate. As we have seen, it is then adapted and tame. A map f from Y to a uniform space is $(\mathcal{S},(pm)^B)$-measurable if and only if $f\circ p$ is $(\mathcal{R},m^B)$-measurable (28.14).

D. Supplements

28.14*. Suppose $(\mathcal{R},M)$ is an upper integral for the positive measure m and $p: (X,\mathcal{R}) \to (Y,\mathcal{S})$ is an $(\mathcal{R},M)$-adequate map.

(i) If $(\mathcal{R},M)$ is essential then so is $(\mathcal{S},pM)$.

(ii) If $(\mathcal{R},M)$ is tight then so is $(\mathcal{S},pM)$: the image of a tight measure is tight.

(iii) If m is tight and $M = \bar{m}^B$ then $pM = (\overline{pm})^B$.

(iv) $p(M_q) = (pM)_q$ for $1 \leq q < \infty$.

28.15. Suppose $m: \mathcal{R} \to F$ is a bounded measure majorized by the finite upper gauge $(\mathcal{R},M)$, and let f be a real-valued $(\mathcal{R},M)$-measurable function. Then $f: X \to \mathbb{R}$ is an $(\mathcal{R},M)$-adequate and tame map (where $Y = \mathbb{R}$, $\mathcal{S} = C_{oo}(\mathbb{R})$). The measure $f(m)$ on the line is bounded. It is called the distribution of f (with respect to m). If the measurable functions (f_k) converge in measure to f, then f is measurable, and the distributions $f_k(m)$ converge vaguely to $f(m)$.

Let (f_n) be a sequence of real- or complex-valued measurable functions. Then (f_n) can be viewed as a measurable (18.3) map from X to $\mathbb{R}^{\mathbb{N}}$ (where $\mathcal{S}$ on $\mathbb{R}^{\mathbb{N}}$ consists of the continuous and bounded functions that depend only on finitely many coordinates). It is $(\mathcal{R},M)$-adequate and tame; the image $(f_n)(m)$ is called the joint distribution of the f_n.

Suppose $(\mathcal{R},M)$ is an upper integral for the positive measure m and f is a positive measurable function. If $(\mathcal{R},M)$ is finite or if f is integrable, then f can be viewed as an $(\mathcal{R},M)$-adequate and tame map to $(\mathbb{R}_+, C_{oo}(\mathbb{R}_+))$. The image of m under f, $f(m)$, is again called the distribution of f.

(i) f is in $\mathcal{L}^p(\mathcal{R},M)$ iff the function $x \to x$ belongs to $\mathcal{L}^p(C_{oo}(\mathbb{R}_+), f(m)^S)$. ($1 \leq p < \infty$; use 27.7.)

(ii) If $f,g: X \to \overline{\mathbb{R}}$ are measurable and such that $|f|(m) = |g|(m)$ on $C_{oo}(\mathbb{R}_+)$, then f and g are rearrangements of each other. Any rearrangement of a p-integrable function is p-integrable ($1 \leq p \leq \infty$).

28.16. (Lifting tight measures). Let X,Y be completely regular spaces, $p\colon X \to Y$ a map, and n a tight measure on $C_b(Y)$. In order that there exist a tight measure m on $C_b(X)$ with image $pm = n$ it is necessary and sufficient that the images $p(K) \subset Y$ of compact sets $K \subset X$ on which the restriction of p is continuous are n^B-dense.

28.17. Suppose $U\colon \mathcal{R} \otimes E \to G$ is a linear map majorized by the weak upper gauge $(\mathcal{R},M)$ and $p\colon (X,\mathcal{R}) \to (Y,\mathcal{S})$ is an $(\mathcal{R},M)$-adequate map. The image, pU, of U under p is then defined by $(pU)(\varphi) = \int \varphi \circ p\, dU$, for $\varphi \in \mathcal{S} \otimes E$. The results 28.4, 28.5 have their analogues for images of linear maps. If U is (weakly) compact so is pU.

28.18*. Let $(X,\mathcal{R})$ and $(Y,\mathcal{S})$ be integration lattices. A <u>morphism</u> from $(X,\mathcal{R})$ to $(Y,\mathcal{S})$ is a map $p\colon X \to Y$ such that $\varphi \circ p \in \mathcal{R}$ for all $\varphi \in \mathcal{S}$.

(i) A morphism is uniformly continuous and $(\mathcal{R},M)$-adequate for every weak upper gauge M on $\mathcal{R}$; if $\mathcal{R}^S$ contains the constants it is also tame.

(ii) If $p\colon X \to Y$ is an $(\mathcal{R},M)$-adequate map, then p is a morphism from $(X,\mathcal{L}^1(\mathcal{R},M))$ to $(Y,\mathcal{S})$ or to $(Y,\mathcal{L}^1(\mathcal{S},pM))$. The converse is not true in general, but is if $\mathcal{S}$ is countably generated.

(iii) If $p\colon (X,\mathcal{R}) \to (Y,\mathcal{S})$ and $q\colon (Y,\mathcal{S}) \to (Z,\mathcal{T})$ are morphisms, then $q \circ p\colon (X,\mathcal{R}) \to (Z,\mathcal{T})$ is a morphism.

(iv) If $X = Y$ and $\mathcal{S} \subset \mathcal{R}$ then the identity map is a morphism; the map $\pi\colon X_0 \to S(\mathcal{R})$ is a morphism from $(X,\mathcal{R})$ to $(S(\mathcal{R}),\hat{\mathcal{R}})$; if $p\colon (X,\mathcal{R}) \to (Y,\mathcal{S})$ is a morphism then it extends to a continuous map $\hat{p}\colon S(\mathcal{R}) \to S(\mathcal{S})$ which is a morphism from $(S(\mathcal{R}),C_{oo}(S(\mathcal{R}))$ to $(S(\mathcal{S}),C_{oo}(S(\mathcal{S})))$.

28.19*. Let $p\colon (X,\mathcal{R}) \to (Y,\mathcal{S})$ be a morphism and $m\colon \mathcal{R} \to F$ a Banach valued measure on $\mathcal{R}$. The <u>image</u> $p(m)$ of m is defined by

$$p(m)(\varphi) = m(\varphi \circ p) \qquad , \text{ for } \varphi \in \mathcal{S}.$$

The map $m \to p(m)$ is linear and decreases variation: $|p(m)| \leq p(|m|)$. It maps $M_F^*(\mathcal{R})$ into $M_F^*(\mathcal{S})$ $(* = S,B)$.

LITERATURE: The treatment is adapted from [5], Ch. VI, IX. Cf. [19]. If $(\mathcal{R},M)$ is tight, then an $(\mathcal{R},M)$-adequate map is called proper in [5].

29. Conditional expectation

Again, a pair of dominated integration lattices, $(X,\mathcal{R})$ and $(Y,\mathcal{S})$, is fixed. $(\mathcal{R},M)$ is a localizable upper integral for the positive measure m, and $p: X \to Y$ is $(\mathcal{R},M)$-adequate and tame.

It will turn out that the image of a measure gm defined by a locally integrable density g, often has a locally $(\mathcal{S},pM)$-integrable derivative,

$$E^p g = d(p(gm))/d(pm),$$

with respect to pm. This derivative, if it exists, is called the conditional expectation of g under p.

A. Admissible functions

Definition. A locally $(\mathcal{R},M)$-integrable map $g: X \to E$ is admissible for p if $g\cdot\varphi\circ p$ is $(\mathcal{R},M)$-integrable for all φ in $\mathcal{S}$.#

This condition says that p is adequate for the upper integral $(\mathcal{R}, \|g\|M)$ (22B , 28A) and is automatically satisfied if, for instance, $\varphi\circ p$ is $\mathcal{R}$-dominated for $\varphi \in \mathcal{S}$. The measure gm has thus an image $p(gm)$ under p. The inequalities

$$|p(gm)| \le p(|gm|) = p(\|g\|m) = \sup\{p((\|g\| \wedge k)m): k = 1, 2, \ldots\}$$

imply that $p(gm)$ is absolutely continuous with respect to $p(m)$. Clearly, $p(gm)$ does not change if g is altered on a locally negligible set. We may thus talk about the measures $g^{\cdot}m$ and $p(g^{\cdot}m)$, for $g^{\cdot}$ an admissible class modulo $\overline{M}$.

Definition. (i) Let $g: X \to E$ be admissible. A conditional expectation of g under p is any locally $(\mathcal{S},pM)$-integrable derivative of $p(gm)$ with respect to pm.

(ii) Let $\dot{g}$ be an admissible class in $L_E^{\ell}(\mathcal{R},M)$. <u>The conditional expectation of</u> $g^{\cdot}$ <u>under</u> p is the class in $L_E^{\ell}(\mathcal{S},pM)$ of any derivative $d(p(gm))/d(p(m))$, provided such exists. It is denoted by $E^p g$ or $E^p(g)$.#

In other words, $E^p g$ is a class in $L_E^{\ell}(\mathcal{S},pM)$ satisfying

$$(1) \qquad \int g\cdot\varphi\circ p\,dm = \int E^p g\cdot\varphi\cdot d(pm) \quad , \text{ for all } \varphi \text{ in } \mathcal{S}.$$

Routine applications of linearity, of Lebesgue's theorem, and of 22.1 yield the following characterization.

<u>29.1. Lemma</u>. $\tilde{g} \in \mathcal{L}_E^{\ell}(\mathcal{S},pM)$ is a conditional expectation of g if and only if

$$\int \tilde{g}f\,d(pm) = \int g\cdot f\circ p\,dm \qquad \text{for all bounded, dominated } (\mathcal{S},pM)\text{-integrable functions } f, \text{ \underline{or}}$$

$$\int \tilde{g}K\,d(pm) = \int g\cdot p^{-1}(K)\,dm \qquad \text{for all dominated } (\mathcal{S},pM)\text{-integrable sets } K.$$

<u>29.2. Lemma</u>. Let $g: X \to E$ be an admissible function, and $\tilde{g}: Y \to E$ an $(\mathcal{S},pM)$-measurable map. Then $\tilde{g}$ is a conditional expectation of g if and only if

$$\int \langle\tilde{g};f\rangle d(pm) = \int \langle g;f\circ p\rangle dm,$$

for all f in $\mathcal{S}\otimes E'$, or for all bounded, dominated, $(\mathcal{S},pM)$-integrable functions f with values in the dual E' of E.

<u>Proof</u>. Suppose $\tilde{g}$ is a conditional expectation of g, and let $\varphi = \sum \varphi_k\xi_k \in \mathcal{S}\otimes E'$. Then

$$\int \langle\tilde{g};\varphi\rangle d(pm) = \sum \langle \int \tilde{g}\varphi_k d(pm);\xi\rangle = \sum \langle \int g\varphi_k\circ p\,dm;\xi_k\rangle = \int \langle g;\varphi\circ p\rangle dm.$$

Let $f\colon Y \to E'$ be a bounded $(\mathcal{S}, pM)$-integrable function which vanishes off the $\mathcal{S}$-dominated set $K \subset Y$. We may assume that $\|f\| < 1/2$. Enlarging K if necessary, we can find a sequence (φ_k) in $\mathcal{S} \otimes E'$ which converges a.e. to f and consists of functions vanishing outside K and having norm not exceeding one. Let $K \leq \psi \in \mathcal{S}$. Since $\|\tilde{g}\|\ \|f - \varphi_k\|$ is majorized by the integrable function $2\psi\|\tilde{g}\|$ and tends to zero a.e.,

$$\left|\int \langle \tilde{g}; f\rangle dm - \int \langle \tilde{g}; \varphi_k\rangle d(pm)\right| \leq (pM)(\|\tilde{g}\|\ \|f - \varphi_k\|) \to 0.$$

Similarly,

$$\left|\int \langle g; f\circ p\rangle dm - \int \langle g; \varphi_k \circ p\rangle dm\right| \leq M(\|g\|\cdot \|f - \varphi_k\| \circ p) \to 0$$

since $\|g\|\cdot\|f - \varphi_k\| \circ p$ tends to zero a.e. and is majorized by the integrable function $2\|g\|\psi\circ p$. This shows that the condition is necessary.

Conversely, if the condition is satisfied, then

$$(pM)(\tilde{g}\psi) = \sup\{\left|\int \langle \tilde{g}\psi; \varphi\rangle d(pm)\right| : \psi \in \mathcal{S} \otimes E',\ \|\varphi\| \leq 1\}$$

$$= \sup\{\left|\int \langle g\psi\circ p; \varphi\circ p\rangle dm\right| : \ldots\} = M(g\psi\circ p) < \infty$$

for all $\psi \in \mathcal{S}$ (21.4), and hence $\tilde{g}$ is locally integrable. For $\varphi \in \mathcal{S}$ and $\xi \in E'$,

$$\langle \int \tilde{g}\varphi d(pm); \xi\rangle = \int \langle \tilde{g}; \varphi\xi\rangle d(pm) = \int \langle g; \xi\varphi\circ p\rangle dm = \langle \int g\varphi dm; \xi\rangle,$$

and therefore $\int \tilde{g}\varphi(pm) = \int g\varphi\circ p dm$; hence $\tilde{g}$ is a conditional expectation of g.#

B. Properties

Several properties of the conditional expectation operator E^p are listed below. For a start note that the admissible maps $g\colon X \to E$ admitting a conditional expectation form a vector space on which E^p is linear.

29.3. If $g: X \to E$ has a conditional expectation then so does $\|g\|$, and furthermore

$$\|E^p g\| \le E^p \|g\|.$$

(We express this behaviour by saying that E^p is <u>contractive</u>.) Indeed, since $(\mathcal{B}, pM)$ is localizable (28.6) and $\|g\|\varphi\circ p$ integrable for all φ in $\mathcal{B}$, $p(\|g\|m)$ has a locally $(\mathcal{B}, pM)$-integrable derivative $E^p\|g\|$ with respect to m. The inequality follows from 28.4 and 22.2:

$$E^p\|g\|\cdot pm = p(\|g\|m) = p(|gm|) \ge |p(gm)| = |E^p g\cdot pm| = \|E^p g\|\cdot pm. \#$$

29.4. If $f \in L_E^{\ell}(\mathcal{B}, pM)$, then $f\circ p$ is admissible and has conditional expectation f.

Indeed, $f\circ p$ is measurable (28.5) and satisfies $\int f\circ p\cdot\varphi\circ p dm = \int f\cdot\varphi d(pm)$ for all $\varphi \in \mathcal{B}$.#

29.5. For any q with $1 \le q < \infty$, every element of $L_E^q(\mathcal{R}, M)$ is admissible and has a conditional expectation. Furthermore, E^p maps $L_E^q(\mathcal{R}, M)$ onto $L_E^q(\mathcal{B}, pM)$ and has operator norm one. As a map from $L^q(\mathcal{R}, M)$ onto $L^q(\mathcal{B}, pM)$, E^p is normal.

Indeed, let $f \in \mathcal{L}_E^q(\mathcal{R}, M)$. Then $f\cdot\varphi\circ p$ is $(\mathcal{R}, M)$-measurable, vanishes off a σ-finite set, and satisfies

$$M(f\cdot\varphi\circ p) \le M_q(f) M_{q'}(\varphi\circ p) < \infty \qquad \text{for all } \varphi \in \mathcal{B}.$$

$(1/q + 1/q' = 1)$. Hence $f\cdot\varphi\circ p$ is integrable for all $\varphi \in \mathcal{B}$; i.e., f is admissible. Suppose now that f is real-valued. Then $p(fm)$ has a locally integrable derivative, $E^p f$, due to the localizability of $(\mathcal{S}, pM)$ (22.6). If $f \in L^q(\mathcal{R},M) \otimes E$, then $E^p f$ exists by linearity, and, from

$$(pM_q)(E^p f) = \sup \{ \int \langle E^p f;\psi\rangle d(pm) : \psi \in \mathcal{S} \otimes E', \ (pM)_{q'}(\psi) \leq 1\}$$

$$= \sup \{ \int \langle f;\psi\circ p\rangle dm : \ldots\}$$

$$\leq \sup \{ \int \langle f;\varphi\rangle dm : \varphi \in \mathcal{L}^1_{E'}(\mathcal{R},M), \ M_{q'}(\varphi) \leq 1\} = M_q(f),$$

E^p is seen to have norm smaller than one on $L^q_E(\mathcal{R},M)$, and hence to be defined by continuity for all of $L^q_E(\mathcal{R},M)$. From 29.4 and $M_q(f\circ p) = (pM)_q(f)$, E^p has operator norm one and is onto. The proof that E^p is normal on $L^q(\mathcal{R},M)$ is left to the reader.#

29.6. If A^∞_E denotes the classes in $L^\infty_E(\mathcal{R},M)$ that have a conditional expectation, then it is closed in $L^\infty_E(\mathcal{R},M)$ and contains $L^\infty(\mathcal{R},M) \otimes E$ and $L^\infty_E(\mathcal{S},pM) \circ p$. Moreover, E^p maps A^∞_E onto $L^\infty_E(\mathcal{S},pM)$ and has operator norm one.

29.7. Let $E \times F \to G$ be as in 21A, suppose $f: X \to E$ has a conditional expectation, and let $g \in L^\ell_F(\mathcal{S},pM)$ be such that $\|f\|\cdot\|g\circ p\|$ is admissible. Then $f\cdot g\circ p$ has conditional expectation

$$E^p(f\cdot g\circ p) = (E^p f)\cdot g \in L^\ell_G(\mathcal{S},pM).$$

<u>29.8. Example</u>. Let $Y = X$ and let $p: X \to Y$ be the identity map. Then p is adequate if and only if $\mathcal{S} \subset \mathcal{L}^1(\mathcal{R},M)$, and tame precisely if the $(\mathcal{S},M)$-integrable sets form an adequate cover (28.1). Note that $pM = M$ and that pm is the restriction of $\int dm$ to $\mathcal{S}$.

The conditional expectation of a function g under the identity map is usually called the conditional expectation of g under $\mathcal{S}$, or under $\mathcal{C}(\mathcal{S},M)$, or under the σ-algebra $T(\mathcal{S},M)$; and is denoted by $E^{\mathcal{S}}g$. It is the unique $(\mathcal{S},M)$-measurable class satisfying

$$\int g\varphi dm = \int E^{\mathcal{S}}g\varphi dm \qquad , \text{ for all } \varphi \in \mathcal{S}, \underline{\text{or}}$$

$$\int gKdm = \int E^{\mathcal{S}}gKdm \qquad , \text{ for all } K \in \mathcal{C}(\mathcal{S},M).$$

C. Supplements

29.9. If $T: E \to F$ is a continuous linear map and $E^p g$ is the conditional expectation of $g \in L^{\ell}_E(\mathcal{R},M)$, then $T \circ E^p g$ is the conditional expectation of $T \circ g$.

29.10. (Jensen's inequality). Suppose $(\mathcal{R},M)$ is a finite upper integral on X and $p: X \to (Y,\mathcal{S})$ is an $(\mathcal{R},M)$-adequate and tame map. Let $\varphi: \mathbb{R} \to \mathbb{R}$ be a convex function and $f \in \mathcal{L}^1(\mathcal{R},M)$ so that $\varphi \circ f$ is integrable. Then

$$E^p(\varphi \circ f) \geq \varphi \circ E^p(f).$$

If f takes its values a.e. in the interval $[a,b]$ then it is sufficient to require that φ be convex on $[a,b]$.

(Hint: φ is the upper envelope of a sequence of affine linear maps $\ell_n: x \to a_n x + b_n$.)

LITERATURE: [1], [11], [13], [16], [19], [31], [33].

§3. Projective limits

30. Projective systems of measures

A. Projective systems of integration lattices

Definition. (i) A projective system of integration lattices is a triple

$$\mathcal{T} = \{ (X_s, \mathcal{R}_s); (s,t); s, t \in T \},$$

where T is an increasingly directed set of indices; the $(X_s, \mathcal{R}_s)$ are integration lattices labeled by $s \in T$; and the (st) are morphisms (28.18) from $(X_t, \mathcal{R}_t)$ to $(X_s, \mathcal{R}_s)$, defined for $t \geq s$ in T and satisfying

$$(ss) = \text{identity and } (rs) \circ (st) = (rt), \text{ for } r \leq s \leq t \text{ in } T.$$

(ii) A <u>projective</u> <u>sublimit</u>[1] of $\mathcal{T} = ((X_s, \mathcal{R}_s); (st); s, t \in T)$ is any integration lattice $(Y, \mathcal{S})$ together with a family $\rho_s : (Y, \mathcal{S}) \to (X_s, \mathcal{R}_s)$ of morphisms such that

$$\rho_s = (st) \circ \rho_t \qquad\qquad , \text{ for } s \leq t \text{ in } T.$$

(iii) A <u>projective</u> <u>limit</u> for $\mathcal{T}$ is a projective sublimit $((X_\infty, \mathcal{R}_\infty); (\infty s); s \in T)$ having the following universal property: whenever $((Y, \mathcal{S}); \rho_s; s \in T)$ is a projective sublimit then there is a unique morphism $\rho_\infty : (Y, \mathcal{S}) \to (X_\infty, \mathcal{R}_\infty)$ such that

$$(st) \circ \rho_t = \rho_s \qquad\qquad , \text{ for all } s \leq t \text{ in } T.$$

We write $((X_\infty, \mathcal{R}_\infty); (\infty s); s \in T) = \lim_{\leftarrow} \mathcal{T}.\#$

---

[1] Often: coherent system: [5], Ch. IX.

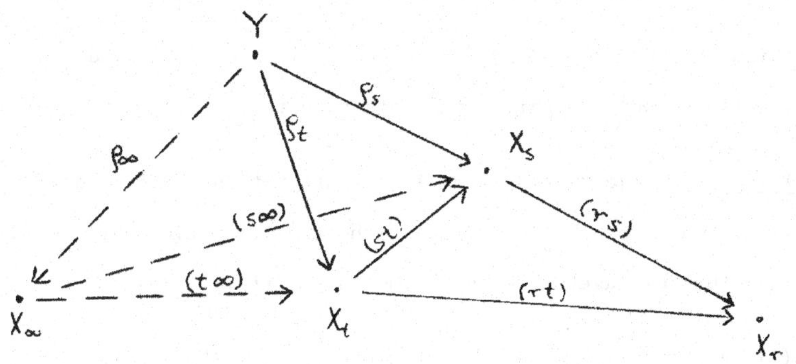

Figure: Projective limits. The dotted line maps can be found so as to make the diagram commutative.

A projective system $\mathcal{T}$ always has a projective limit $(X_\infty, \mathcal{R}_\infty)$, unique in the obvious sense. However, X_∞ may be void. To see this, let us call a thread on T an element $x = (x_t)_{t \in T}$ in the cartesian product $\Pi\{X_t : t \in T\}$ satisfying

$$(st)(x_t) = x_s \qquad , \text{ for } s \leq t \text{ in } T.$$

Let X_∞ denote the set of all threads on T, and define

(1) $$(t\infty): X_\infty \to X_t \text{ by } (t\infty)(x) = x_t \qquad , \text{ for } t \in T,\ x = (x_t)_{t \in T} \in X_\infty.$$

Finally, write

(2) $$\mathcal{R}_\infty = \cup\{\mathcal{R}_t \circ (t\infty) : t \in T\}.$$

Evidently, $\{(X_\infty, \mathcal{R}_\infty), (t\infty), t \in T\}$ is a projective sublimit of $\mathcal{T}$. If $\{(Y, \mathcal{B}), \rho_t, t \in T\}$ is any other projective sublimit, then

$$\rho_\infty(y) = (\rho_t(y))_{t \in T} \qquad , y \in Y,$$

defines a morphism $\rho_\infty : (Y,\mathcal{B}) \to (X_\infty, \mathcal{R}_\infty)$ such that $(t\infty)\circ\rho_\infty = \rho_t$; and, conversely, ρ_∞ is the only morphism with this property. To summarize:

30.1. Proposition. A projective system $\mathcal{T}$ of integration lattices has a projective limit, which can be identified with the integration lattice (2) on the set of all threads together with the morphisms (1).#

Definition. (i) A projective system $\mathcal{T}$ of integration lattices is epimorphic if (st) maps X_t onto X_s for all $s \leq t$ in T.

(ii) $\mathcal{T}$ has an epimorphic projective limit if $(t\infty)$ maps X_∞ onto X_t for all $t \in T$.

(iii) $\mathcal{T}$ is full if every thread defined on an increasingly directed subset S of T can be extended to a thread on all of T. The projective limit of a full system is called a full projective limit.#

If $\mathcal{T}$ has an epimorphic limit it is clearly epimorphic, and if it is full it has an epimorphic limit.

30.2. Example. If $\mathcal{T}$ is epimorphic and $T = \mathbb{N}$, then $\mathcal{T}$ is full. Indeed, if $S \subset T$ is cofinal in T then a thread on S defines a thread on T; and if S has a maximal element m, then a thread on S can be extended to all of T by induction in $n \geq m$. The same reasoning applies if T but has a countable cofinal subset.#

30.3. Example. If all the X_t are compact in the $\mathcal{R}_t$-topology and if $\mathcal{T}$ is epimorphic, then $\mathcal{T}$ has a full projective limit, and X_∞ is compact in the $\mathcal{R}_\infty$-topology.

Indeed, let $S \subset T$ be increasingly directed and let $x' : S \to \Pi\{X_s : s \in S\}$ be a thread. If $t \in T$ is given, choose a point $x_t \in X_t$ such that

$(st)x_t = x'_s$. Such exists since $\cap\{(st)^{-1}(x'_s) : t \geq s \in S\}$ is not empty, due to the compactness of X_t. Define a map $x^t : T \to \Pi\{X_t : t \in T\}$ by putting $x^t_s = x'_s$ for $s \in S$, $x^t_r = (rt)x_t$ for $r \leq t$ in T, and assigning arbitrary values elsewhere. The net (x^t) has a converging subnet (from Tychonoff's theorem) whose limit is clearly a thread on T and extends x'.#

B. Epimorphic systems of measures

Suppose now that we are given a projective system $\mathcal{T}$ as above, and for each $t \in T$ a Banach valued measure $m_t : \mathcal{R}_t \to E$ such that

(3) $$(st)(m_t) = m_s \quad , \text{ for } s \leq t \text{ in T (28.19).}$$

Definition. The system $\mathcal{I} = \{(X_t, \mathcal{R}_t), m_t, (st); T\}$ satisfying (1) is called a projective system of elementary integrals, and it is called epimorphic or full if $\mathcal{T}$ is.#

It is henceforth assumed that $\mathcal{T}$ has an epimorphic projective limit $((X_\infty, \mathcal{R}_\infty), (t\infty))$.

Definition. The projective limit, $\lim\limits_{\leftarrow} \mathcal{I}$, is the triple,

$$\{(X_\infty, \mathcal{R}_\infty, m_\infty), (t\infty), t \in T\},$$

where $\{(X_\infty, \mathcal{R}_\infty), (t\infty), t \in T\} = \lim\limits_{\leftarrow} \mathcal{T}$ and where $m_\infty : \mathcal{R}_\infty \to E$ is the measure given by

(4) $$m_\infty(\varphi) = m_t(\varphi_t) \quad , \text{ for } \varphi \in \mathcal{R}_\infty \text{ with } \varphi = \varphi_t \circ (t\infty),\ \varphi_t \in \mathcal{R}_t \text{ say.\#}$$

Equation (3) guarantees that $m_\infty(\varphi)$ is actually independent of the $t \in T$ used on the right of (4). Clearly, $m_\infty : \mathcal{R}_\infty \to E$ is linear, and each m_t is the image of m_∞ under $(t\infty)$. It will now be investigated under which circumstances m_∞ has finite variation, or is *-continuous.

If m_∞ has finite variation then so does m_t, because $(t_\infty)|m_\infty|$ majorizes m_t (28.4).

<u>30.4. Proposition</u>. The measure m_∞ has finite variation if and only if all the m_t do and

$$\text{(5)} \qquad \sup\{|m_t|(\varphi\circ(st)) : s \le t \text{ in } T\} < \infty,$$

for each $s \in T$ and $\varphi \in \mathcal{R}_{s+}$. In this instance,

$$\text{(6)} \qquad |m_\infty|(\varphi) = \sup\{|m_t|(\varphi_t) : t \in T,\ \varphi_t \in \mathcal{R}_{t+},\ \varphi = \varphi_t\circ(t\infty)\},$$

for $\varphi \in \mathcal{R}_{\infty+}$.

<u>Proof</u>. The necessity is obvious. If the expression (5) is finite then the right hand side of (6) is finite for all $\varphi \in \mathcal{R}_{\infty+}$ and depends additively on φ; hence it has a unique linear extension to all of $\mathcal{R}_\infty$ (3.2); and this is the variation of m_∞.#

<u>30.5. Remark</u>. Let $\mathcal{F},\mathcal{T}$ be as above, and let $\{(Y,\mathcal{S}),\rho_t, t\in T\}$ be a projective sublimit of $\mathcal{T}$. The system

$$\mathcal{F}' = \{\rho_t : (Y,\mathcal{S}) \to (X_t,\mathcal{R}_t,m_t); t \in T\}$$

is called a <u>promeasure</u> on $(Y,\mathcal{S})$. For any $\varphi \in \mathcal{S}$ of the form $\varphi = \psi\circ\rho_t$ $(t\in T, \psi\in\mathcal{R}_t)$ we may define

$$n(\varphi) = m_t(\psi).$$

Then

$$\bigcup \{ \mathscr{R}_t \circ \rho_t : t \in T \} = \mathscr{S}'$$

is an integration sublattice of $\mathscr{S}$; and n is a measure on $\mathscr{S}'$, of finite variation if (5) holds. The system $\{ \rho_t : (Y, \mathscr{S}', n) \to (X_t, \mathscr{R}_t, m_t); t \in T \}$ or, variously, $(Y, \mathscr{S}', n)$, is called a projective sublimit of $\mathscr{S}'$.

As an example of particular interest consider a locally convex Hausdorff space F and let T denote the family of all closed subspaces of finite codimension. T is ordered by $s \leq t$ if $t \subseteq s$. For $t \in T$ let $X_t = F/t$. Then X_t is a finite dimensional vector space. For $s \leq t$ there is the canonical linear map $(st): X_t \to X_s$ which sends $\xi + t$ to $\xi + s$ $(\xi \in F)$. Take for $\mathscr{R}_t$ the uniformly continuous and bounded functions on X_t. Then $\mathscr{T} = \{ (X_t, \mathscr{R}_t), (st), s \leq t$ in $T \}$ is an epimorphic (but not full) projective system, and $\mathscr{S} = \{ (F, \mathscr{S}), \xi \to \xi + t, t \in T \}$ is an epimorphic projective sublimit. Here $\mathscr{S}$ can be taken to be the uniformly closed integration lattice spanned by the bounded continuous functions that are finite infima of affine functions. A projective system of (positive, tight) measures on $\mathscr{T}$, or, variously, its projective sublimit n on $\mathscr{S}$, is called a promeasure on F. It is of great interest to determine when n is *-continuous or tight (Cf. [5]).#

C. Kolmogoroff's theorem

The following result is fundamental in the theory of probability (33B) and stochastic processes.

<u>30.6. Theorem</u>. (Kolmogoroff). Suppose that $\mathcal{S}$ is a full projective system of elementary integrals (as above) such that $\mathcal{R}_s$ separates the points of X_s for all $s \in T$ and m_∞ has finite variation.

If all the $(\mathcal{R}_t, m_t)$ are tight then $(\mathcal{R}_\infty, m_\infty)$ is S-continuous.

<u>Proof</u>. The requirement that $\mathcal{R}_s$ separate the points of X_s is needed to insure that a compact set in X_s is closed, which in turn implies the finite intersection property.

First, each m_t is extended to the dominated Borel functions $\mathcal{R}_t^B$. $((X_t, \mathcal{R}_t^B, m_t), (st), T)$ forms a projective system with full limit $(X_\infty, \mathcal{R}'_\infty, m_\infty)$, where $\mathcal{R}'_\infty = \cup\{\mathcal{R}_t^B \circ (t\infty) : t \in T\}$. The new measure, m_∞, also has finite variation. For the S-continuity, it suffices to show that m_∞ is σ-additive on the sets in $\mathcal{R}'_\infty$ (3.12, 6.3, 3.15). Let (A^n) be a decreasing sequence of sets in $\mathcal{R}'_\infty$ with void intersection. It has to be shown that $|m_\infty|(A^n)$ tends to zero. By way of contradiction, assume that there is an $a > 0$ such that $|m_\infty|(A^n) > 2a$ for all n.

To simplify the notation, write $K^\uparrow = (s\infty)^{-1}(K)$, for $K \subset X_s$. Define, by induction, an increasing sequence $(s(n))$ in T and a collection of compact sets $K_n \subset X_{s(n)}$ with the following properties:

(a) there is an $A_n \in \mathcal{R}^B_{s(n)}$ such that $A^n = A_n^\uparrow$;

(b) $|m_\infty|(A^n) \leq |m_n|(A_n) + a2^{-n}$;

(c) $K_n \subset A_n \cap (s(n-1)s(n))^{-1}(K_{n-1})$; and

(d) $|m_\infty|(A^n - K_n^\uparrow) \leq 2a(1 - 2^{-n})$.

For $n = 1$, this is done as follows. There is an $s'(1) \in T$ such that $A^1 = A_1^\uparrow$ for some $A_1 \in \mathcal{R}^B_{s'(1)}$; and every $t \in T$ exceeding $s'(1)$ has the same property. From (4), an $s(1) \geq s'(1)$ can be chosen so that (b) is also satisfied. As for (d), there is a compact subset $K_1 \subset A_1$ such that $|m_{s(1)}|(A_1 - K_1) \leq a/2$, since $|m_{s(1)}|$ is tight. When combined with (b), this gives $|m_\infty|(A^1 - K_1^\dagger) \leq a2^{-1} + a/2 = a = 2a(1 - 2^{-1})$.

Suppose now that $s(k)$ and K_k have been chosen according to the specifications for all $k < n$. By essentially repeating the procedure for $n = 1$, pick an $s(n) \geq s(n-1)$ such that (a) and (b) hold. Writing $K_n' = (s(n-1)s(n))^{-1}(K_{n-1})$, choose a compact set $K_n \subset A_n \cap K_n'$ such that

$$|m_n|(A_n \cap K_n' - K_n) < a2^{-n}.$$

From this,

$$|m_\infty|(A^n - K_n^\dagger) = |m_\infty|(A^n \setminus K_{n-1}^\dagger) + |m_\infty|(A^n \cap K_{n-1}^\dagger - K_n^\dagger)$$

$$\leq |m_\infty|(A^{n-1} - K_{n-1}^\dagger) + |m_n|(A_n \cap K_n' - K_n) + a2^{-n}$$

$$\leq 2a(1 - 2^{-n+1}) + a2^{-n} + a2^{-n} = 2a(1 - 2^{-n}).$$

The sets $K_n^\dagger \subset X_\infty$ are decreasing and non-void as they have positive measure. Define

$$K^n = \cap\{(s(n)s(m))K_m : n \leq m \in \mathbb{N}\}.$$

Each K^n is a compact non-void subset of $X_{s(n)}$. Moreover, $(s(n)s(m))$ maps K^m onto K^n, for $m \geq n$.

Now, define inductively a thread x' on $S = \{s(n): n = 1, 2, \ldots\} \subset T$ so that $x'_{s(n)} \in K^n$ for all n. Since the projective system at hand is full, x' has an extension to a thread x on all of T. The element x of X_∞ evidently lies in the intersection of all the A^n; this is the desired contradiction.#

30.7. Corollary. Suppose $\mathcal{F}$ is an epimorphic system of tight elementary integrals as above such that $\mathcal{R}_s$ separates the points of X_s for all $s \in T$ and m_∞ has finite variation.

If T has a countable cofinal subset then $(\mathcal{R}_\infty, m_\infty)$ is tight.

Proof. Notice that $\mathcal{F}$ is full (30.2) and adopt the notation of the proof above. There is a countable cofinal subset S of T which is order-isomorphic with the natural numbers.

Let a set A in $\mathcal{R}'_\infty$ and an $\varepsilon > 0$ be given. There are an $s(1) \in S$ and a compact set $K_1 \subset X_{s(1)}$ such that $K_1^\uparrow \subset A$ and $|m_\infty|(A - K_1^\uparrow) < \varepsilon 2^{-1}$. Continuing inductively, we establish the existence of a cofinal sequence $(s(n))$ in S and a sequence (K_n) of compact sets $K_n \subset X_{s(n)}$ such that

$$K_n^\uparrow \subset K_{n-1}^\uparrow \quad \text{and} \quad |m_\infty|(K_{n-1}^\uparrow - K_n) < \varepsilon 2^{-n}.$$

Clearly $K = \cap K_n^\uparrow$ is compact in X_∞, in the $\mathcal{R}_\infty$-topology, and is $(\mathcal{R}_\infty, |m_\infty|^S)$-integrable (30.6). Moreover, $|m_\infty|^S(A - K) \leq \varepsilon$. Since the sets in $\mathcal{R}'_\infty$ form an adequate cover, the compact sets are $(\mathcal{R}_\infty, |m_\infty|^S)$-dense. From 24.1, $|m_\infty|$ is tight.#

If T does not have a countable cofinal subset then the following theorem gives a condition when m_∞ is tight. Its simple proof is left to the reader.

30.8. Theorem. (Prokhorov). Suppose $\mathcal{F}$ is a full system of tight elementary integrals as above such that $\mathcal{R}_s$ separates the points of X_s for all $s \in T$ and m_∞ has finite variation. Then $(\mathcal{R}_\infty, |m_\infty|)$ is tight, provided the following condition holds:

(P) For every set A in $\mathcal{R}'_\infty$ and $\varepsilon > 0$ there is a compact set $K \subset A$ such that

$$\sup\{ |m_t|^B((t\infty)(A - K)) : (t\infty)(A) \in \mathcal{R}_t^B \} \leq \varepsilon.$$

D. Supplements

30.9. The condition that $\mathcal{F}$ be full is necessary in Kolmogoroff's theorem. (Remove one point from a countable product of unit intervals to construct an example.) It is sufficient to require $\mathcal{R}_s$ separate all those points of X_s that $\mathcal{R}_\infty \circ (s\infty)$ separates.

30.10. Kolmogoroff's theorem also holds if E is a Banach space over the proper L-space R and if m_∞ has finite R-variation.

30.11. If T is linearly ordered and has no countable cofinal subset, then if the m_t are all S-continuous so is m_∞.

30.12. Let S be an increasingly directed and cofinal subset of T. Then, in an obvious sense,

$$\varprojlim\{(X_t,\mathcal{R}_t,m_t),(st),s \leq t \text{ in } T\} = \varprojlim\{(X_s,\mathcal{R}_s,m_s),(rs),r \leq s \text{ in } S\}.$$

LITERATURE: [5], Ch. IX, [13].

31. Martingale theorems

These afford a means to study the projective limit $(\mathcal{R}_\infty, m_\infty)$ from the "local" angle of a given upper integral $(\mathcal{R}_\infty, N)$.

A. Martingales

The following situation will be considered in the sequel: X is a set and T an increasingly directed set of indices. For each $t \in T$, $\mathcal{R}_t$ is a dominated integration lattice on X such that

$$\mathcal{R}_s \subset \mathcal{R}_t \qquad , \text{ for } s \leq t \text{ in } T,$$

and $m_t : \mathcal{R}_t \to E$ is an S-continuous measure of finite variation such that

$$m_t | \mathcal{R}_s = m_s \qquad , \text{ for } s \leq t \text{ in } T.$$

This is a special full projective system where all the (st) are the identity map. It is of a general nature in that if a projective system $\mathcal{I} = ((X_t, \mathcal{R}^t, m^t);(st);T)$ with epimorphic limit $((X_\infty, \mathcal{R}_\infty, m_\infty);(t\infty))$ is given, then

$$((X_\infty, \mathcal{R}_t = \mathcal{R}^t \circ (t\infty); m_t = m_\infty | \mathcal{R}_t), T)$$

is a system as considered here and has most of the relevant properties of $\mathcal{I}$. (Note, though, that the $\mathcal{R}_t$ need not separate the points even if the $\mathcal{R}^t$ do (Cf. 30.6).)

We shall suppress the index ∞ and write $(\mathcal{R}, m)$ for the projective limit of the system; here $\mathcal{R}$ is the union of the $\mathcal{R}_t$ and m is the unique measure that coincides with m_t on $\mathcal{R}_t$, for all $t \in T$. Also, we note that the integration lattice $\mathcal{R}$ is itself dominated. It will henceforth be assumed that m has <u>finite variation</u>. According to 30.4, this is equivalent to

$$|m|(\varphi) = \sup\{|m_t|(\varphi): t \in T, \varphi \in \mathcal{R}_t\} < \infty$$

for all $\varphi \in \mathcal{R}_+$.

In the following, we also have at our disposal a fixed <u>localizable</u> upper integral $(\mathcal{R},N)$. The identity map from $(X,\mathcal{R})$ to $(X,\mathcal{R}_s)$ is automatically $(\mathcal{R},N)$-adequate; and it will, in addition, be assumed to be <u>tame</u>. This is to ensure that the $(\mathcal{R}_s,N)$-integrable sets form an $(\mathcal{R},N)$-adequate cover, and thus that an $(\mathcal{R}_s,N)$-measurable function is $(\mathcal{R},N)$-measurable (28.5); it will be automatically true if the $\mathcal{R}_s$ contain the constants, or if $\mathcal{R}_s^S$ does. The unique positive measure on $\mathcal{R}_s$ which coincides with N on $\mathcal{R}_{s+}$ is henceforth denoted by n_s, and, similarly, $n = N|\mathcal{R}$. Note that $((\mathcal{R}_s,n_s), s \in T)$ is a projective system with limit $(\mathcal{R},n)$.

Lastly, we assume not only that $m_t \ll n_t$ for all $t \in T$ but, more restrictively, that m_t has a locally $(\mathcal{R}_t,N)$-integrable derivative g_t with respect to n_t:

(1) $$m_t = g_t n_t\ , \quad \text{and} \quad g_t \in \mathcal{L}_E^{\ell}(\mathcal{R}_t,N).$$

The problem studied below is to decide when m has a derivative $g = dm/dn$, and, if so, in which sense (g_t) converges to g. (Of course, m cannot be expected to have a derivative in general, as it would then be S-continuous.)

If $s \leq t$ in T, then g_s is a conditional expectation of g_t under $\mathcal{R}_s$, and from 29.3,

$$E^{\mathcal{R}_s}\|g_t\| \geq \|g_s\|.$$

Definition. Let $\mathcal{R}$, N, T, $\mathcal{R}_t$ $(t \in T)$ be as above. The triple $(\mathcal{R}_s, g_s, s \in T)$ is an (increasing) martingale if the g_s are locally $(\mathcal{R}_s, N)$-integrable functions such that g_s is a conditional expectation of g_t under $\mathcal{R}_s$, for $s \le t$ in T. If the g_s take values either in $\mathbb{R}$ or in a Banach lattice, then the triple is a submartingale if

$$g_s \le E^{\mathcal{R}_s} g_t \qquad \text{locally almost everywhere for } s \le t \text{ in } T.$$

A martingale or submartingale is locally p-integrable, real-valued, positive, etc. if all the g_s are.#

If $(\mathcal{R}_s, g_s, s \in T)$ is a martingale then the system $m_s = g_s n_s$ is called the associated projective system of measures. There is an evident 1-1 correspondence of martingales with projective systems of measures that have locally integrable derivatives. From these considerations and from 30.4, we get the following proposition. The details are left to the reader.

31.1. Proposition. (i) Let $(\mathcal{R}_t, g_t, t \in T)$ be a submartingale such that

$$m(\varphi) = \lim\left(\int g_t \varphi dn: t \in T\right) \quad \left(= \sup\left\{\int g_t \varphi dn_t : \varphi \in \mathcal{R}_t\right\}\right)$$

is finite for all $\varphi \in \mathcal{R}_+$. Then m is a measure on $\mathcal{R}$ and is denoted by $m = \underset{\leftarrow}{\lim}(g_t n_t)$; both $(\mathcal{R}_t, |g_t|, T)$ and $(\mathcal{R}_t, g_{t+}, T)$ are positive submartingales; and m has finite variation if and only if $m_+ = \underset{\leftarrow}{\lim}(g_{t+} n_t)$ or $|m| = \underset{\leftarrow}{\lim}(|g_t| n_t)$ (and then both) exist.

(ii) Let $(\mathcal{R}_t, g_t, t \in T)$ be a Banach-valued martingale. Then $(\mathcal{R}_t, \|g_t\|, t \in T)$ is a submartingale and $m = \underset{\leftarrow}{\lim}(g_t n_t)$ has variation

$$|m| = \underset{\leftarrow}{\lim}(\|g_t\| n_t)$$

if this limit exists.#

B. Convergence in mean

The first result on martingales establishes a connection between the convergence in mean of the g_t and the existence of a derivative dm/dn.

31.2. Theorem. Let $(\mathcal{R}_t, g_t, t \in T)$ be an E-valued martingale such that $m = \underset{\leftarrow}{\lim}(g_t n_t)$ has finite variation, and suppose that each g_t is locally $(\mathcal{R}_t, N_p)$-integrable $(1 \leq p < \infty)$.

Then (g_t) converges locally in p-mean to a locally $(\mathcal{R}, N_p)$-integrable function g, i.e.,

$$\lim\{N_p(\varphi(g - g_t)) : t \in T\} = 0 \qquad \text{for all } \varphi \in \mathcal{R},$$

if and only if m has a locally $(\mathcal{R}, N_p)$-integrable derivative dm/dn; and in this instance

$$g = dm/dn.$$

Proof. Assume that (g_t) converges to g locally in p-mean, and let $\varphi \in \mathcal{R}$, $\varphi \in \mathcal{R}_r$ say. Since $m(\varphi) = \int \varphi g_t dn$ for $r \leq t \in T$,

$$|m(\varphi) - (gn)(\varphi)| = |\int (g_t - g)\varphi dn| \leq N_p((g_t - g)\varphi) N_{p'}([\varphi \neq 0]),$$

where p' is the conjugate of p (21A). The right hand side can be made arbitrarily small by choosing t sufficiently large, and so $m(\varphi) = (gn)(\varphi)$.

Conversely, assume that $m = gn$, where $g \in \mathcal{L}_E^{\ell}(\mathcal{R}, N_p)$. Let a $K \in \mathcal{C}_0(\mathcal{R}, N)$ and an $\varepsilon > 0$ be given. By enlarging K if necessary, we may assume that $K \in \mathcal{C}_0(\mathcal{R}_r, N)$. There is a $\varphi \in \mathcal{R} \otimes E$, $\varphi \in \mathcal{R}_s \otimes E$ say, such that $N_p(gK - \varphi K) < \varepsilon$. For $t \geq r, s$ in T,

$$N_p(gK - g_tK) \leq N_p(gK - \varphi K) + N_p(\varphi K - g_tK) \leq 2\varepsilon,$$

since $\varphi K - g_tK$ is equal to $E^{\mathcal{R}_t}(\varphi K - gK)$, which has norm smaller than $E^{\mathcal{R}_t}\|\varphi K - gK\|$, and hence p-norm smaller than ε.#

C. Maximal martingale theorems

In this subsection we consider a fixed submartingale $(\mathcal{R}_t, g_t, t \in T)$ with countable and linearly ordered index set T. For fixed $r, t \in T$ we write

$$g_r^t = \sup\{g_s : r \leq s \leq t,\ s \in T\},$$

$$g_r^* = \sup\{g_s : r \leq s \in T\}, \quad \text{and}$$

$$g^* = \sup\{g_s : s \in T\}.$$

These functions are called maximal functions of the submartingale, and the following statement concerning their distribution is the key to all subsequent results.

31.3. Theorem. (Maximal distributional martingale theorem). Let $(\mathcal{R}_t, g_t, T)$ be a submartingale with countable and linearly ordered index set T.

(i) For $r < t$ in T and $\varphi \in \mathcal{R}_r$,

$$\int \varphi[g_r^t \geq \varphi]\,dn \leq \int g_t[g_r^t \geq \varphi \neq 0]\,dn.$$

(ii) If $g_t \geq 0$ for all $t \in T$ then

$$N(K[g_r^* \geq \xi]) \leq \xi^{-1} \sup_{r \leq t \in T} N(Kg_t[g_r^* \geq \xi])$$
$$\leq \xi^{-1} \sup_{r \leq t \in T} N(Kg_t)$$

for any $\xi > 0$, $r \in T$, and $K \in \mathcal{C}(\mathcal{R}_r, N)$.

(iii) If, additionally, the g_t are all integrable, then

$$\overline{N}([g^* \geq \xi]) \leq \xi^{-1} \sup_{t\in T} \int g_t[g^* \geq \xi]dn$$
$$\leq \xi^{-1} \sup_{t\in T} N(g_t).$$

Proof. (i): For all $t \in T$ define

$$C_t = [g_t \geq \varphi] \quad \cup \{[g_s \geq \varphi] : s < t\}.$$

Since g_s is $(\mathcal{R}_t, N)$-measurable for $t \geq s$ and since T is countable, C_t is $(\mathcal{R}_t, N)$-measurable. The C_s, $r \leq s \leq t$, are mutually disjoint, countable in number, and have union $[g_r^t \geq \varphi]$; and so

$$\int \varphi[g_r^t \geq \varphi]dn = \sum_{r\leq s\leq t} \int \varphi C_s dn \leq \sum_{r\leq s\leq t} \int g_s[\varphi \neq 0]C_s dn$$
$$\leq \sum_{r\leq s\leq t} \int g_t[\varphi \neq 0]C_s dn = \int g_t[g_r^t \geq \varphi \neq 0]dn.$$

(ii): Replacing $\mathcal{R}_r$ by $\mathcal{L}_0^1(\mathcal{R}_r, N)$ and φ by ξK,

$$\int K[g_r^t \geq \xi]dn \leq \xi^{-1} \int Kg_t[g_r^t \geq \xi]dn \leq \xi^{-1} \int Kg_t[g_r^* \geq \xi]dn.$$

The statement follows upon taking the supremum over $t \in T$. Finally, (iii) is obtained from (ii) by taking the supremum over $K \in \mathscr{C}(\mathcal{R}_r, N)$ and $r \in T$.#

31.4. Corollary. Suppose $(\mathcal{R}_t, g_t, T)$ is a submartingale and fix a $\xi > 0$, an $r \in T$ and a $K \in \mathscr{C}(\mathcal{R}_r, N)$. Then for any $k \in (0,1)$,

$$\int K[g_r^* \geq \xi]dn \leq \xi^{-1}(1-k)^{-1} \sup_{t\in T} \int Kg_t[g_t \geq k\xi]dn.$$

<u>Proof</u>. Let $r < t \in T$ and put

$$\tilde{g}_t = \begin{cases} g_t & \text{on } [g_t \geq k\xi] \\ 0 & \text{on } [g_t < k\xi] \end{cases}$$

and $\tilde{g}_s = E^{\mathcal{R}_s}\tilde{g}_t$ for $r \leq s \leq t$. Then $(\mathcal{R}_s, \tilde{g}_s, r \leq s \leq t)$ is a positive martingale and

$$g_s \leq \tilde{g}_s + k\xi .$$

Therefore

$$[\sup_{r\leq s\leq t} g_s \geq \xi] \subset [\sup_{r\leq s\leq t} \tilde{g}_s \geq \xi(1-k)] ,$$

and by the theorem applied to $(\mathcal{R}_s, \tilde{g}_s, t \geq s \in T)$,

$$\int [\sup_{r\leq s\leq t} g_s \geq \xi] K dn \leq \{\xi(1-k)\}^{-1} \int \tilde{g}_t K dn = \{\xi(1-k)\}^{-1} \int g_t K [g_t \geq k\xi] dn.$$

The claim follows upon taking the supremum over all $t \in T$.#

<u>31.5. Theorem</u>. (Maximal theorem for L^p, $p > 1$). Let $1 < p \leq \infty$, and suppose $(\mathcal{R}_t, g_t, T)$ is a positive submartingale with countable and linearly ordered index set T. Then for any $r \in T$ and $K \in \mathscr{C}(\mathcal{R}_r, N)$,

$$N_p(g_r^* K) \leq \frac{p}{p-1} \sup_{r<t\in T} N_p(g_t K).$$

In particular,

$$\bar{N}_p(g^*) \leq \frac{p}{p-1} \sup\{\bar{N}_p(g_t) : t \in T\}.$$

<u>Proof</u>. If $N_p(g_t K) = \infty$ for some $t \in T$ there is nothing to prove. If this number is finite we calculate as follows. The function $\xi \to \xi^p$ on $\mathbb{R}_+$ has positive and continuous derivative $p\xi^{p-1}$ and therefore (27.7), with $k \in (0,1)$,

$$N(K(g_r^t)^p) = \int_0^\infty p\xi^{p-1} N(Kg_r^t \geq \xi) d\xi \leq p \int_0^\infty \xi^{p-2}(1-k)^{-1}\{\int Kg_t [g_t \geq k\xi] dn\} d\xi .$$

Since the function $(x,\xi) \to \xi^{p-2} K(x) g_t(x) [g_t(x) \geq k\xi]$ is measurable on $X \times \mathbb{R}_+$ (27.6), we may interchange the order of integration and arrive at

$$N(K(g_r^t)^p) \leq p(1-k)^{-1} \int Kg_t \{\int \xi^{p-2} [Kg_t \geq k\xi] d\xi\} dn = \frac{pk^{1-p}}{(p-1)(1-k)} \int (Kg_t)^p dn < \infty$$

Hence $g_r^t \in \mathcal{L}^p$ if $Kg_t \in \mathcal{L}^p$. To obtain the inequality we calculate similarly, using 31.3 instead of 31.4:

$$\begin{aligned} N(K(g_r^t)^p) &\leq p \int_0^\infty \xi^{p-2} \{\int Kg_t [g_r^t \geq \xi] dn\} d\xi \\ &= p \int Kg_t \{\int \xi^{p-2} [g_r^t \geq \xi] d\xi\} dn \\ &= \frac{p}{p-1} \int Kg_t (g_r^t)^{p-1} dn \\ &\leq \frac{p}{p-1} N_p(Kg_t) N_{p'}((g_r^t)^{p-1}) = \frac{p}{p-1} N_p(Kg_t) N(K(g_r^t)^p)^{1/p'}, \end{aligned}$$

since $(p-1)p' = p$. As $1 - 1/p' = 1/p$, we get

$$N_p(K(g_r^t)) \leq \frac{p}{p-1} N_p(Kg_t),$$

and taking the supremum over $t \in T$, the inequality of the statement follows.#

31.6. Theorem. (Maximal martingale theorem for L^1). Suppose $(\mathcal{R}_t, g_t, T)$ is a positive submartingale with countable and linearly ordered index set T. Then

$$N(g_r^* K) \leq k^{-1} N(K) + (1-k)^{-1} \sup_{r<t\in T} N(Kg_t (\log g_t)_+)$$

for any $K \in \mathcal{C}(\mathcal{R}_r, N)$ and $0 < k < 1$.

Proof. Fix $K \in \mathcal{C}(\mathcal{R}_r, N)$, $k \in (0,1)$, and $r < t \in T$. Then, from 31.4,

$$\begin{aligned}\int Kg_r^t dn &= \int_0^\infty N([Kg_r^t \geq \xi])d\xi \\ &\leq \int_0^{1/k} N(K)d\xi + (1-k)^{-1} \int_{1/k}^\infty \xi^{-1} \Big\{ \int_{1/k}^\infty Kg_t [Kg_t \geq k\xi] dn \Big\} d\xi \\ &= k^{-1} N(K) + (1-k)^{-1} \int Kg_t \int_0^\infty \xi^{-1} (1/k,\ 1/k \vee Kg_t/k)(\xi) d\xi dn \\ &= k^{-1} N(K) + (1-k)^{-1} \int Kg_t (\log (Kg_t)_+) dn. \#\end{aligned}$$

D. Pointwise convergence

We consider again a martingale $(\mathcal{R}_t, g_t, T)$ with values in a Banach space E and with countable and linearly ordered index set T, and assume that $m = \lim\limits_{\leftarrow} g_t n_t$ has finite variation. Recall that $(\mathcal{R}_t, \|g_t\|, T)$ is a positive submartingale and that $\lim\limits_{\leftarrow} \|g_t\| n_t$ is the variation of $\lim\limits_{\leftarrow} g_t n_t$. In preparation for Doob's martingale theorem we establish two lemmas.

31.7. Lemma. Let $(\mathcal{R}_s, g_s, s \in T)$ be a positive submartingale, with linearly ordered and countable index set T, such that $m = \lim\limits_{\leftarrow} g_t n_t$ exists and is disjoint from n.

Then (g_t) converges to zero locally a.e.

Proof. Without changing premises or conclusion we may replace $\mathcal{R}_s$ by its full closure for all $s \in T$, and $m_s = g_s n_s$ and n_s by their natural extensions. We may thus assume that $\mathcal{R}_s$ is full for $s \in T$ $(s \neq \infty!)$. The point in doing so is that we have now sufficiently many sets in $\mathcal{R}$. Let K be one of them: $K \in \mathcal{R}_r$ with $r \in T$, say. Let $0 < \delta$ and $0 < \varepsilon < 1$ be given. There are disjoint sets K', K'' in $\mathcal{R}$ with union K such that $n(K') + m(K'') < \varepsilon\delta/2$ (3.8), and there is an $s \geq r$ in T such that K', K'' belong to $\mathcal{R}_s$. Using 31.3,

$$N([\sup_{s<t} g_t > \varepsilon K]K) = N([\sup_{s<t} g_t > \varepsilon K]K') + N([\sup_{s<t} g_t > \varepsilon K]K'')$$

$$\leq \varepsilon\delta/2 + 1/\varepsilon \cdot N([\sup_{s<t} g_t > \varepsilon K'']\varepsilon K'') \leq \delta/2 + 1/\varepsilon \cdot \sup_{s<t} N(g_t K'')$$

$$= \delta/2 + 1/\varepsilon\, m(K'') \leq \delta/2 + 1/\varepsilon \cdot \varepsilon \cdot \delta/2 = \delta.$$

This shows that the set $\bigcap_{s\in T} K \cap [\sup_{s<t} g_t > k^{-1}] = A_k$ is negligible for each k. Hence so is the set $\bigcup\{A_k : k = 1, 2, \ldots\}$ of points in K at which g_t does not converge to zero.#

<u>31.8. Lemma</u>. Let $(\mathcal{R}_t, g_t, t \in T)$ be an E-valued martingale with countable linearly ordered index set T. If there is a locally $(\mathcal{R},N)$-integrable function g such that $g_t = E^{\mathcal{R}_t} g$ locally a.e. for all $t \in T$, then (g_t) converges to g pointwise locally a.e. (and locally in N-mean (31.2)).

<u>Proof</u>. Again, we assume that $\mathcal{R}_t = \mathcal{L}_0^1(\mathcal{R}_t, N)$ for all $t \in T$. Let an $\varepsilon > 0$ and a set $K \in \mathcal{R}$ be given, $K \in \mathcal{R}_r$ say. There is a $\varphi \in \mathcal{R} \otimes E$ such that $N(Kg - K\varphi) < \varepsilon^2$, say $\varphi \in \mathcal{R}_s \otimes E$ with $r \leq s \in T$. If $s \leq t, u$ in T, then $\varphi = E^{\mathcal{R}_t}\varphi = E^{\mathcal{R}_u}\varphi$, so that

$$\|(g_t - g_u)K\| \leq \|(E^{\mathcal{R}_t} g - E^{\mathcal{R}_t}\varphi)K\| + \|(E^{\mathcal{R}_u}\varphi - E^{\mathcal{R}_u} g)K\| \leq KE^{\mathcal{R}_t}\|g - \varphi\| + KE^{\mathcal{R}_u}\|g - \varphi\|,$$

and therefore

$$\bigcup_{s\leq t,u} K\cap[\|g_t - g_u\| > 2\varepsilon] \subset \bigcup_{s<t} [KE^{\mathcal{R}_t}\|g-\varphi\| > K\varepsilon] = (1/\varepsilon)\varepsilon K[\sup_{s<t} E^{\mathcal{R}_t}\|g-\varphi\| \geq \varepsilon K] \text{ a.e.}$$

Now, $(\mathcal{R}_t, E^{\mathcal{R}_t}\|g-\varphi\|, t\in T)$ is a positive submartingale, and, from 31.3, the set on the right hand side has measure smaller than

$$(1/\varepsilon) \sup_{s\leq t} N(KE^{\mathcal{R}_t}\|g-\varphi\|) \leq (1/\varepsilon)N(K\|g-\varphi\|) \leq \varepsilon.$$

This proves that $(g_t(x))$ is a Cauchy net for locally almost all $x \in X$; hence (g_t) converges locally a.e. to a limit f. To see that $f = g$ locally a.e., recall that (g_t) converges to g locally in mean (31.2).#

<u>31.9. Theorem</u>. (Doob's martingale theorem). Let $(\mathcal{R}_t, g_t, t\in T)$ be an E-valued martingale with countable linearly ordered index set T, and suppose that $m = \underleftarrow{\lim}(g_t n_t)$ has finite variation. Let m' denote the part of m absolutely continuous with respect to n (17.3).

Then m' has a locally $(\mathcal{R},N)$-integrable derivative dm'/dn if and only if (g_t) converges locally a.e.; and in this instance,

$$dm'/dn = \lim g_t \qquad \text{locally a.e.}$$

<u>Proof</u>. Assume that $m' = gn$ where $g \in \mathcal{L}_E^{\ell}(\mathcal{R},N)$, and set

$$f_t = E^{\mathcal{R}_t} g \text{ and } h_t = g_t - f_t \qquad \text{, for all } t\in T.$$

$(\mathcal{R}_t, h_t, t\in T)$ is a martingale and $p = \underleftarrow{\lim}(h_t n_t : t\in T)$ has finite variation not exceeding $|m| + \|g\|n$. For all $\varphi \in \mathcal{R}$,

$$p(\varphi) = \lim \int \varphi(g_t - f_t)dn = m(\varphi) - m'(\varphi).$$

Therefore, p is disjoint from n and so $(\|h_t\|)$ converges to zero locally a.e. (31.7). From 31.8, (f_t) converges to g locally a.e., and hence (g_t) converges to g locally a.e.

Conversely, assume that (g_t) converges locally a.e. to a function g. Then g is measurable and, in fact, locally $(\mathcal{R},N)$-integrable. Indeed, for $\varphi \in \mathcal{R}$,

$$\lim\inf N(\|g_t\|\varphi) = |m|(|\varphi|) < \infty,$$

and so, by Fatou's lemma 8.11,

$$\lim\inf(\|g_t\||\varphi|) = \|g\||\varphi| \in \mathcal{L}^1(\mathcal{R},N).$$

To see that $g = dm'/dn$, consider the martingale $f_t = g_t - E^{\mathcal{R}_t}g$. Once it has been shown that the limit

$$q = \underset{\leftarrow}{\lim}(f_t n_t : t \in T) = m - gn$$

is disjoint from n, the proof will be finished, since then evidently $m' = gn$.

For every η in E', $(\mathcal{R}_t, \langle f_t;\eta\rangle)$ is a real-valued martingale, and the part $\langle q;\eta\rangle'$ in (n) of the limit

$$\underset{\leftarrow}{\lim}(\langle f_t;\eta\rangle n_t : t \in T) = \langle q;\eta\rangle$$

has a locally integrable derivative $d\langle q;\eta\rangle'/dn$, since $(\mathcal{R},N)$ is localizable. From the above, (f_t) converges to zero locally a.e. and $(\langle f_t;\eta\rangle)$ converges to $d\langle q;\eta\rangle'/dn$ locally a.e. Hence $\langle q;\eta\rangle' = 0$ for all $\eta \in E'$. From the Hahn-Banach theorem, the part q' of q in $(n)_E$ is zero.#

31.10. Corollary. Suppose $(\mathcal{R}_t, g_t, T)$ is a real-valued martingale with countable and linearly ordered index set T, and $m = \varprojlim g_t n_t$.

(i) If $1 < p \leq \infty$ and $\sup\{N_p(g_t) : t \in T\} < \infty$ then $m \ll n$ and $g_t \to dm/dn$ a.e. and in p-mean.

(ii) If $\sup\{N(g_t(\log g_t)_+) : t \in T\} < \infty$ then $m \ll n$ and $g_t \to dm/dn$ a.e. and locally in mean.

Proof. (i): We know that $g_t \to dm'/dn$ a.e. From 31.5, $\sup\{|g_t| : t \in T\}$ belongs to $\mathcal{L}^p$. From Lebesgue's theorem, $g_t \to dm'/dn$ in p-mean, and therefore $m = m' \in (n)$ (31.2). (ii) follows similarly from 31.6.#

E. Descending martingales

We study now the behavior of the martingale $(\mathcal{R}_t, g_t, T)$ to the left, i.e., as t decreases. Again, T is assumed to be linearly ordered and countable. For simplicity's sake we assume additionally that the $\mathcal{R}_t$ contain the constants and that the g_t are real-valued $(t \in T)$. We put

$$\mathcal{R}_0 = \cap \{\mathcal{L}^1(\mathcal{R}_s, N) : s \in T\}$$

and let m_0, n_0 denote the restrictions of $m_t = g_t n_t$, n_t, respectively to $\mathcal{R}_0$. Clearly

$$g_0 = E^{\mathcal{R}_0} g_t = dm_0/dn_0$$

is an $(\mathcal{R}_0, N)$-integrable derivative of m_0 with respect to n_0.

31.11. Theorem. As t decreases in T, (g_t) converges to g_0 a.e. and in mean. If g_t is p-integrable for some $t \in T$ and some $1 \leq p < \infty$, then (g_t) converges to g_0 in p-mean.

<u>Proof</u>. We may assume that $g_0 = 0$ and $m_0 = 0$; the general version of the statement is obtained by applying this special one to the martingale $(g_t - g_0)$. Fix $t \in T$, $\xi > 0$ real, and put

$$G_s = [\sup_{r<s} g_r \geq \xi] \;, \quad G = [\limsup_{s \to 0} g_s \geq \xi] .$$

Then G_s decreases to G as $s \to 0$. From 31.3,

$$\int G dn \leq \int G_s dn \leq \xi^{-1} \int g_s G_s = \xi^{-1} \int g_s (G_s - G) dn + \xi^{-1} \int g_s G dn$$

$$= \xi^{-1} \int g_t (G_s - G) dn + \xi^{-1} \int g_s G dn \; .$$

The last equation is true since (g_s) is a martingale and $G_s - G$ $(\mathcal{R}_s, N)$-integrable. As $G \in \mathcal{R}_0$ and $g_s n = 0$ on $\mathcal{R}_0$, the last integral vanishes. The integral before tends to zero as $s \to 0$ from the monotone convergence theorem. Hence G is negligible and so

$$\limsup g_s = 0 \qquad \text{a.e.}$$

Applying the same reasoning to the martingale $(-g_s)$ we obtain $\liminf(g_s) = 0$ a.e., and therefore $\lim(g_s) = 0$ a.e. as claimed.

To see that (g_s) converges in mean, we consider the submartingale $(|g_s|)$ and compute as follows. For $\xi > 0$ and $s < t$ in T,

$$\int |g_s| [|g_s| \geq \xi] dn \leq \int |g_t| [|g_s| \geq \xi] dn$$

$$= (|g_t| N)([|g_s| \geq \xi]) \leq (|g_t| N)([\sup_{s<t} |g_s| \geq \xi]) \; .$$

From 31.3, $N([\sup_{s<t} |g_s| \geq \xi])$ is smaller then $\xi^{-1} N(|g_t|)$ and so converges to zero as $\xi \to \infty$. Since $|g_t| N$ is absolutely continuous with respect to N (7.20), the first integral tends to zero as $\xi \to \infty$, uniformly for

all $s < t$. Hence $(|g_s|)_{s<t}$ is uniformly integrable, and by 18.20 converges in mean to its pointwise limit $g_0 = 0$.

Finally, if $g_t \in \mathcal{L}^p(\mathcal{R}_t, N)$ for some $1 < p < \infty$, then $\sup\{|g_s|: s < t\}$ is p-integrable (3.15). From Lebesgue's theorem, $g_s \to g_0$ in p-mean.#

F. Supplements

Henceforth $(\mathcal{R}, N)$ is a finite upper integral on X and $\{\mathcal{R}_t : t \in T\}$ is an increasingly directed family of integration lattices with union $\mathcal{R}$ and containing the constants.

31.12. If $(\mathcal{R}_t, f_t)$ and $(\mathcal{R}_t, g_t)$ are two martingales (submartingales) then so is $(\mathcal{R}_t, af_t + bg_t)$, for $a, b \geq 0$. $(\mathcal{R}_t, f_t \vee g_t)$ is a submartingale.

31.13. $(\mathcal{R}_t, g_t)$ is a <u>supermartingale</u> if

$$E^{\mathcal{R}_s} g_t \leq g_s \qquad \text{a.e. for } s < t \text{ in } T.$$

This amounts to saying that $(\mathcal{R}_t, -g_t)$ is a submartingale. The results 31.1, 31.3 - 31.7, 31.12 can easily be formulated as results on supermartingales.

31.14. A submartingale or supermartingale $(\mathcal{R}_t, g_t)$ is a martingale if and only if $\int g_t dn$ is independent of t.

31.15. Let (g_t) be a supermartingale with $T = \mathbb{N}$. If (and only if) there exists a submartingale (f_t) with $g_t \geq f_t$ a.e. for all $t \in T$, there is a martingale (g_t^o) such that $(g_t - g_t^o)$ is a positive supermartingale with limit zero a.e. and in mean. In this instance, $g_t^o \geq f_t$ a.e. for all $t \in \mathbb{N}$.

LITERATURE: [33], [31], [13]. Proofs are from [6], [7], [8].

32. Infinite products of positive measures

A. Generalities

Let I be any set, and for each $i \in I$ let $(X_i, \mathcal{R}_i, m_i)$ be a positive elementary integral such that $\mathcal{R}_i$ contains the constants and $m_i(X_i) = 1$. We shall define their product and show that it is a $*$-continuous elementary integral if all the $(\mathcal{R}_i, m_i)$ are.

Let T denote the family of all finite subsets of I, ordered by inclusion, and let $s \leq t$ in T. Then $(X_t, \mathcal{R}_t, m_t)$ denotes the product of the $(X_i, \mathcal{R}_i, m_i)$ with $i \in t$, and (st) the natural projection of X_t onto X_s. (st) is clearly a morphism from $(X_t, \mathcal{R}_t)$ to $(X_s, \mathcal{R}_s)$, and satisfies

$$(st)(m_t) = m_s .$$

Indeed, let $\varphi \in \mathcal{R}_s$ and set $r = t - s \in T$. Then $\varphi \circ (st) = \varphi \otimes X_r$ in the notation of section 27A, and

$$((st)(m_t))(\varphi) = m_t(\varphi \circ (st)) = m_t(\varphi \otimes X_r) = m_s(\varphi) m_r(X_r) = m_s(\varphi).$$

It is here that the assumption $m_i(X_i) = 1$ is needed. To summarize: $((X_s, \mathcal{R}_s), m_s, (st), T)$ is a projective system of elementary integrals and is evidently full. The product of the $(X_i, \mathcal{R}_i, m_i)$ is defined as the projective limit of this system.

Definition. $\Pi\{(X_i, \mathcal{R}_i, m_i) : i \in \underline{T}\} = \lim\limits_{\leftarrow}(X_s, \mathcal{R}_s, m_s; T)$.#

The underlying set of the product is the cartesian product, $X = \Pi\{X_i : i \in I\}$, of the X_i; its integration lattice is

$$\mathcal{R} = \mathcal{R}_\infty = \cup\{\mathcal{R}_t \circ (t\infty) : t \in T\},$$

where $(t\infty)$ is the natural projection from X onto X_t (which is a morphism). $\mathcal{R}$ contains the constants; the projective limit, $m = m_\infty$, of the m_t is positive on $\mathcal{R}$; and $m(X) = 1$. We write

$$\mathcal{R} = \Pi\{\mathcal{R}_i : i \in I\}.$$

Here are a few observations: $\mathcal{R}$ is the integration lattice spanned by the vector space $\otimes \mathcal{R}_i$ of functions

$$\sum_k \varphi^k_{i_1} \otimes \cdots \otimes \varphi^k_{i_n} : x = (x_i)_{i\in I} \to \sum \varphi^k_{i_1}(x_{i_1}) \cdots \varphi^k_{i_n}(x_{i_n}),$$

where

$$\{i_1, \ldots, i_n\} \in T\ , \quad \varphi^k_{i_1} \in \mathcal{R}_{i_1}, \ldots, \varphi^k_{i_n} \in \mathcal{R}_{i_n},$$

and the sum (over k) is finite. $\otimes \mathcal{R}_i$ is uniformly dense in $\Pi\, \mathcal{R}_i$.

For $x = (x_i)_{i\in I} \in X$ call x_i the ith coordinate of x, and observe that $x_i = (\{i\}\infty)(x)$. A function $\varphi: X \to \mathbb{R}$ is said not to depend on the jth coordinate x_j if $\varphi(x) = \varphi(y)$ for all pairs of points $x,y \in X$ that differ only in the jth coordinate. Any function $\varphi \in \mathcal{R}$ depends only on finitely many coordinates: it is of the form $\psi \circ (s\infty)$ for some $s \in T$ and some $\psi \in \mathcal{R}_s$ and so depends only on the coordinates in s.

Let I be the disjoint union of two non-empty subsets K and L. Then $\Pi\{X_i;\ i \in I\}$ is in one-to-one correspondence with $\Pi\{X_i;\ i \in L\} \times \Pi\{X_i;\ i \in K\}$ via the map $\pi: ((x_i)_{i\in L}, (x_i)_{i\in K} \to (x_i)_{i\in I}$.

Write $(X^K,\mathcal{R}^K,m^K) = \Pi\{(X_i,\mathcal{R}_i,m_i);\ i\in K\}$. For $\varphi \in \mathcal{R} = \mathcal{R}^I$, $\varphi\circ\pi$ is clearly in $\mathcal{R}^K \times \mathcal{R}^L$ and $\varphi \to \varphi\circ\pi$ is an isomorphism of $\mathcal{R}^I$ with $\mathcal{R}^K \times \mathcal{R}^L$. Its transpose maps $m^K \times m^L$ onto $m = m^I$. More precisely, let $\varphi \in \mathcal{R}^I$. Then φ depends only on finitely many coordinates x_i, $i \in t$ for some $t \in T$; the map $(x_i)_{i\in L} \to \varphi((x_i)_{i\in I})$ is in $\bar{\mathcal{R}}^L$ for all $(x_i)_{i\in K} \in X^K$;

$$(x_i)_{i\in K} \to \int \varphi((x_i)_{i\in K},(x_i)_{i\in L})dm^L((x_i)_{i\in L}) \text{ is in } \mathcal{R}^K; \text{ and}$$

$$m(\varphi) = \int\left(\int \varphi((x_i)_{i\in K},(x_i)_{i\in L})dm^L((x_i)_{i\in L})\right)dm^K((x_i)_{i\in K}).$$

These facts are easily established for $\varphi\in \otimes\{\mathcal{R}_i,\ i\in I\}$ and can be extended to $\varphi \in \Pi\{\mathcal{R}_i;\ i \in I\}$ by continuity. The details are left to the reader. The integrals are extensions by uniform continuity (cf. 4.7).

B. *-continuity

<u>32.1. Theorem</u>. $(\mathcal{R},m)$ is a *-continuous elementary integral if and only if all the $(\mathcal{R}_i,m_i)$ are $(* = S \text{ or } B)$.

<u>Proof</u>. Let $i \in I$. The map $(\{i\}\infty)$ is a morphism from $(X,\mathcal{R})$ to $(X_i,\mathcal{R}_i)$ under which m has image m_i. If m is *-continuous then so is m_i (28A). The condition is therefore necessary. That it is sufficient is more difficult to show. We do it for the case $* = B$, and so all the m_i are assumed B-continuous. We may well-order I: that is, we may assume that I is an ordinal δ. Let Φ be a decreasingly directed subset of $\mathcal{R}$. We shall show that if $\inf\{m(\varphi);\ \varphi \in \Phi\} > 0$ then $\inf \Phi \neq 0$. By way of contradiction, assume that $\inf m(\Phi) = a > 0$. For $0 < \beta \leq \delta$, let

$$(X^\beta,\mathcal{R}^\beta,m^\beta) = \Pi\{(X_\alpha,\mathcal{R}_\alpha,m_\alpha);\ \alpha < \beta\}, \text{ and}$$

$$(X^{\beta'},\mathcal{R}^{\beta'},m^{\beta'}) = \Pi\{(X_\alpha,\mathcal{R}_\alpha,m_\alpha);\ \beta \leq \alpha < \delta\}.$$

For each $\varphi \in \Phi$ and each $\beta \leq \delta$, define $\varphi_\beta \in \mathcal{R}^\beta$ by

$$\varphi_\beta((x_\alpha)_{\alpha<\beta}) = \int \varphi((x_\alpha)_{\alpha<\beta}, (x_\alpha)_{\beta\leq\alpha<\delta})dm^{\beta'}((x_\alpha)_{\beta\leq\alpha<\delta}).$$

We have $\varphi_0 = m(\varphi) \geq a$, and $\varphi_\delta = \varphi$. We shall show by induction the existence of a point $(a_\alpha)_{0\leq\alpha<\delta}$ of X such that

$$(P_\beta): \varphi_\beta((a_\alpha)_{\alpha<\beta}) \geq a \quad \text{for all} \quad \beta \leq \delta \quad \text{and} \quad \varphi \in \Phi.$$

Once this is done, we are finished, as then

$$\varphi((a_\alpha)_{\alpha<\delta}) \geq a \quad \text{for all} \quad \varphi \in \Phi.$$

Suppose $(a_\alpha)_{\alpha<\gamma}$ have been chosen so that (P_β) holds for all $\beta \leq \gamma$. If $\gamma = \delta$ we are finished, and therefore we assume that $\gamma < \delta$. Consider the functions

$$\psi: x \to \varphi_{\gamma+1}((a_\alpha)_{\alpha<\gamma}, x) \qquad (x \in X_\gamma, \ \varphi \in \Phi).$$

They belong to $\mathcal{R}_\gamma$, and their integrals,

$$\int \psi dm_\gamma = \varphi_\gamma((a_\alpha)_{\alpha<\gamma}),$$

are at least as large as a. The infimum of the ψ, as φ ranges over Φ, is therefore not smaller than a everywhere, from the B-continuity of m. Hence there is an $a_\gamma \in X_\gamma$ such that $\psi(a_\gamma) = \varphi_{\gamma+1}((a_\alpha)_{\alpha\leq\gamma+1}) \geq a$. This argument applies also to the case $\gamma = 0$. The proof is finished.#

C. Some limit theorems

If $(X,\mathcal{R},m) = (X_1,\mathcal{R}_1,m_1) \times \cdots \times (X_n,\mathcal{R}_n,m_n)$ is a finite product and if f is an $(\mathcal{R},m^S)$-integrable function, then its integral can be calculated iteratively, by Fubini's theorem:

$$\int fdm = \int \cdots (\int f(x_1,\ldots,x_n)dm_1(x_1)) \cdots dm_n(x_n).$$

It is natural to ask what the analogue of this statement is, for infinite products. This subsection is devoted to this and related questions. For simplicity's sake, we shall assume that the infinite product is countable and, accordingly, consider a sequence $(X_k,\mathcal{R}_k,m_k)$ of positive elementary *-integrals of mass one, and let $(X,\mathcal{R},m)$ denote their product. We shall need the following notation.

For $n = 0, 1, 2,\ldots$ put

$$(X_{(n)},\mathcal{R}_{(n)},m_{(n)}) = (X_1,\mathcal{R}_1,m_1) \times \cdots \times (X_n,\mathcal{R}_n,m_n) \quad \text{and}$$

$$(X_{(n')},\mathcal{R}_{(n')},m_{(n')}) = \Pi\{(X_k,\mathcal{R}_k,m_k): k = n+1, n+2,\ldots\} .$$

From the above, $(X,\mathcal{R},m) = (X_{(n)},\mathcal{R}_{(n)},m_{(n)}) \times (X_{(n')},\mathcal{R}_{(n')},m_{(n')})$ in the sense that there is a canonical isomorphism,

$$(x_k)_{k\in \mathbb{N}} \to ((x_k)_{k=1,\ldots,n},\ (x_k)_{k=n+1,\ldots}),$$

from one product with the other. Next, let $((n)\infty)$ and $((n')\infty)$ denote the canonical maps from $(X,\mathcal{R},m)$ to $(X_{(n)},\mathcal{R}_{(n)},m_{(n)})$ and to $(X_{(n')},\mathcal{R}_{(n')},m_{(n')})$, respectively, and put

$$\mathcal{R}^{(n)} = \mathcal{R}_{(n)}\circ((n)\infty) \quad \text{and} \quad \mathcal{R}^{(n')} = \mathcal{R}_{(n')}\circ((n')\infty).$$

Then $(\mathcal{R}^{(n)})$ is an increasing family of integration sublattices of $\mathcal{R}$ with union $\mathcal{R}$; and $(\mathcal{R}^{(n')})$ is a decreasing family whose intersection consists of the constants on X.

Lastly, if f is an $(\mathcal{R},m^*)$-integrable function, put

$$f_{(n)}(x_1,\ldots,x_n) = \int f(x_1,\ldots,x_n,\ (t_k)_{k=n+1,\ldots})dm_{(n')}((t_k)_{k=n+1,\ldots}) \ ,$$

$$f_{(n')}(x_{n+1},\ldots) = \int f(t_1,\ldots,t_n,\ x_{n+1},\ldots)dm_{(n)}(t_1,\ldots,t_n)$$

$$= \int \{\cdot[\int f(t_1,\ldots,t_n,\ x_{n+1},\ldots)dm_1(t_1)\}\cdot]dm_n(t_n) \ , \quad \text{and}$$

$$f^{(n)} = f_{(n)}\circ((n)\infty) \quad \text{and} \quad f^{(n')} = f_{(n')}\circ((n')\infty).$$

The functions $f_{(n)}$ and $f_{(n')}$ are integrable functions on $X_{(n)}$ and $X_{(n')}$, respectively, while $f^{(n)}$ and $f^{(n')}$ are integrable functions on X.

<u>32.2. Lemma</u>. $f^{(n)}$ is a conditional expectation of f under $\mathcal{R}^{(n)}$; and $f^{(n')}$ is a conditional expectation of f under $\mathcal{R}^{(n')}$. Hence $(\mathcal{R}^{(n)};f^{(n)};n = 1, 2,\ldots)$ is an increasing martingale, for each integrable function f, and $(\mathcal{R}^{(n')};f^{(n')};\ n = 1, 2,\ldots)$ is a decreasing martingale.

<u>Proof,</u> for (n). From Fubini's theorem, $f_{(n)}$ is $(\mathcal{R}_{(n)},m_{(n)}{}^*)$-integrable on $X_{(n)}$ and so $f^{(n)}$ is $(\mathcal{R}^{(n)},m^*)$-integrable on X (Note that $((n)\infty)m^* \leq m_{(n)}{}^*$ and use 28.5). In particular, $f^{(n)}$ is $(\mathcal{R}^{(n)},m^*)$-measurable. If $\varphi \in \mathcal{R}^{(n)}$, then

$$\int f\varphi dm = \int (\int f(x,t)\varphi(x)dm_{(n')}(t))dm_{(n)}(x)$$

$$= \int f_{(n)}(x)\varphi(x)dm_{(n)}(x) = \int f^{(n)}(x)\varphi(x)dm(x),$$

so that $f^{(n)}$ is, indeed, a conditional expectation of f under $\mathcal{R}^{(n)}$.#

From simple applications of the martingale theorems 31D we obtain the following result.

32.3. Theorem. (Jessen). Let $(\mathcal{R},M)$ be an upper integral for m and $f \in \mathcal{L}^p_E(\mathcal{R},M)$ $(1 \le p < \infty$, E a Banach space).

(i) $f = \lim_{n\to\infty} f^{(n)}$ a.e. and in p-mean.

(ii) $\int f dm = \lim_{n\to\infty} f^{(n')}$ a.e. and in p-mean.

Proof. (i) is a consequence of 31.2 and 31.8 in conjunction with the fact that $(\mathcal{R}^{(n)}, f^{(n)}, n \in \mathbb{N})$ is a martingale.

For (ii), observe that $(\mathcal{R}^{(n')}, f^{(n')}, n \in \mathbb{N})$ is a martingale if is given the order opposite to the usual one. Let an $\varepsilon > 0$ be given. There is a $\varphi \in \mathcal{R} \otimes E$, $\varphi \in \mathcal{R}^{(n)} \otimes E$ say, so that $M_p(f - \varphi) < \varepsilon$. For $k > n$, $\varphi^{(k')}(x) = \int \varphi(t,x)dm_k(t)$ is constant and equal to $\int \varphi dm$. As the conditional expectation is contractive in p-mean (29.5),

$$M_p(f^{(k')} - \varphi^{(k')}) = M_p(f^{(k')} - \int \varphi dm) < \varepsilon,$$

and hence

$$M_p(f^{(k')} - \int fdm) < 2\varepsilon$$

for all $k \ge n$: $f^{(k')}$ converges to $\int fdm$ in p-mean.

To prove convergence pointwise a.e., put

$$g^{(j')} = f^{(j')} - \int fdm \qquad , \ j = 1, 2, \ldots,$$

and observe that $(\mathcal{R}^{(j')}, g^{(j')}, k \ge j \ge i)$ is a martingale for all i,k in . We apply 31.3, with $\varphi = \varepsilon$, to the submartingale $(\mathcal{R}^{(j')}, \|g^{(j')}\|, k \ge j \ge i)$ and obtain

$$\varepsilon \int [\sup_{k\ge j\ge i} |g^{(j')}| \ge \varepsilon] dm \le \sup_{k\ge j\ge i} \int |g^{(j')}| dm = \int |g^{(i')}| dm,$$

hence $\int [\sup_{j\geq i} |g^{(j')}| \geq \varepsilon] dm \leq \frac{1}{\varepsilon} \int |g^{(i')}| dm$. As $\int |g^{(i')}| dm =$ $M(f^{(i')} - \int fdm)$ converges to zero as $i \to \infty$,

$$\lim_{i\to\infty} \int [\sup_{j\geq i} |g^{(j')}| \geq \varepsilon] dm = 0$$

for all $\varepsilon > 0$, which shows that $\limsup_{j\to\infty} |g^{(j')}| = 0$ a.e. Hence $\lim f^{(j')} = \int fdm$ a.e., as claimed.#

From 31.11 we know that the functions $f^{(n')}$ converge a.e. to the conditional expectation of $f^{(n')}$ under

$$\mathcal{L}^{(\infty')} = \bigcap \{\mathcal{L}^1(\mathcal{R}^{(n')}, M) : n = 1, 2, \ldots\}.$$

Comparison with 32.3 (ii) shows that $\mathcal{L}^{(\infty')}$ consists a.e. of constants only. Indeed, if $f \in \mathcal{L}^{(\infty')}$, then

$$f = f^{(n')} \underset{n\to\infty}{\to} \int fdm \qquad \text{a.e.}$$

In particular, a set A in

$$\mathcal{B}^{(\infty')} = \mathcal{B}(\mathcal{L}^{(\infty')}) = \bigcap \{\mathcal{B}(\mathcal{R}^{(n')}, M) : n = 1, 2, \ldots\},$$

being a.e. constant, satisfies either $\int Adm = 0$ or $\int Adm = 1$.

Consider an $A \in \mathcal{B}^{(\infty')}$. For every n there is an $\mathcal{R}^{(n')}$-Baire set $A^{n'}$ equivalent to A. Since the $\mathcal{R}^{(n')}$ are decreasing, the set

$$\tilde{A} = \limsup\{A^{n'} : n \to \infty\}$$

is $\mathcal{R}^{(n')}$-Baire for all n and equivalent to A. Let us call two points $x = (x_1, x_2, \ldots)$ and $y = (y_1, y_2, \ldots)$ of X ultimately equal if $x_n = y_n$

for all n exceeding some $m \in \mathbb{N}$; and write $x \sim y$. Clearly $\sim$ is an equivalence relation on X; and the set $\tilde{A}$ is saturated under $\sim$. Conversely, assume $A \subset X$ is saturated under $\sim$ and $(\mathcal{R}, M)$-integrable. Since the functions

$$A^{(n)}: (x_1, \ldots) \to \int A(t_1, \ldots, t_n, x_{n+1}, \ldots) dm_{(n)}(t_1, \ldots, t_n)$$

are all equal to A, A is $(\mathcal{R}^{(n')}, m^*)$-integrable for all n and so belongs to $\mathfrak{G}^{(\infty')}$. To summarize:

<u>32.4. Corollary</u>. (Zero-One law). The sets in $\mathfrak{G}^{(\infty')}$ are exactly the $(\mathcal{R}, m^*)$-integrable sets saturated under $\sim$, modulo negligible sets; and each of them has measure either zero or one.#

For $k = 1, 2, \ldots$ let $\mathcal{R}^k$ denote the functions in $\mathcal{R}$ that depend only on the kth coordinate, i.e., $\mathcal{R}^k = \mathcal{R}_k \circ (\infty\{k\})$; and let A^k be an $(\mathcal{R}^k, M)$-integrable subset of X.

<u>32.5. Corollary</u>. (Borel-Cantelli lemma). If

$$B = \limsup A^k = \bigcap_n \bigcup_{k \geq n} A^k$$

then

$$\int B dm = \begin{cases} 0 & \text{if } \sum \int A^k dm < \infty \\ 1 & \text{if } \sum \int A^k dm = \infty. \end{cases}$$

<u>Proof</u>. A point $x \in X$ belongs to B if and only if there exist arbitrarily high indices k such that $x \in A^k$. Hence B is evidently saturated under $\sim$ and so has integral either zero or one. Since

$$M(B) \leq M(\cup\{A^k: k \geq n\}) \leq \Sigma\ \{\int A^k dm: k \geq n\}$$

for all n, we have $M(B) = \int B dm = 0$ if $\Sigma \int A^k dm$ is finite.

Assume then that this sum is infinite. Since $1 + a \leq (1 - a)^{-1}$ for $0 \leq a \leq 1$, we have

$$\sum_{k=n}^{m} \int A^k dm \leq \prod_{k=n}^{m} (1 + \int A^k dm) \leq \prod_{k=n}^{m} (1 - \int A^k dm)^{-1}$$

for all $m \geq n$, and thus

$$\prod_{k=n}^{\infty} (1 - \int A^k dm) = 0 \quad \text{, for all } n.$$

From 32.3, we obtain

$$M(X - B) = \int \bigcup_n \bigcap_{k \geq n} (X - A^k) dm$$

$$= \lim_n \int X_1 \times \cdots \times X_{n-1} \times (X_n - A^n) \times (X_{n+1} - A^{n+1}) \times \cdots dm$$

$$= \lim_n \prod_{k=n}^{\infty} (1 - \int A^k dm) = 0. \#$$

D. Supplement

32.6. Let $(X, \mathcal{R}, m)$ be the product of the $*$-integrals $(X_i, \mathcal{R}_i, m_i)_{i \epsilon}$ and let $f_i: X_i \to \mathbb{R}_+$ be $(\mathcal{R}_i, m_i^*)$-integrable functions. Then $f = \otimes\{f_i: i \in \mathbb{N}\}$ is integrable if and only if $\Pi\{\int f_i dm_i: i \in \mathbb{N}\}$ exists, and in this instance, this product equals $\int f dm$.

LITERATURE: Standard Texts, e.g. [19], [21].

33. Probability

A. Introduction

The usual probabilistic model for a physical system P consists of a probability space $(\Omega, \mathcal{A}, m)$; that is a set Ω of points, interpreted as the possible states of the system, together with a σ-algebra $\mathcal{A}$ on Ω, interpreted as the events, and a positive, σ-additive measure m on $\mathcal{A}$. If $A \in \mathcal{A}$ then m(A) is interpreted as the probability that the state of the system belongs to A. It is measured as the frequency with which a large number of identical systems is found to be in a state belonging to A. A measurable function f on Ω is interpreted as an observable taking random values, and $\int f dm$ as the expected or average value of f.

This is all quite intuitive, except for the σ-additivity of m. The latter is usually justified by the success of the theory. Here is how one can arrive at this model from first assumptions not involving any continuity requirements.

Let $\mathcal{F}$ denote a family of real-valued observables of the system P; these are (prescriptions of) measurements which will produce random values if the state of the system is not known precisely.

Let $\mathcal{R}(\mathcal{F})$ denote the closure of $\mathcal{F}$ under composition with bounded continuous functions: If $\underline{X} = (X_1, \ldots, X_n)$ is an n-tuple of observables in $\mathcal{R}$ and $\varphi: \mathbb{R}^n \to \mathbb{R}$ a continuous bounded function, then $\varphi(\underline{X}) = \varphi(X_1, \ldots, X_n)$ is another observable; when X_i has the observed value x_i $(1 \leq i \leq n)$ then $\varphi(\underline{X})$ takes the value $\varphi(x_1, \ldots, x_n)$. Clearly, there is a smallest family $\mathcal{R} = \mathcal{R}(\mathcal{F})$ closed under composition as above and containing $\mathcal{F}$; and $\mathcal{R}$ is closed under taking linear combinations, finite suprema, and infima with 1. (1 is the observable which always takes the value 1, whichever the state of the system.)

A probabilistic model of P requires the existence of an expected or average value, $E(X)$, for every $X \in \mathcal{R}$. It is evidently reasonable to assume that E is linear and positive on $\mathcal{R}$, and that $E(1) = 1$. Note that $\mathcal{R}(\mathcal{F})$ is not viewed as a family of functions on some ideal set of states of P. The construction, in purely mathematical terms, of a set on which $\mathcal{R}(\mathcal{F})$ appears as functions--but which is not a set of ideal "states"--is now to follow.

Let $\underline{X}$ be an n-tuple in $\mathcal{F}$ and put $\mathbb{R}_{\underline{X}} = \mathbb{R}^n$. The map

$$\varphi \to E(\varphi(\underline{X})) \qquad , \quad \varphi \in C_b(\mathbb{R}_{\underline{X}}),$$

is a positive and tight measure $m_{\underline{X}}$ on $\mathbb{R}_{\underline{X}}$. Moreover, if $\underline{Y}$ is an m-tuple of observables from among the $\{X_1,\dots,X_n\}$ $(m \leq n)$, then $m_{\underline{Y}}$ is evidently the image of $m_{\underline{X}}$ under the canonical map from $\mathbb{R}_{\underline{X}}$ onto $\mathbb{R}_{\underline{Y}}$. We obtain a projective system $(\mathbb{R}_{\underline{X}}, m_{\underline{X}})$ of tight measures, which is full. Indeed, the projective limit of the $\mathbb{R}_{\underline{X}}$ is nothing but $\mathbb{R}^{\mathcal{F}}$. According to Kolmogoroff's theorem (30.6), the projective limit m of the $m_{\underline{X}}$ is an S-measure on the projective limit $\mathcal{R}_\infty$ of the $C_b(\mathbb{R}_{\underline{X}})$.

The triple $(\mathbb{R}^{\mathcal{F}}, \mathcal{R}_\infty, m)$ so constructed from $\mathcal{F}$ and E is called the <u>stochastic model</u> of $(\mathcal{F}, E)$. It is a probability space as described at the start of the section. There is an evident 1-1-correspondence of the observables in $\mathcal{R}$ with the functions in $\mathcal{R}_\infty$ on $\mathbb{R}^{\mathcal{F}}$; this can be extended to an interpretation of the $(\mathcal{R}_\infty, m^S)$-measurable functions as observables of the system. Whence the name random variable for the $(\mathcal{R}_\infty, m^S)$-measurable functions. If $f \in \mathcal{L}^1(\mathcal{R}_\infty, m^S)$ then $\int f dm$ is interpreted as the average or expected value of f.

The difference from the intuitive probabilistic model set out in the beginning is that $\mathbb{R}^{\mathcal{F}}$ cannot usually be interpreted as the set of states of the system. For instance, if $\mathcal{F}$ is chosen too small, then different states--distinguishable by other measurements, not in $\mathcal{F}$--will produce the same values for all observables in $\mathcal{F}$; on the other hand, if X and $2X$ belong to $\mathcal{F}$ then there is a point $\omega \in \mathbb{R}^{\mathcal{F}}$ with $X(\omega) = (2X)(\omega) = 1$. Evidently ω does not correspond to any state of the system. The lack of an interpretation for $\mathbb{R}^{\mathcal{F}}$ is inessential in those questions that deal with the distributions and expectations of the observables $X \in \mathcal{F}$, i.e., with the directly observable probabilistic behaviour of the system.

B. Sequences of random variables

Henceforth $(\Omega, \mathcal{A}, m)$ is a fixed probability space, for instance $(\mathbb{R}^{\mathcal{F}}, T(\mathcal{R}_\infty, m^S), m)$ where $(\mathbb{R}^{\mathcal{F}}, \mathcal{R}_\infty, m)$ is the stochastic model for the observables $\mathcal{F}$ of a physical system. An $(\mathcal{S}(\mathcal{A}), m^S)$-measurable real-valued function X is usually called a <u>random variable</u> and

$$E(X) = \int X(\omega) dm(\omega)$$

its <u>expectation</u>.

In studying sequences (f_n) of random variables it is often helpful to reduce $(\Omega, \mathcal{A}, m)$ to a smaller probability space by constructing the stochastic model of the family $\{f_n\}$: For any finite subfamily $F \subset \{f_n\}$ let m_F denote its joint distribution on $\mathbb{R}_F = \mathbb{R}^k$ (where $F = \{f_{n(1)}, \ldots, f_{n(k)}\}$). The $(\mathbb{R}_F, m_F, C_b(\mathbb{R}_F))$ form a projective system, and its project limit $(\Omega', \mathcal{R}', m')$, which is tight (30.7), is called the <u>stochastic model</u> for the sequence (f_n). Evidently, m' is the image, or joint distribution, of (f_n) as a measurable (18.17) map from Ω to

$\Omega' = \mathbb{R}^{\mathbb{N}}$. On the stochastic model, each f_n corresponds to the nth coordinate function:

$$f_n = X_n \circ (f_k) \quad \text{with} \quad X_n : (x_1, x_2, \ldots) \in \Omega' \to x_n .$$

The (e.g., convergence) behaviour of the f_n translates into the behaviour of the coordinate functions on $\Omega' = \mathbb{R}^{\mathbb{N}}$.

C. Independent random variables

Definition. The family $\mathcal{F}$ of random variables is said to be independent if the joint distribution of any finite subfamily $\{f_1, \ldots, f_n\} \subset \mathcal{F}$ is the product of the distributions of the f_i($1 \leq i \leq n$):

$$(f_1, \ldots, f_n)(m) = f_1(m) \times \cdots \times f_n(m).\#$$

From the above, the sequence (f_n) of random variables is independent if $m' = (f_n)(m)$ is the infinite product

$$m' = \Pi\{ f_n(m) : n = 1, 2, \ldots \}.$$

To give a probabilistic interpretation of this condition, consider Borel sets $A_i \subset \mathbb{R}$ $(1 \leq i \leq n)$. From Fubini's theorem $A_1 \times \cdots \times A_n \in \mathbb{R}^n$ is $m_{(n)}$-integrable and

$$E[(f_1, \ldots, f_n) \in A_1 \times \cdots \times A_n] = m_{(n)}(A_1 \times \cdots \times A_n)$$

$$= f_1(m)(A_1) \ldots f_n(m)(A_n) = E[f_1 \in A_1] \ldots E[f_n \in A_n].$$

In other words, the expectation that f_i takes its value in A_i for $1 \leq i \leq n$ is the product of the expectations that $f_i \in A_i$, for $1 \leq i \leq n$.

33.1. Theorem. (Law of large numbers). Let (f_n) be a sequence of independent and square integrable random variables with expectation zero. If

$$\sum_{i=1}^{\infty} \frac{1}{i^2} E(f_i^2) < \infty$$

then

$$\lim_{n\to\infty} \frac{1}{n} \sum_{i=1}^{n} f_i(\omega) = 0 \qquad \text{for almost all } \omega \in \Omega.$$

Proof. Let $(\Omega', \mathcal{R}, m')$ be the stochastic model of (f_n). By assumption, the coordinate functions X_i $(i = 1, 2, \ldots)$ have expectation zero and are square integrable:

$$\int X_i dm' = 0 , \quad X_i \in \mathcal{L}^2(m') , \quad \text{and} \quad \sum \frac{1}{i^2} \int X_i^2 dm' < \infty.$$

We have to show that

$$Y_n = \frac{1}{n} \sum_{i=1}^{n} X_i$$

converges to zero m'-a.e. on Ω'. From (33.3) below, it is sufficient to show that the sequence

$$Z_n = \sum_{i=1}^{n} \frac{1}{i} X_i$$

converges a.e. Now, Z_n is clearly a square-integrable martingale with respect to the increasing sequence of integration sublattices $\mathcal{R}^{(n)} \subset \mathcal{R}$ of functions that depend only on the first n coordinates. Indeed, since $\int X_n dm = 0$,

$$E^{\mathcal{R}^{(n)}} Z_{n+t} = Z_n \qquad , \text{ for } n, t \in \mathbb{N} .$$

Since

$$\int X_i X_j dm' = \int x_i x_j d\{(X_i, X_j)(m')\} = \int x_i x_j d\{X_i(m') \times X_j(m')\}$$

$$= \int x_i d(X_i(m')) \int x_j d(X_j(m')) = \int X_i dm' \int X_j dm' = 0$$

for $i \neq j$,

$$\int Z_n^2 dm' = \sum_{i=1}^{n} \int \frac{1}{i^2} X_i^2 dm \leq \sum_{i=1}^{\infty} \frac{1}{i^2} E(X_i^2).$$

From 31.10(i), (Z_n) converges a.e.#

<u>33.2. Corollary</u>. If (f_n) is a sequence of independent, square-integrable, and identically distributed random variables, then

$$\lim_{n\to\infty} \frac{1}{n} \sum_{i=1}^{n} f_n(\omega) = E(f_1) \qquad \text{a.e.}$$

<u>Proof</u>. To say that the f_n are identically distributed means $f_i(m) = f_j(m)$ for $i, j \in \mathbb{N}$. Hence $\int f_i dm = \int x\, d(f_i m)(x) = \int x\, d(f_j m)(x) = \int f_j dm$ and, similarly, $E((f_i - Ef_i)^2) = E((f_j - Ef_j)^2)$. Hence

$$\frac{1}{n} \sum_{i=1}^{n} (f_i - E(f_i)) = \frac{1}{n} \sum_{i=1}^{n} f_i - E(f_i) \to 0 \qquad \text{a.e.\#}$$

This corollary shows that the mathematical model for probability set up in 33A is consistent with the interpretation of $E(f)$ as the limit of the arithmetic means of repeated independent measurements of f: Let $P_1, P_2, \ldots$ be infinitely many systems indistinguishable from P and let f_n be the random observable "measurement of f at P_n." Then, with probability one, $E(f)$ is the limit of the arithmetic means $\frac{1}{n} \sum_{i=1}^{n}$ (observed valued of f_i), as n tends to infinity.

D. Supplements

33.3. If (a_i) is a sequence of complex numbers such that $\sum i^{-1}a_i$ exists then $\lim_{n\to\infty} \frac{1}{n} \sum_{i=1}^{n} a_i = 0$.

33.4. Suppose $\{f_{ij}: i \in I, j \in J\}$ are independent random variables, and for $i \in I$ let φ_i be a bounded Borel function on $\mathbb{R}^{n(i)}$. Then $\{\varphi_i(f_{ij(1)},\ldots,f_{ij(n(i))}): i \in I\}$ is an independent family.

LITERATURE: [1], [16], [31]. The law of large numbers is more adequately done in the realm of ergodic theory. Cf., e.g. [4], [18]. There are other stochastic models of interest for $(\mathcal{F},E)$, which are not full projective limits of the finite joint distributions. See, e.g. [5, Ch. IX], [16], [31], [33].

V. LIFTINGS AND DERIVATIVES

This chapter contains a generalization of the Radon-Nikodym theorem to the case of vector-valued measures; the determination of the dual of L_E^p and, more generally, of all continuous linear maps from L_E^p to another Banach space; and the disintegration of measures. The principal tool in these investigations is the notion of a lifting, to be developed in the first two sections. It is not attempted to give a complete presentation of all known integral representation results; this chapter rather contains an exposition of the most common methods used to obtain them.

§1. Liftings

34. Liftings and dense topologies

Henceforth $(\mathcal{R},M)$ is a strong and essential upper gauge on the set X, $\mathcal{R}$ is dominated, and M does not vanish on $\mathcal{R}_+$.

A. Definition. A lifting for $(\mathcal{R},M)$ is a map T from $L^\infty(\mathcal{R},M)$ to $\mathcal{L}^\infty(\mathcal{R},M)$ having the following properties.

(L1) $Tf^\cdot \in f^\cdot$, for $f^\cdot \in L^\infty$.

(L2) $T(rf^\cdot + sg^\cdot) = rTf^\cdot + sTg^\cdot$, for $r,s \in \mathbb{R}$, $f^\cdot, g^\cdot \in L^\infty$.

(L3) $T(f^\cdot g^\cdot) = Tf^\cdot Tg^\cdot$.

(L4) $T1^\cdot = 1 = X.\#$

In other words, a lifting is an algebra homomorphism from L^∞ to $\mathcal{L}^\infty$ which is a right inverse for the natural quotient map: $\mathcal{L}^\infty \to L^\infty$. Or else: a lifting picks a representative from every class in L^∞ in a linear and multiplicative manner. One may view a lifting as an algebra automorphism T from $\mathcal{L}^\infty$ to itself satisfying $f \doteq g \Rightarrow Tf = Tg \doteq f$. Accordingly, we shall write Tf or $Tf^\cdot$ interchangeably. The value of Tf at $x \in X$ is denoted by $T_x f$. Here are a few consequences of the definition:

(L5) If $0 \le f^{\bullet} \in L^{\infty}$, then $Tf^{\bullet} \ge 0$ everywhere.

Indeed, $f^{\bullet}$ is a square and so is Tf.

(L6) $T(f^{\bullet} \vee g^{\bullet}) = Tf^{\bullet} \vee Tg^{\bullet}$; $T(f^{\bullet} \wedge g^{\bullet}) = Tf^{\bullet} \wedge Tg^{\bullet}$; $T|f| = |Tf|$ (for $f^{\bullet}, g^{\bullet} \in L^{\infty}$).

Indeed, the map $f^{\bullet} \to T_x f^{\bullet}$ from L^{∞} to $\mathbb{R}$ is a character of the algebra L^{∞} and thus also of the Riesz space L^{∞} (20.5). Conversely and for the same reason, the multiplicativity (L3) of T is implied by (L2) and any one of the equations in (L6).

(L7) $|T_x f| \le M_{\infty}(f)$, for $f \in \mathcal{L}^{\infty}$, $x \in X$.

Indeed, we infer from (L5) and the equation $-M_{\infty}(f)1^{\bullet} \le f^{\bullet} \le M_{\infty}(f)1^{\bullet}$ that $-M_{\infty}(f) \le T_x f^{\bullet} \le M_{\infty}(f)$. A lifting has therefore norm one from L^{∞} to $\mathcal{L}^{\infty}$, the latter being equipped with the pointwise supremum norm.

The role liftings play is partly illuminated by the following propositions.

<u>34.1. Proposition</u>. Let T be a lifting for $(\mathcal{R},M)$ and put $\mathcal{D} = \mathcal{D}_T = T(L^1 \cap L^{\infty})$. Then $\mathcal{D}$ is an integration lattice and algebra dense in $\mathcal{L}^p(\mathcal{R},M)$, for $1 \le p < \infty$. The upper gauge $(\mathcal{D},M)$ is B-continuous and has support X. A function f in $\mathcal{D}^B_{\uparrow}$ is $(\mathcal{R},M)$-measurable, and is $(\mathcal{R},M)$-integrable if $M(f) < \infty$.

<u>Proof</u>. The first statement is evident. For the second, let $\mathcal{D}_+ \supset \Phi \uparrow^B \psi \in \mathcal{D}^B_{\uparrow}$ (8C), and let $K = TK \in \mathcal{C}(\mathcal{R},M)$ and $k \in \mathbb{N}$. Then

$$\mathcal{D}_+ \supset K\Phi \wedge k \uparrow^B K\psi \wedge k \in \mathcal{D}^B_{\uparrow} \qquad \text{(L3,L6)}.$$

If $\psi_1^{\bullet}$ denotes the supremum of $\{K\Phi \wedge k\}^{\bullet}$ in L^1 (8.13), then

$$T\psi_1^{\bullet} \geq \sup\{T(K\varphi \wedge k) : \varphi \in \Phi\} = K\psi \wedge k,$$

due to the monotonicity of T. Since

$$M(\psi_1^{\bullet}) \geq M(K\psi \wedge k) \geq \sup M(K\Phi \wedge k) = M(\psi_1^{\bullet}),$$

$\sup M(\Phi) \geq M(K\psi \wedge k)$. Taking the supremum over all $K \in \mathfrak{C}(\mathfrak{R},M)$ and all integers k, one gets

$$\sup M(\Phi) \geq M(\psi).$$

Since the reverse inequality is obvious, $\sup M(\Phi) = M(\psi)$, as desired. $(\mathfrak{D},M)$ has support X, since there is no negligible function in $\mathfrak{D}$ apart from zero. Hence no $\mathfrak{D}$-open set has measure zero except the empty set. The last statement of the proposition is a consequence of the B-continuity of $(\mathfrak{D},M)$ (8.15).#

<u>34.2. Corollary</u>. If $(\mathfrak{R},M)$ admits a lifting, it must be strictly localizable. (See 16.4. We shall see in the next section that this condition is also sufficient.)#

The following results show that a lifting can be used to fix measurable functions having other than real values, within their classes.

<u>34.3. Proposition</u>. Put $\mathfrak{D}^{\infty} = \mathfrak{D}_T^{\infty} = T(L^{\infty})$.

(i) $\mathfrak{D}^{\infty}$ is an integration lattice and algebra, is uniformly closed, and contains the constants. Any real-valued function uniformly continuous in the $\mathfrak{D}^{\infty}$-uniformity on X is already in $\mathfrak{D}^{\infty}$ (and is negligible only if it is zero).

34A

(ii) For any $(\mathcal{R},M)$-measurable function f with values in a compact metric space Y there exists a unique $\mathcal{D}^\infty$-uniformly continuous function Tf, necessarily measurable, with values in Y, which equals f almost everywhere. (Note here that $g: X \to Y$ is $\mathcal{D}^\infty$-uniformly continuous if and only if

$$\varphi\circ g \in \mathcal{D}^\infty, \text{ i.e., } \varphi\circ g = T(\varphi\circ g)$$

for all uniformly continuous functions φ on Y.)

Proof. (i): From $|Tf - f_n| = |T(f - f_n)| = T|f - f_n| \leq M_\infty(f - f_n)$ for $f_n \in \mathcal{D}^\infty$, $f \in \mathcal{L}^\infty$, it follows that $\mathcal{D}^\infty$ is uniformly closed. The rest of (i) is clear (Cf. 5D).

(ii): If f is a measurable step function s whose steps have width $A_i \in T(\mathcal{R},M)$ then f equals almost everywhere a measurable step function Ts whose steps have width $T(A_i)$ (where i ranges over the number of steps). Ts is clearly $\mathcal{D}^\infty$-uniformly continuous. An arbitrary measurable function f is everywhere the uniform limit of measurable step functions (s_n) (19.6). The sequence (Ts_n) converges uniformly to a $\mathcal{D}^\infty$-uniformly continuous function Tf equal almost everywhere to f. If g is a second such function then

$$(\varphi\circ f \doteq)\ \varphi\circ Tf = \varphi\circ g \in \mathcal{D}^\infty$$

for all uniformly continuous functions φ on Y. As the latter separate points, Tf equals g everywhere, and so Tf is unique.#

Notice that, in the proof of (ii), use has been made only of the effect of T on the measurable sets: the argument used has the following corollary.

34.4. Corollary. Let $T: \mathcal{L}^{\infty}(\mathcal{R},M) \to \mathcal{L}^{\infty}(\mathcal{R},M)$ be a lifting. Its restriction Θ_T to $T(\mathcal{R},M)$ satisfies

(L'1) $\Theta_T(A) = \Theta_T(B) \doteq A$, for $A \doteq B$ in $T(\mathcal{R},M)$;

(L'2) $\Theta_T(A \cup B) = \Theta_T(A) \cup \Theta_T(B)$, for A,B in $T(\mathcal{R},M)$;

(L'3) $\Theta_T(A \cap B) = \Theta_T(A) \cap \Theta_T(B)$, for A,B in $T(\mathcal{R},M)$;

(L'4) $\Theta_T(\emptyset) = \emptyset$, $\Theta_T(X) = X$.

Conversely, given a map Θ from $T(\mathcal{R},M)$ to itself satisfying (L'1) through (L'4), there is a unique lifting T of $(\mathcal{R},M)$ such that $\Theta = \Theta_T$. (For this reason, such Θ is also called a lifting for $T(\mathcal{R},M)$.)

Proof. (L'2) and (L'3) result from (L6). Any given Θ can be extended to measurable real-valued step functions as above, and to L^{∞} by continuity in M_{∞}.#

34.5. Example. Let $Y = \overline{\mathbb{R}}$. From the proposition, a function Tf can be picked uniquely in each class of measurable functions $f: X \to \overline{\mathbb{R}}$. Furthermore,

$$T(f \vee g) = Tf \vee Tg, \quad T|f| = |Tf|, \quad \text{and}$$

$$T(rf + sg) = rTf + sTg, \quad T(fg) = Tf \cdot Tg \qquad (r,s \in \overline{\mathbb{R}},\ f,g \in \mathcal{L}_{\overline{\mathbb{R}}}(\mathcal{R},M))$$

wherever both sides are defined. For instance, if $T_x f \cdot T_x g$ is defined then both T(fg) and $Tf \cdot Tg$ are $\mathfrak{D}^{\infty}$-continuous in x and must coincide as there is no negligible $\mathfrak{D}^{\infty}$-open set containing x. The map $f^{\bullet} \to T_x f^{\bullet}$ is a numerical character from L^p to $\overline{\mathbb{R}}$ in the sense of 5C, for each $x \in X$, $1 \leq p \leq \infty$.#

<u>Definition</u>. Let Y be a uniform space. A map $f: X \to Y$ is <u>scalarly measurable</u> if $\varphi \circ f$ is measurable for every uniformly continuous real-valued function φ on Y. $\mathcal{M}_Y^\infty(\mathcal{R}, M)$ denotes the family of scalarly measurable maps from X to Y having relatively compact image $f(X)$ in Y. Two maps $f, g \in \mathcal{M}_Y^\infty(\mathcal{R}, M)$ are called <u>scalarly equivalent</u> if

$$\varphi \circ f \doteq \varphi \circ g \ (M)$$

for every uniformly continuous function φ on Y.#

<u>34.6. Proposition</u>. Every map $f \in \mathcal{M}_Y^\infty(\mathcal{R}, M)$ is scalarly equivalent to a unique $\mathcal{D}^\infty$-uniformly continuous map $Tf \in \mathcal{M}_Y^\infty(\mathcal{R}, M)$; i.e.,

$$\varphi \circ Tf = T(\varphi \circ f)$$

for every uniformly continuous map φ on Y.

<u>Proof</u>. For φ in the set UC of uniformly continuous real-valued functions on the compact set $\overline{f(X)}$ and for $x \in X$, put

$$t_x(\varphi) = T_x(\varphi \circ f).$$

t_x is clearly a character of the algebra UC. There is a point $T_x f \in \overline{f(X)} \subset Y$ such that

$$t_x(\varphi) = T_x(\varphi \circ f) = \varphi(T_x f) \qquad , \text{ for all } \varphi \in UC.$$

The map $Tf: x \to T_x f$ is scalarly measurable by its very definition. Since $\varphi \circ Tf$ is in $\mathcal{D}^\infty$ for $\varphi \in UC$, Tf is $\mathcal{D}^\infty$-uniformly continuous in the UC-uniformity on $\overline{f(X)}$, which coincides with the given uniformity of $\overline{f(X)}$ due to compactness.#

B. Dense topologies

With a lifting T, there is associated a topology τ_T with the property that every real-valued measurable function is a.e. τ_T-continuous. Conversely, a topology with this (and a second) property gives rise to a lifting.

Definition. A dense topology for the upper gauge $(\mathcal{R},M)$ is a topology τ consisting of $(\mathcal{R},M)$-measurable sets and satisfying

(T1): τ contains no non-void negligible set, and

(T2): Every measurable set contains an equivalent τ-open set.#

34.7. Example. Let T be a lifting for $(\mathcal{R},M)$ and define

$$\tau_T = \{T(A)\setminus N: A \in T(\mathcal{R},M),\ N \text{ negligible}\}.$$

Then every set $A \in T(\mathcal{R},M)$ contains the equivalent set $A \cap T(A) \in \tau_T$; and if $T(A)\setminus N \doteq \emptyset$ then $T(A) \doteq \emptyset$ and $T(A) = \emptyset$, so that there is no negligible non-void set in τ_T. It only needs to be shown that τ_T is a topology. Clearly, τ_T is closed under finite intersections. To see that it is closed under arbitrary unions, let $\{T(A_i)\setminus N_i: i \in I\}$ be any family in τ_T, with union U. Since $(\mathcal{R},M)$ is strictly localizable (34.2), it is localizable (21.8), and there is a supremum $A^{\bullet}$ in L^{∞} of the classes $\{A_i^{\bullet}: i \in I\}$. Clearly, $U \subset TA^{\bullet}$. For any integrable subset K of $TA^{\bullet}$, there is a countable subfamily $\{A_i: i \in I_K \subset I\}$ which covers K adequately. Hence $\{T(A_i)\setminus N_i: i \in I_K\}$ covers K adequately; and so $K\setminus U$ is negligible. Therefore U is measurable and equivalent to $TA^{\bullet}$. Finally, $U = TA^{\bullet}\cap U$ is in τ_T.

τ_T is called the dense topology associated with T.

Note that τ_T is hyperstonean: the closure of an open set $T(A) \setminus N$ is $T(A)$, which is again open.#

The properties of a lifting have not all been used in establishing that τ_T is a dense topology. The following will do:

<u>Definition</u>. A <u>predensity</u> for $(\mathcal{R},M)$ is a map Θ from $T(\mathcal{R},M)$ into itself satisfying

(D1) $\Theta(A) = \Theta(B) \doteq A$, for $A \doteq B$ in $T(\mathcal{R},M)$;

(D2) $A \subset B$ implies $\Theta(A) \subset \Theta(B)$, for A,B in $T(\mathcal{R},M)$;

(D3) If $A_1 \cap \ldots \cap A_k \doteq \emptyset$ in $T(\mathcal{R},M)$, then $\Theta(A_1) \cap \ldots \cap \Theta(A_k) = \emptyset$;

(D4) $\Theta(X) = X$, $\Theta(\emptyset) = \emptyset$.

The <u>dense topology associated with</u> Θ is given by

$$\tau_\Theta = \{\Theta(A) \setminus N : A \in T(\mathcal{R},M),\ N \text{ negligible}\}.\#$$

(D3) implies that τ_Θ is closed under finite intersections and contains no non-void negligible set. The fact that τ_Θ is closed under arbitrary unions is shown exactly as in the example above, provided it is known that $(\mathcal{R},M)$ is strictly localizable. To see this, pick a maximal family $P^\bullet$ of mutually disjoint nonzero classes of integrable sets and put $P = \{\Theta(K) : K^\bullet \in P^\bullet\}$. Evidently, P is an $(\mathcal{R},M)$-adequate partition. Predensities are of technical interest only; they and the associated dense topologies will be used in the existence proof for liftings (35.1). The interest in dense topologies is illuminated by the following result.

34.8. Proposition. Let τ be an $(\mathcal{R},M)$-dense topology. A function on X with values in $\overline{\mathbb{R}}$ (or in a compact metric space) is measurable if and only if it is τ-continuous almost everywhere.

Proof. From 19.5(iv), the condition is sufficient. A measurable step function with steps $A_i \in T(\mathcal{R},M)$ is evidently τ-continuous on the union of the τ-open kernels of the A_i, a set of negligible complement. An arbitrary measurable function is the uniform limit of measurable step functions s_n (19.6) and hence is τ-continuous at all points $x \in X$ except those where one of the s_n is not.#

C. Connection between dense topologies and liftings

It will now be shown that a dense topology τ in turn gives rise to a lifting. For $x \in X$, let I_x denote the ideal in L^∞ of classes $f^\cdot$ that contain a function f τ-continuous and zero at x. I_x is a proper ideal; for if $f \in 1^\cdot$ were continuous and zero at x then $[f < 1/2]$ would be a non-void negligible set in τ; and so $1^\cdot$ does not belong to I_x. For each $x \in X$, choose a non-trivial character T_x of L^∞ annihilating I_x (20.8).

Consider a class $f^\cdot \in L^\infty$, a representative $f \in f^\cdot$, and a point $x \in X$ in which f is τ-continuous. Clearly $f^\cdot - f(x)1^\cdot$ belongs to I_x, and so $0 = T_x(f^\cdot - f(x)1^\cdot) = T_xf^\cdot - f(x)$. Therefore, $T_xf^\cdot = f(x)$ for all x in which f is τ-continuous; in particular, $Tf^\cdot \in f^\cdot$ (34.8). Since the algebraic properties (L2) through (L4) are satisfied by definition, T is a lifting for $(\mathcal{R},M)$. It is called a lifting subordinate to τ.

T depends on the choice of characters T_x. However, some information about τ is reflected by T:

(1) $$T(A) \supset A \quad \text{for} \quad A \in \tau.$$

Indeed, A is continuous at all points of A, and is thus reproduced by T in these points.

(2) If A is both closed and open then $T(A) = A$.

Indeed, $T(A) \supset A$, $T(\complement A) \supset \complement A$ and $T(A) \cap T(\complement A) = \emptyset$. To summarize:

<u>34.9. Proposition</u>. (Oxtoby-Dixmier). The following are equivalent.

(i) There exists a lifting for $(\mathcal{R},M)$.

(ii) There exists a dense topology for $(\mathcal{R},M)$.

Furthermore, given a dense topology τ there exists a lifting T satisfying (1) and (2) above.

D. Strong liftings

Of particular interest are liftings T of $(\mathcal{R},M)$ that reproduce $\mathcal{R}$: $T\varphi = \varphi$ for $\varphi \in \mathcal{R}$. There is no hope that such exist if the support of $(\mathcal{R},M)$ differs from X, for then there is a negligible non-zero function in $\mathcal{R}$, which could not possibly be reproduced. We shall ask less:

<u>Definition</u>. Let T be a lifting for $(\mathcal{R},M)$, and $Y \subset X$. Then T is <u>strong on</u> Y if

(SL) $$T\varphi = \varphi \quad \text{on } Y \text{ for all } \varphi \text{ in } \mathcal{R}.$$

T is simply <u>strong</u> if it is strong on $\underline{X} = \operatorname{supp}(\mathcal{R},M)$.#

Suppose $M(|\varphi|) = 0$ implies $\varphi = 0$ for $\varphi \in \mathcal{R}$ and $(\mathcal{R},M)$ admits a strong lifting T. Then $(\mathcal{R},M)$ must be B-continuous, from 34.1. Similarly, if a lifting T for $(\mathcal{R},M)$ is strong on $Y \subset X$, then $(\mathcal{R},YM)$ is B-continuous. In other words, strong liftings are only defined for B-continuous upper gauges, so that it makes sense to talk about $\text{supp}(\mathcal{R},M)$.

It is an open question of importance whether B-continuity is sufficient for a strong lifting to exist.

If T is strong then the associated dense topology τ_T is finer than the $\mathcal{R}$-topology on $\underline{X}$. Indeed, a function in $\mathcal{R}$ is then τ_T-continuous on $\underline{X}$. Conversely, if τ is a dense topology finer than the $\mathcal{R}$-topology on $\underline{X}$ then there is a strong lifting for $(\mathcal{R},M)$. To see this, let S be a lifting subordinate to τ. Every function $\varphi \in \mathcal{R}$ is τ-continuous in every point of the τ-open kernel $\overset{\circ}{\underline{X}}$ of $\underline{X}$, and is thus reproduced there by S: S is strong on $\overset{\circ}{\underline{X}}$. Changing S at all points x of the negligible set $\underline{X} - \overset{\circ}{\underline{X}}$ by redefining S_x to be a character of L^∞ which extends the character $\varphi \to \varphi(x)$ of $\mathcal{R}$ (20.9), we obtain a strong lifting for $(\mathcal{R},M)$. To summarize:

<u>34.10. Proposition</u>. A B-continuous upper gauge $(\mathcal{R},M)$ admits a strong lifting if and only if there is a dense topology which is finer than the $\mathcal{R}$-topology on $\text{supp}(\mathcal{R},M)$.#

E. Supplements

34.11*. If τ is an $(\mathfrak{R},M)$-dense topology and $A \in T(\mathfrak{R},M)$, let $\underset{o}{A}$ denote the essential τ-open kernel of A, i.e.,

$$\underset{o}{A} = \bigcup \{U \in \tau : U \dot{\subset} A \ (M)\}.$$

The map $\Theta_T : A \to \underset{o}{A}$ from $T(\mathfrak{R},M)$ to itself is a density, i.e., it satisfies $(L'1)$, $(L'3)$ and $^o(L'4)$ (and therefore is in particular a predensity).

34.12*. Let $(\mathfrak{R},M)$ be an essential upper integral, T a lifting for $(\mathfrak{R},M)$ and $m = M|\mathfrak{R}$. If n is a scalar measure absolutely continuous with respect to m, then $g = dn/dm$ can be calculated as follows. Let P be a fixed T-invariant partition such that $gL \in \mathcal{L}^1$ for $L \in P$; and let $\mathcal{P}$ denote the family of all T-invariant $(\mathfrak{R},M)$-adequate partitions Q such that every set in P is the finite union of sets in Q. Then $\mathcal{P}$ is increasingly directed under refinement; and for $Q \in \mathcal{P}$

$$g_Q = \sum \{K \int K dn / \int K dm : K \in Q\}$$

is a conditional expectation of g under the δ-ring spanned by Q and converges loc. a.e. to g as Q increases in $\mathcal{P}$. If $g \in \mathcal{L}^\infty$ and $g = Tg$, then g_Q converges uniformly to g on every T-invariant integrable set.

34.13. Suppose τ is a dense topology for $(\mathfrak{R},M)$ and let $\mathcal{V}(x)$ denote the τ-neighborhoods of $x \in X$. Select an ultrafilter $\mathcal{U}(x)$ finer than the section filter on $\mathcal{V}(x)$ for each x, and for a measurable function $f: X \to \overline{\mathbb{R}}$ define

$$T_x(f) = \lim\{\int fU dm / \int U dm : U \in \mathcal{U}(x)\}.$$

Then T is a lifting subordinate to τ. If τ is finer than the $\mathfrak{R}$-topology on $\underline{X}$, i.e., if τ is strong, then so is T.

34.14. Suppose T is a lifting for $(\mathfrak{R},M)$ and let τ denote the associated dense topology. Then (i) A subset is negligible if and only if it is nowhere dense and closed. (ii) For $f \in \mathcal{L}^\infty$ and s a real number, $(Tf)_+ = T(f_+)$, $[Tf > s] \subset T[f > s]$, $[Tf < s] \subset T[f < s]$, $[Tf \geq s] \supset T[f \geq s]$, and $[Tf \leq s] \supset T[f \leq s]$.

LITERATURE: [22] and the literature cited there. Concerning 34.12, Cf. [20]. Cf. also [28].

35. Existence of liftings

A. The existence theorem

35.1. Theorem. (Ionescu-Tulcea). Let $(\mathcal{R},M)$ be an essential upper integral, and let $\mathcal{R}_0 \subset \mathcal{R}$ be an integration sublattice of $\mathcal{R}$ such that the $(\mathcal{R}_0,M)$-integrable sets form an $(\mathcal{R},M)$-adequate cover.

Then any lifting S for $(\mathcal{R}_0,M)$ can be extended to a lifting T for $(\mathcal{R},M)$, i.e.,

$$Tf^{\cdot} = Sf^{\cdot} \qquad , \text{ for } f^{\cdot} \in L^{\infty}(\mathcal{R}_0,M).$$

Proof. The assumption on $\mathcal{R}_0$ implies, and is needed to insure, that an $(\mathcal{R}_0,M)$-measurable function is $(\mathcal{R},M)$-measurable (28.5). Let $\mathcal{T}$ denote the set of pairs $(\mathcal{L},\tau)$, where $\mathcal{L}$ is an integration sublattice of $\mathcal{L}^1(\mathcal{R},M)$ equal to $\mathcal{L}^1(\mathcal{L},M)$ and containing $\mathcal{R}_0$; and where τ is an $(\mathcal{L},M)$-dense topology finer than τ_S. $\mathcal{T}$ is not void: it contains the couple $(\mathcal{L}^1(\mathcal{R}_0,M),\tau_S) = (\mathcal{L}_0,\tau_0)$.

$\mathcal{T}$ is ordered as follows: $(\mathcal{L}_1,\tau_1) \leq (\mathcal{L}_2,\tau_2)$ if $\mathcal{L}_1 \subset \mathcal{L}_2$ and $\tau_1 \subset \tau_2$. It will be shown below that this order on $\mathcal{T}$ is inductive and that a maximal element $(\mathcal{L},\tau)$ has $\mathcal{L} = \mathcal{L}^1(\mathcal{R},M)$. This will prove the theorem: Any lifting T subordinate to τ reproduces the open and closed sets of τ_0 (34.9), which are exactly the sets SA with $A \in T(\mathcal{R}_0,M)$ (34.7). Hence T agrees with S on $T(\mathcal{R}_0,M)$ and thus on $\mathcal{L}^{\infty}(\mathcal{R}_0,M)$ (34.4).

To see that $\mathcal{T}$ is inductively ordered, consider a chain c in $\mathcal{T}$. Two cases are distinguished.

(a) c has no countable cofinal subset. In this case we put $\mathcal{L}' = \bigcup \{\mathcal{L}: (\mathcal{L},\tau) \in c\}$ and let τ' be the topology spanned by $\bigcup\{\tau: (\mathcal{L},\tau) \in c\}$. Since every $(\mathcal{L}',M)$-integrable function is the limit of a sequence in $\mathcal{L}'$, it lies in one of the $\mathcal{L}$ with $(\mathcal{L},\tau) \in c$ and thus in $\mathcal{L}'$; and so $\mathcal{L}'$ equals $\mathcal{L}^1(\mathcal{L}',M)$. A similar reason shows that τ' is $(\mathcal{L}',M)$-dense. Obviously $(\mathcal{L}',\tau')$ is an upper bound for the chain c.

(b) If there is a countable, cofinal, increasing sequence $(\mathcal{L}_n,\tau_n)$ in c, put $\mathcal{L}' = \mathcal{L}^1(\cup \mathcal{L}_n, M)$. The definition of τ' is more complicated: For every n choose a fixed lifting T_n of $(\mathcal{L}_n,M)$ subordinate to τ_n (34.9). For any $(\mathcal{L}',M)$-measurable set A, put

$$\Theta(A) = [\lim T_n E^n A = 1],$$

where $E^n A$ is a conditional expectation of A under $\mathcal{L}_n$. Let us show that Θ is a predensity for $(\mathcal{L}',M)$. $(\mathcal{L}_n, T_n E^n A)$ is a countable increasing martingale and thus converges a.e. to A (31.8). Therefore $\Theta(A) \doteq A$ and (D1) follows. The monotonicity of E^n and T_n yield (D2) immediately. To show (D3), let $A_1,\ldots,A_k \in T(\mathcal{L}',M)$ be such that $A_1 \cap \ldots \cap A_k \doteq 0$. Then $A_1 + \ldots + A_k \lesssim k-1$ and so $T_n E^n A_1 + \ldots + T_n E^n A_k \leq k-1$: The $T_n E^n A_i$ cannot all converge to one in the same point, and therefore $\Theta(A_1) \cap \ldots \cap \Theta(A_k) = \emptyset$. (D4) is evident.

Now, τ' is defined to be the dense topology associated with Θ. It is clearly $(\mathcal{L}',M)$-dense; and so it has only to be shown finer than all the τ_n. If $A \in \tau_n$, then $E^m A_n \doteq A_n$ and $T_m E^m A_n = T_m A_n \supset A_n$ for all $m \geq n$ (34.9). Hence $\Theta(A) \supset A$ and $A = \Theta(A) \cap A \in \tau'$; and thus $\tau_n \subset \tau'$ for all integers n: $(\mathcal{L}',\tau')$ is, indeed, an upper bound for the chain c.

$\mathcal{T}$ is now known to be inductively ordered and so has a maximal element $(\mathcal{L},\tau)$. It must still be shown that $\mathcal{L} = \mathcal{L}^1(\mathcal{R},M)$. By way of contradiction assume this is not so. Then there is an $(\mathcal{R},M)$-integrable set A not belonging to $\mathcal{L}$ (8.4). Let A_o denote its essential τ-open kernel (34.11), i.e.,

$$\underset{o}{A} = \bigcup \{U \in \tau : M(U \setminus A) = 0\},$$

and consider also the essential τ-open kernel $\underset{o}{A}'$ of its complement A'. Since $\underset{o}{A} \cap \underset{o}{A}' = \emptyset$, we may change A on a negligible set and arrive at $\underset{o}{A} \subset A$, $\underset{o}{A}' \subset A'$. Now, put

$$\mathcal{L}' = \{Af + A'g: f,g \in \mathcal{L}\} \quad \text{and} \quad \tau' = \{UA \cup VA': U,V \in \tau\}.$$

Clearly, $\mathcal{L}'$ is an integration lattice equal to $\mathcal{L}^1(\mathcal{L}',M)$ and properly containing $\mathcal{L}$; indeed

$$A^{\bullet} = \sup\{A^{\bullet}f^{\bullet}: 1^{\bullet} \geq f^{\bullet} \in L^1_+(\mathcal{L},M)\}$$

belongs to $L^1(\mathcal{L}',M)$ but not to $L^1(\mathcal{L},M)$. The $(\mathcal{L}',M)$-measurable sets are

$$T(\mathcal{L}',M) = \{AK \cup A'L: K,L \in T(\mathcal{L},M)\}.$$

Also, $\tau' \subset T(\mathcal{L}',M)$ is a topology finer than τ and $(\mathcal{L}',M)$-dense. The only thing possibly not obvious here is that τ' has no non-void negligible set. But if $UA \cap VA' \in \tau'$ is negligible, then $U \dot{\subset} A'$ and $V \dot{\subset} A$, $U \subset \underset{o}{A}'$ and $V \subset \underset{o}{A}$, and so $UA = VA' = \emptyset$ by construction. Hence $(\mathcal{L}',\tau')$ is an element of $\mathcal{T}$ properly exceeding $(\mathcal{L},\tau)$, in contradiction to the maximality of the latter.#

<u>35.2. Theorem</u>. (von Neumann, Maharam, Ionescu-Tulcea). For any strong and essential upper gauge $(\mathcal{R},M)$ the following are equivalent:

(i) There exists a lifting for $(\mathcal{R},M)$;

(ii) There exists a dense sublattice $\mathcal{D}$ of $\mathcal{L}^1(\mathcal{R},M)$ so that $(\mathcal{D},M)$ is B-continuous;

(iii) There exists a dense topology for $(\mathcal{R},M)$;

(iv) $(\mathcal{R},M)$ is strictly localizable.

Moreover, if any infinite $(\mathcal{R},M)$-adequate partition P is given, then there is a lifting T such that $TK = K$ for all $K \in P$.

Proof. Only (iv) $\Rightarrow$ (i) needs to be proved (34.1, 34.9, 16.4). For any K in the given partition P, choose an upper integral $(\mathcal{R}|K,N_K)$ (on K) equivalent to the finite upper gauge $(\mathcal{R}|K,M_K)$ (15.8), and let $(\mathcal{R}|K)_0$ denote the integration lattice (on K) consisting of the scalar multiples of the identity. Further, let S_K denote the lifting for $((\mathcal{R}|K)_0,N_K)$ which sends rK to rK for $r \in \mathbb{R}$. Due to the theorem above, there is a lifting T_K for $(\mathcal{R}|K,N_K)$ which extends S_K. T_K is also a lifting for $(\mathcal{R}|K,M_K)$ (18.1). The desired lifting T will be obtained by piecing together the liftings T_K, $K \in P$.

A function f in $\mathcal{L}^\infty(\mathcal{R},M)$ has $(\mathcal{R}|K,M_K)$-measurable restriction $f|K$ to K (18.15), and we may define

$$T_x f = T_{Kx}(f|K) \qquad , \text{ for } x \in K \in P,\ f \in L^\infty(\mathcal{R},M);$$

For every x in the negligible set $X - \cup P$, select a non-zero character T_x of $L^\infty(\mathcal{R},M)$. The function Tf is then defined on all of X and is $(\mathcal{R},M)$-measurable (18.9). T is evidently a lifting for $(\mathcal{R},M)$. If P is infinite, then the functions in $L^\infty(\mathcal{R},M)$ that vanish outside a finite union of sets in P form a proper ideal I of $L^\infty(\mathcal{R},M)$. For $x \in X - \cup P$, the character T_x may then be chosen to annihilate I (20.8). This results in $TK = K$ for $K \in P$.#

B. Existence of strong liftings

35.3. Proposition. If $\mathcal{R}$ is countably generated then every upper gauge $(\mathcal{R},M)$ admits a strong lifting.

Proof. Since $\mathfrak{R}$ is strictly localizable (16.1), there is a lifting S for $(\mathfrak{R},M)$. Let $\mathfrak{R}^c$ be a countable uniformly dense subset of $\mathfrak{R}$ (1.3), and let $\underline{X}$ be the support of $(\mathfrak{R},M)$ (16B). The set N of points $x \in \underline{X}$ at which $S_x\varphi \neq \varphi(x)$ for some $\varphi \in \mathfrak{R}^c$, is $(\mathfrak{R},M)$-negligible. For every $x \in N$ choose a character T_x of L^∞ that extends the character $\varphi \to \varphi(x)$ of $\mathfrak{R}$ (20.9), and put

$$T_x f = \begin{cases} S_x f & \text{for } x \notin N \\ T_x f & \text{for } x \in N \end{cases} \quad , \text{ for } f \in L^\infty .$$

Evidently T is a lifting for $(\mathfrak{R},M)$ and reproduces $\mathfrak{R}^c$ on $\underline{X}$. Since a lifting is uniformly continuous (34A), it reproduces all of $\mathfrak{R}$ on $\underline{X}$.#

35.4. Proposition. Again, let $\mathfrak{R}$ be an arbitrary integration lattice. The set $M_0^L(\mathfrak{R})$ of measures $m \in M^B(\mathfrak{R})$ such that $(\mathfrak{R},\bar{m}^B)$ admits a strong lifting, is a band.

The set $M^L(\mathfrak{R})$ of measures $m \in M^B(\mathfrak{R})$ that are majorized by an upper integral admitting a strong lifting is a band, and $M_0^L(\mathfrak{R}) \subset M^L(\mathfrak{R})$.

Proof. We use criterion 2.21. Let $n \in M^L(\mathfrak{R})$ be majorized by the upper integral $(\mathfrak{R},N)$, let T be a strong lifting for $(\mathfrak{R},N)$, and let $m \ll n$. In showing that $m \in M^L(\mathfrak{R})$, we may assume that $n = N|\mathfrak{R}_+$ and that m is positive.

Let g be a positive locally $(\mathfrak{R},N)$-integrable derivative for m (22.6) and set $M = gN$ and $G = T[g \neq 0]$. $(\mathfrak{R},M)$ is an upper integral for $(\mathfrak{R},m)$ and

$$Sf = T(Gf) \quad , \ f \in L^\infty(\mathfrak{R},M)$$

defines a strong lifting for $(\mathfrak{R},M)$, if altered suitably on the M-negligible set $X - G$. If $N = \bar{n}^B$ then $M = \bar{m}^B$ (22.8), and the corresponding result for $M_0^L(\mathcal{R})$ follows. The details are left to the reader.

Next, let D be a family of mutually disjoint positive measures in $M_0^L(\mathcal{R})$ with sum $n \in M^B(\mathcal{R})$, and for $m \in D$ let $(\mathcal{R}, M_m)$ be an upper integral for $(\mathcal{R}, m)$ with strong lifting S_m. Clearly, $\sum \{M_m : m \in D\} = N$ is an upper integral for n and is B-continuous and therefore strictly localizable. Let P be an $(\mathcal{R}, N)$-adequate partition consisting of dominated sets (16.4). For each $K \in P$ there are at most countably many elements in the family D_K of measures $m \in D$ having $\int K dm > 0$. Every set $K \subset P$ is the disjoint union of countably many dominated $(\mathcal{R}, N)$-integrable sets K_m, $m \in D_K$, having the property that $\int K dm = \int K_m dm$ for $m \in D_K$. Replacing P by $\{K_m : m \in D_K, K \in P\}$ and then D by $\{Km: K \in P, m \in D\}$, we may assume that there is a one-to-one correspondence $m \longleftrightarrow K_m$ of D with P such that $\int X dm = \int K_m dm$ for all $m \in D$. Redefining N accordingly, we may define

$$Tf = \sum \{K_m S_m f : m \in D\} \qquad \text{for } f \in L^\infty(\mathcal{R}, N).$$

Routine arguments, which are left to the reader, show that T becomes a strong lifting if suitably altered on $X - \cup \{K_m \cap TK_m : m \in D\}$, and that $N = \bar{n}^B$ if $M_m = \bar{m}^B$ for all $m \in D$.#

C. Supplements

35.5. A locally compact space is said to have the strong lifting property (sℓp) if for every positive Radon measure m, m^B admits a strong lifting.

(i) If every compact space has the sℓp, then so does every locally compact space.

(ii) If $[0,1]^a$ has the sℓp for every cardinal a, then every compact space has the sℓp.

(iii) If 2^a has the sℓp then so does $[0,1]^a$.

(iv) A closed subspace of a space with the sℓp has the sℓp.

Problem. Does the product of two spaces with the sℓp have the sℓp?

LITERATURE: The proof 35.1 and 35.2 is essentially that of Ionescue-Tulcea [22]. For liftings commuting with the action of a group see [23].

36. Disintegration of measures

As a typical application of liftings it will now be shown that a measure m on $(X,\mathcal{R})$ with image $n = pm$ on $(Y,\mathcal{S})$ can be decomposed into the integral of a field λ of measures on $(X,\mathcal{R})$ over n:

$$m = \int \lambda dn.$$

The precise data considered are the following. $(\mathcal{R},M)$ is an essential upper integral for the positive measure m on $(X,\mathcal{R})$; $(Y,\mathcal{S})$ is another integration lattice, which will be assumed to separate the points of Y; and p is an $(\mathcal{R},M)$-adequate and tame map from X into Y. Resuming the notation of section 28, we denote by n the image of m under p and write $N = pM$.

A. Definition. An $(\mathcal{S},N)$-integrable field λ on Y of measures on $\mathcal{R}$ is called an (integrable) disintegration of m along p if

$$m = \int \lambda dn. \qquad \#$$

The following lemma is the first step toward establishing the existence of disintegrations. It is given mostly because all known disintegration results are obtained from variants of its proof.

36.1. Lemma. Suppose $(\mathcal{S},N)$ has a lifting T and set $\mathcal{D} = T(L' \cap L^\infty)$. Then there is an $(\mathcal{S},N)$-integrable disintegration of m,

$$\lambda: Y \to I_+(\mathcal{R}),$$

such that $\langle\lambda;\varphi\rangle \in \mathcal{D}$ for all $\varphi \in \mathcal{R}$.

Proof. For every $y \in Y$ and $\varphi \in \mathfrak{R}$, put

$$\langle \lambda_y ; \varphi \rangle = T_y E^p \varphi .$$

Since both E^p and the character T_y are linear and positive, λ_y is a positive measure. Since E^p is contractive in 1-norm and in ∞-mean (29.3, 29.6), $E^p \varphi$ belongs to $\mathcal{L}^1(\mathcal{S},N) \cap \mathcal{L}^\infty(\mathcal{S},N)$, and so $\langle \lambda ; \varphi \rangle$ belongs to $\mathfrak{D}$. Finally

$$\int \langle \lambda ; \varphi \rangle dn = \int TE^p \varphi dn = \int \varphi dm = m(\varphi) . \#$$

The measures λ_y will, in general, not even be S-continuous, and so the integration theory of fields of measures developed in section 25 will not be applicable. The situation turns out to be satisfactory, though, if $(\mathfrak{R},M)$ is tight.

36.2. Lemma. Let T, $\mathfrak{D}$, λ be as in the lemma above. If X is compact in the $\mathfrak{R}$-topology then λ_y is B-continuous for all $y \in Y$.

If, additionally, p is uniformly continuous and T is strong, then

$$\operatorname{supp} \lambda_y \subset p^{-1}(y)$$

for all y in the support of n.

Proof. The first statement follows from the fact that a Radon measure is B-continuous (4.11); indeed, λ_y has an extension to a Radon measure on X, since $\mathfrak{R}$ is uniformly dense in $C(X)$ (4.10, 5.3).

For the second statement, let $\varphi \in \mathfrak{R}$ and $\psi \in \mathcal{S}$. Then $\varphi \cdot \psi \circ p$ belongs to $C(X)$, and so, by continuity,

$$\langle \lambda_y ; \varphi \cdot \psi \circ p \rangle = T_y E^p (\varphi \cdot \psi \circ p) = T_y (\psi E^p \varphi) = \psi(y) \cdot TE^p \psi = \psi(y) \langle \lambda_y ; \varphi \rangle$$

for all $y \in \operatorname{supp}(\mathcal{S},N)$. On taking the infimum over all $\psi \in \mathcal{S}_+$ that equal one in y,

$$\langle \lambda_y; \varphi p^{-1}(y) \rangle = \langle \lambda_y; \varphi \rangle.$$

The assumption that $\mathcal{S}$ separates the points of Y is needed here. This shows that λ_y has support in $p^{-1}(y)$ for all y in $\operatorname{supp}(\mathcal{S},N) = \operatorname{supp} n$.#

36.3. Theorem. Suppose $(\mathcal{R},M)$ is essential and tight; then $(\mathcal{S},N)$ is B-continuous (28.11) and hence admits a lifting T. Put $\mathcal{D} = T(L' \cap L^\infty)$.

(1) There is a $(\mathcal{D},N)$-B-adequate field λ of tight measures on $\mathcal{R}$ such that

$$m = \int \lambda dn \quad \text{and} \quad \int X d\lambda_y = 1 \qquad \text{for all } y \in Y.$$

(2) If T is strong then λ_y is concentrated on $p^{-1}(y)$ for all $y \in Y$, in the sense that

$$(*) \qquad \lambda_y^B(X - p^{-1}(y)) = 0.$$

Definition. A disintegration $m = \int \lambda dn$ into B-continuous measures is said to be strong if it satisfies (*).

Proof of 36.3. (1) Let Q be a T-invariant $(\mathcal{S},N)$-adequate partition of Y. For every $L \in Q$, $p^{-1}(L)$ is $(\mathcal{R},M)$-integrable (28.5) and so can be $(\mathcal{R},M)$-adequately covered by a countable collection P_L of disjoint, non-negligible, compact sets on which p is uniformly continuous. From 28.2, $P = \cup \{P_L : L \in Q\}$ is an $(\mathcal{R},M)$-adequate partition. Set

$$\langle\lambda^K;\varphi\rangle = TE^P(\varphi K) \qquad , \text{ for } \varphi\in\mathcal{R},\ K\in P,$$

and define

$$\langle\lambda;\varphi\rangle = \sum\{\langle\lambda^K;\varphi\rangle : K\in P\}.$$

The sum exists, since its finite partial sums are majorized in absolute value by $TE^P|\varphi|$; this shows at the same time that $\langle\lambda_y;\varphi\rangle \le 1$ for $1 \ge \varphi \in \mathcal{R}_+$, so that

$$0 \le \int X d\lambda \le 1 \qquad \text{everywhere.}$$

Since λ_y^K is tight for all K and all y, and the tight measures form a band (24.2), λ_y is tight. Evidently, $\langle\lambda;\varphi\rangle \in \mathcal{D}_\uparrow^B$ for all $\varphi\in\mathcal{R}_+$, and so λ is $(\mathcal{D},N)$-B-adequate. From

$$\int \langle\lambda;\varphi\rangle dn = \sum\{\int\langle\lambda^K;\varphi\rangle dn : K\in P\}$$

$$= \sum\{\int E^P K\varphi dn : K\in P\} = \sum\{\int K\varphi dm : K\in P\} = m(\varphi),$$

the field λ is an integrable disintegration of m.

If T is strong, then $\operatorname{supp}\lambda_y^K \subset K\cap p^{-1}(y)$ for all $y\in\operatorname{supp}(\mathcal{S},N)$ and all $K\in P$ (35.2). From

$$\langle\lambda^K;\varphi\rangle = TE^P(\varphi\cdot K\cdot L\circ p) = L\langle\lambda^K;\varphi\rangle \qquad , \text{ for } K\in P_L,$$

we get

$$\lambda_y = \sum\{\lambda_y^K : K\in P_L\} \qquad , \text{ for } y\in L\in Q,$$

and

$$\lambda_y = 0 \qquad , \text{ for } y\in X - \cup Q.$$

Hence

$$\lambda_y^B(X - p^{-1}(y)) = \sum \lambda_y^{KB}(X - p^{-1}(y)) = 0$$

in either case.

To satisfy the last requirement, that $\int X d\lambda \equiv 1$, λ will have to be altered on a negligible set. This will be done after a short discussion of the integration theory of λ.

B. The integration theory of $\int \lambda dm$

From 26.3, the new upper integral

$$M' = \int \overline{\lambda}^B d\overline{N}$$

on $\mathcal{R}$ is B-continuous, and hence smaller than m^B (9.1). Since the compact sets are $(\mathcal{R},m^B)$-dense in X (24.1), they form an $(\mathcal{R},M')$-adequate cover. If f is an $(\mathcal{B},N)$-measurable function then the compact subsets of X on which $f \circ p$ is uniformly continuous are $(\mathcal{R},M)$-dense (28.11), hence $(\mathcal{R},m^B)$-dense in all compact sets, and so form an $(\mathcal{R},M')$-adequate cover. Therefore $f \circ p$ is $(\mathcal{R},M')$-measurable (18.9). In particular, if $\psi \in \mathcal{B}$ then $\psi \circ p$ is $(\mathcal{R},M')$-measurable. As $(\mathcal{R},M')$ is regular (25.6), $\psi \circ p$ is $(\mathcal{R},M')$-integrable, and therefore p is $(\mathcal{R},M')$-adequate. Both $(\mathcal{R},M)$ and $(\mathcal{R},M')$ are upper integrals for m and thus coincide on the sets $p^{-1}(L)$, $L \in Q$, which are integrable for both. From Fubini's theorem,

$$N(L) = M(p^{-1}(L)) = M'(p^{-1}(L)) = \int L\left(\int p^{-1}(L) d\lambda\right) dn \leq \int L\left(\int X d\lambda\right) dn \leq N(L).$$

Therefore, $\int X d\lambda = 1$ a.e. on L.

At points $y \in Y$ where $\int X d\lambda_y < 1$, λ_y may be replaced by a point measure on the fibre $p^{-1}(y)$. The field λ so obtained has all the properties claimed in theorem 36.3. This finishes the proof of 36.3 and yields the following further information.

36.4. Corollary. Suppose $m = \int \lambda dn$ is a disintegration as in the theorem. Then

$$M' = \int \overline{\lambda}^B d\overline{N}$$

defines a B-continuous upper integral, $(\mathcal{R},M')$, for m having the following properties: the map p is $(\mathcal{R},M')$-adequate and tame, and $pM' = N$; the compact sets on which p is uniformly continuous constitute an $(\mathcal{R},M')$-adequate cover; and a function f on Y is $(\mathcal{S},N)$-measurable if and only if $f \circ p$ is $(\mathcal{R},M')$-measurable; if f is a Banach-valued $(\mathcal{R},M')$-integrable (or $(\mathcal{R},m^B)$-integrable) function, then f is $(\mathcal{R},\overline{\lambda}_y^B)$-integrable for almost all $y \in Y$, $\int f d\lambda$ is $(\mathcal{S},N)$-integrable, and

$$\int f dm = \int (\int f d\lambda) dn;$$

if f is an $(\mathcal{R},M')$-measurable function then f is $(\mathcal{R},\lambda_y)$-measurable for almost all $y \in Y$, and if f is Banach-valued then $\langle \lambda^B; f \rangle$ is $(\mathcal{S},N)$-measurable.#

36.5. Example. Let X be a polish (24.7) or a Souslin (24.8) space, $\mathcal{R}$ an integration lattice on X which separates the points of X, and m a positive S-measure on $\mathcal{R}$. Since $\mathcal{R} \subset UC_b(X)_\uparrow^B$ and $UC_b(X)$ is countably generated, m is B-continuous (3.16).

Furthermore, let f: X → Y be a surjective map (e.g., the quotient map

X → X/R, where R is an equivalence relation on X). Let us assume that

there is a countably generated integration lattice $\mathcal{S}$ on Y separating

the points of Y and satisfying $\varphi \circ f \in \mathcal{L}^1(\mathcal{R}, m^S)$ for all $\varphi \in \mathcal{S}$. (We say

that Y (or R) is countably separated with respect to m.)

The map f: X → Y is then scalarly measurable and, since Y is

separable and metrizable in the $\mathcal{S}$-uniformity, is actually measurable

(19.10). Hence f is $(\mathcal{R}, m^S)$-adequate. Write n = f(m). Since $(X, \mathcal{R})$

is radonian (24.7, 24.8, 24.4), m is tight. Hence so is n (28.12).

Since $\mathcal{S}$ is countably generated, there exists a strong lifting for

$(\mathcal{S}, n^B)$ (35.3), and therefore a strong disintegration of m into tight

measures (36.3)

$$ m = \int \lambda dn. \qquad\qquad \# $$

C. Supplements

36.6. Let X = Y × A be the product of two compact spaces, and let
n be the image on Y, under the natural projection p: X → Y, of the
positive Radon measure m on X. Then there exists an $\tilde{n}^B$-integrable disintegration $m = \int \lambda dn$ such that $\int X d\lambda_y = 1$ and supp $\lambda_y \subset p^{-1}(y)$ for
all y ∈ Y. (Hint: choose a lifting T for $(Y, \tilde{n}^B)$, put $\lambda_y(\varphi \otimes \psi) = \varphi(y) T_y E^p \psi$
for $\varphi \in C(Y)$ and $\psi \in C(A)$, and extend by linearity and uniform
continuity. [22].)

36.7. Let X be a compact space, m a positive Radon measure on X,
and M an upper integral for m. There exists a strong lifting T for
M if and only if for all compact spaces Y, continuous maps p from Y
to X and measures n on Y such that p(n) = m, there is a strong disintegration $n = \int \lambda dm$ along p. (Hint: take for Y the spectrum of
$L^\infty(M)$ and for p the natural map: $Y → X_s$; cf. 20.12.)

36.8. Let $(X,\mathcal{R},m)$ and $(Y,\mathcal{S},n)$ be two tight positive elementary integrals and $p: X \to Y$ a uniformly continuous map such that the pre-image of an $\bar{n}^B$-negligible set is $\bar{m}^B$-negligible. Then there exists a field $\lambda: X \to M_+^T(\mathcal{S})$ such that $\langle\lambda;\varphi\rangle$ is $(\mathcal{S},\bar{n}^B)$-integrable for all $\varphi \in \mathcal{R}$ (and all $\varphi \in \mathcal{L}^1(\mathcal{R},m^B)$) and

$$\int \langle\lambda;\varphi\rangle dn = \int \varphi dm.$$

Moreover, if $(\mathcal{S},\bar{n}^B)$ admits a strong lifting, the λ_y can be chosen to be concentrated on $p^{-1}(y)$, for each $y \in Y$.

(Hint: For $K \subset X$ compact, $p(Km) \ll n$; and $m = \sup\{Km: K \text{ compact}\}$. This observation reduces the problem to the case $pm \ll n$. Simplify further by assuming that X and Y are locally compact (24.1).)

36.9. Let E be a separable Banach space. A <u>support function</u> is a sublinear and continuous real-valued function q on E. Define $\|q\| = \sup\{q(\xi): \xi \in E_1\}$. Next let $(X,\mathcal{R},M)$ be a strictly localizable upper integral and $q: x \to q_x$ a weakly measurable ($\langle q;\xi\rangle: x \to q_x(\xi)$ is measurable for all $\xi \in E$) family of support functions such that $\|q\|: x \to \|q_x\|$ has $M_1(\|q\|) < \infty$ and $M_\infty(\|q\|) < \infty$. Define the weak integral Q by

$$Q(\xi) = \int q_x(\xi) dm(x) \qquad (\xi \in E,\ m = M|\mathcal{R}).$$

<u>Theorem</u>. (Strassen). Q is a support function and $\|Q\| \le M_1(\|q\|)$. If $\eta \in E'$ is majorized by Q ($\eta(\xi) \le Q(\xi)$ for $\xi \in E$), then there exists a field $\zeta: X \to E'$ of continuous linear maps $\zeta_x \in E'$ with the following properties:

(i) $\langle\zeta;\xi\rangle: x \to \zeta_x(\xi)$ is measurable and smaller than $\langle q;\xi\rangle$, for each $\xi \in E$.

(ii) $\zeta(X)$ is relatively $\sigma(E',E)$-compact.

(iii) $\eta(\xi) = \int\langle\zeta_x;\xi\rangle dm(x)$, for $\xi \in E$.

(Hint: Extend η, Q to a linear form and support function $\tilde{\eta},\tilde{Q}$ on L_E^∞, respectively, so that $\tilde{\eta} \le \tilde{Q}$. Show that $\tilde{\eta}$ is continuous in 1-mean on $L_E^\infty \cap L_E^1$, and produce ζ from 34.6.)

The condition that E be separable can be replaced by the following one: there is a negligible set $A \subset X$ with $T\langle q;\xi\rangle \le \langle q;\xi\rangle$ on A for all $\xi \in E$.

36.10. Let $(X,\mathcal{R})$ and $(Y,\mathcal{S})$ be dominated integration lattices, M an upper integral for $m \in M_+^*(\mathcal{R})$, $p: X \to Y$ an $(\mathcal{R},M)$-adequate and tame map such that $(\mathcal{S},pM)$ is strictly localizable. Let $f: X \to E$ be a measurable map with values in a locally compact space E. For $\varphi \in C_{00}(E)$ define

$$\langle\lambda_y,\varphi\rangle = T_y E^p(\varphi \circ f) \qquad , y \in Y.$$

The field λ of Radon measures on E is $(\mathcal{S},pM)$-integrable and tame, and $\int \lambda d(pm) = f(m)$. λ is called the <u>conditional distribution</u> of f under p. (T is a lifting for $(\mathcal{S},pM)$.)

LITERATURE: [22]. Cf. also [11], [20].

§2. Integral representations of linear maps

To avoid repetitions, the following data are fixed for the next two sections. $\mathcal{R}$ is a dominated integration lattice, and m is a positive scalar measure on $\mathcal{R}$; $(X,\mathcal{R},M)$ is an essential and strictly localizable upper integral for m; and E, F, G are three Banach spaces; finally, $n: \mathcal{R} \to F$ is a measure of finite variation and absolutely continuous with respect to m.

37. Radon-Nikodym theorems

It will now be investigated under which circumstances there is a derivative of n with respect to m.

A. Weakly locally integrable derivatives

37.1. Proposition. Let T be any lifting for $(\mathcal{R},M)$ and let V be any norming subspace of F' (4D). There is then a V-scalarly measurable map g from X to V' having the following properties.

(i) $\|g\|$ is locally integrable and $\|g\| = d|n|/dm$.

(ii) $\langle n(\varphi);v\rangle = \int \langle g;v\rangle \varphi dm$, for all $v \in V$ and $\varphi \in \mathcal{R}$.

(iii) Let $\mathcal{R}_1[M] = \{\varphi \in \mathcal{R}: M(|\varphi|) \leq 1\}$. For every $K \in T(\mathcal{R},M)$, $g(TK)$ is contained in the V-weak closure (in V') of

$$\int \mathcal{R}_1[M]Kdn = \int \mathcal{R}_1[M]d(Kn).$$

(iv) There is a negligible set $B \subset X$ such that

$$T\langle g;v\rangle = \langle g;v\rangle \text{ on } X-B \qquad , \text{ for all } v \in V.$$

Proof. Fix $\gamma = d|n|/dm \in \mathcal{M}^{\infty}_{\overline{\mathbb{R}}}(\mathcal{R},M)$ by $T\gamma = \gamma$ (22.6, 34.3). Then γ is finite except in the points of a negligible set B. For any $v \in V$, fix

$$\gamma^v = d\langle n;v\rangle/dm \in \mathcal{M}^{\infty}_{\overline{\mathbb{R}}}(\mathcal{R},M) \text{ by } T\gamma^v = \gamma^v.$$

Since $|\langle n;v\rangle| \leq \|v\||n| = \|v\|\gamma m$, we have

$$|\gamma^v(x)| \leq \|v\|\gamma(x) \qquad , \text{ for all } x \in X.$$

In particular, γ^v is finite except possibly in the points of B. For every $x \in X - B$, the map $v \to \gamma^v(x)$ defines a linear form $g(x)$ on V of norm not exceeding $\gamma(x)$, by

$$\langle g(x);v\rangle = \gamma^v(x) \qquad (34.5).$$

For $x \in B$ define $g(x) = 0$.

From its very definition,

$$\langle n(\varphi);v\rangle = \langle n;v\rangle(\varphi) = \int \langle g(x);v\rangle\varphi dm$$

for all $v \in V$ and $\varphi \in \mathcal{R}$. Since

$$\|g(x)\| = \sup\{|\langle g(x);v\rangle| \wedge n: v \in V_1; n \in \mathbb{N}\} \leq \gamma(x),$$

$\|g\|$ belongs to $\mathfrak{D}_\uparrow$ and so is measurable and locally integrable (34.1); and since $\|g\|m$ majorizes n and thus $|n|$ (3B), $\|g\| = \gamma$ a.e.: $\|g\| = d|n|/dm$.

To show (iii), let $K \in T(\mathcal{R},M)$ and let $v \in V$ belong to the polar of $\int \mathcal{R}_1 K dn$, i.e.,

$$\langle\int \mathcal{R}_1 K dn;v\rangle = \int \mathcal{R}_1 K\langle g;v\rangle dm \in (-\infty,1].$$

Then $\sup\{\int \varphi K\langle g;v\rangle dm: \varphi \in \mathcal{R}, M(\varphi) \leq 1\} \leq 1$, and so $K\langle g;v\rangle \leq 1$ a.e. (21.3). Hence $TK\langle g;v\rangle \leq 1$ everywhere. That is, $g(TK)$ is contained in the bipolar of $\int \mathcal{R}_1 K dn$, as claimed.#

Definition. We say that a function $g: X \to V'$ with properties (i), (ii) above is a V'-valued V-scalarly locally integrable derivative of n with respect to m.

We ask ourselves now when g can be chosen to have its values in F. The following definition describes sufficient conditions on n for this to be the case.

Definition. (i) $WC(n,M,V)$ denotes the family of $(\mathcal{R},M)$-measurable sets K for which the measure Kn is V-weakly M-compact; and n is said to be almost V-weakly M-compact if $WC(n,M,V)$ is $(\mathcal{R},M)$-dense.

(ii) $WC(n,V)$ denotes the family of $\mathcal{R}$-Baire sets (6.16) K for which the measure Kn is V-weakly $|n|$-compact; and n is said to be almost V-weakly $|n|$-compact if $WC(n,V)$ is $(\mathcal{R},n^S)$-dense.

(If $V = F'$ in (i) or (ii), it is suppressed in the notation.)

(iii) $C(n,M)$ and $C(n)$ are the families of $(\mathcal{R},M)$-measurable or $\mathcal{R}$-Baire sets K, respectively, for which Kn is (norm,M)-compact or (norm, $|n|$)-compact, respectively; and n is said to be almost M-compact (almost $|n|$-compact), if $C(n,M)$ (or $C(n)$) is dense for $(\mathcal{R},M)$ (or for $(\mathcal{R},|n|^S)$, respectively).#

The use of the word "almost" is in the spirit of 19B. It will turn out below (37.7) that the almost weakly $|n|$-compact measures and the almost $|n|$-compact measures coincide and are exactly those measures that have a locally integrable derivative with respect to any (strictly localizable) measure with respect to which they are absolutely continuous. We start by investigating the consequences of the weaker conditions (i) and (ii) of the definition.

37.2. Lemma. (i) The measure n is almost V-weakly M-compact if and only if it is almost V-weakly $|n|$-compact, and in this instance

$$|n| = \sup\{K|n|: K \in \mathcal{K}_{\downarrow}^{S},\ Kn \text{ is V-weakly } |n|\text{-compact}\}.$$

(ii) n is almost M-compact if and only if it is almost $|n|$-compact.

Proof. Put $N = \gamma M$ with $\gamma = d|n|/dm$, and set $\mathcal{R}_1[M] = \{\varphi \in \mathcal{R}: M(|\varphi|) \leq 1\}$. Let $L \in \mathcal{B}(\mathcal{R}^S)$ be a dominated Baire set and put $L_0 = L \cap [\gamma = 0]$, $L_1 = L \cap [\gamma \neq 0]$. We may assume that L_0, L_1 are, again, Baire sets (18.19).

$\Rightarrow$: The sets $K \in \mathcal{B}(\mathcal{R},M) \cap WC(n,M,V)$ for which γK is bounded and bounded away from zero and Kn is V-weakly M-compact are $(\mathcal{R},M)$-dense, and so $(\mathcal{R},N)$-dense in L_1. For each of them, let $\tilde{K}$ denote an equivalent Baire set contained in L (18.19). From $N(L-K) = \int (L-\tilde{K})dn$, the sets $\tilde{K}$ are $|n|^S$-dense in L. Since the norms $\tilde{K}M$ and KN are equivalent, $\int \tilde{K}\mathcal{R}_1[N]dn$ is relatively V-weakly compact in F. Hence n is almost V-weakly $|n|$-compact.

$\Leftarrow$: The Baire sets $K \subset L_1$ for which γK is bounded and bounded away from zero and Kn is V-weakly N-compact are $(\mathcal{R},N)$-dense in L_1 (9.1), and so are $(\mathcal{R},M)$-dense in L_1 (22.4). Since the sets $K \cup L_0$ are $(\mathcal{R},M)$-dense in L, and

$$\int (K \cup L_0)\mathcal{R}_1[M]dn = \int K\mathcal{R}_1[M]dn$$

is relatively V-weakly compact in F, $WC(n,M,V)$ is an $(\mathcal{R},M)$-adequate cover dense in every dominated Baire set. As $WC(n,M,V)$ is closed under taking integrable subsets, it is dense.

The equality in (i) follows trivially, and (ii) is proved literally the same way as (i).#

37.3. Corollary. If n is almost V-weakly M-compact or almost V-weakly $|n|$-compact, then there is a V-scalarly locally integrable derivative g having values in F and satisfying (i) through (iv) of 37.1.

Proof. Let P be an adequate partition in $WC(n,M,V)$. For $K \in P$, $g(TK) \in F$ by 37.1(iv). If we enlarge B by $X - \cup \{K \cap TK: K \in P\}$ and set g equal to zero in all points of the larger set, then g satisfies the description of this corollary.#

37.4. Corollary. Let E be a Banach space and $n: \mathcal{R} \to E'$ a measure of finite variation, absolutely continuous with respect to m.

Then there exists a (E-)scalarly locally integrable E'-valued derivative of n with respect to m.

Proof. n is E-weakly $|n|$-compact since the unit ball of E' is E-weakly compact.#

B. Locally integrable derivatives

It will now be investigated when g can be chosen to be locally (Bochner) integrable.

Let us see that an additional requirement on n is necessary. If $g = dn/dm$ is a locally $(\mathcal{R},M)$-integrable derivative with values in F, then the family $U(g)$ of sets on which g is uniformly continuous is dense. For each $K \in U(g)$, the map

$$\varphi \to \int \varphi K dn = \int \varphi K g dm \qquad (\varphi \in \mathcal{R})$$

from $\mathcal{R}$ to F is strongly M-compact; indeed, from 10.5, the image of the M-unit ball of $\mathcal{R}$ is contained in the multiple $\int K dm \cdot \overline{ce}(g(K))$ of the closed convex equilibrated hull of $g(K)$, which is relatively norm-compact (5.7 and 10.4). We are therefore led to require that n be almost (norm,M)-compact. It is actually sufficient to require that n be almost weakly M-compact. We set out considering the case that n is weakly M-compact. Then $\int \mathcal{R}_1[M]dn$ is bounded in F, and so n is majorized by a multiple of M. We may extend n to a weakly compact map from the Banach space $L^1(\mathcal{R},M)$ to F, and find the following situation.

37.5. Theorem. (i) Let $(\mathcal{R},M)$ be a strictly localizable upper integral and $n: L^1(\mathcal{R},M) \to F$ a weakly compact linear map of norm $\|n\|$. There is a derivative $g = dn/dm \in \mathcal{L}_F^\infty(\mathcal{R},M)$ satisfying

(a) $M_\infty(g) = \|n\|$ and

(b) $g(X)$ is relatively weakly compact in F.

(ii) Let $M_{FW}^\infty(\mathcal{R},M)$ denote the subspace of L_F^∞ of those classes that contain a function with relatively weakly compact image. Then M_{FW}^∞ is a closed subspace of L_F^∞, and the map $g \to gm$ is an isometry of M_{FW}^∞ with the space $L_0(L^1,F)$ of weakly compact linear maps from L^1 to F.

Proof. Let $g: X \to F$ be a scalarly integrable derivative of n with respect to m having the properties (i) through (iv) of 37.1. Since the variation of n is majorized by a multiple of m, $\|g\| \in \mathcal{L}^\infty$. From 37.1(iii) $g(X)$ is relatively weakly compact in F.

It has to be shown that g is measurable. To do this, we assume at first that M is finite. It will suffice to show that, then, $n(\mathcal{R})$ lies in a separable subspace F^c of F. Indeed, if this is so, let $\eta \in F'$ be such that $\langle F^c;\eta\rangle = 0$. Then

$$0 = \langle \int \varphi dn;\eta\rangle = \int \langle g;\eta\rangle \varphi dm \qquad , \text{ for all } \varphi \in \mathcal{R},$$

and hence $\langle g;\eta\rangle = 0$ a.e. Since $T\langle g;\eta\rangle = \langle g;\eta\rangle$ except on a negligible set B that is independent of η, $\langle g,\eta\rangle = 0$ on $X - B$. That is $g(X - B) \subset F^c$, and so g is almost separably-valued. From 19.9, g is then measurable.

In order to show that $n(\mathcal{R})$ is separable, it is sufficient to prove that the image $n(\mathcal{R}^1)$ of the sup-norm unit ball $\mathcal{R}^1$ of $\mathcal{R}$ is relatively norm-compact in F. Indeed, a norm-compact subset spans a separable subspace, and $n(\mathcal{R})$ is contained in the subspace spanned by $n(\mathcal{R}^1)$.

Since F is a metric space, it will suffice to show that any sequence $(\xi_k) = (n(\varphi_k))$, $\varphi_k \in \mathcal{R}^1$, has a convergent subsequence. Let $\mathcal{R}^c$ denote the integration lattice spanned by the φ_k, and let F^c denote the separable subspace of F generated by $n(\mathcal{R}^c)$ (1.3). The image of the M-unit ball of $\mathcal{R}^c$ is relatively weakly compact in F^c, since F^c is weakly closed in F (Theorem of Hahn-Banach; it is here that the assumption $V = F'$ is needed). From 37.3, there exists a scalarly $(\mathcal{R}^c,M)$-integrable and F^c-valued derivative of $n^c = n|\mathcal{R}^c$ with respect to $m^c = m|\mathcal{R}^c$; and since F^c is separable, it is even $(\mathcal{R}^c,M)$-measurable:

$$g^c = dn^c/dm^c \qquad (19.9).$$

Since $\|g^c\| \leq \|g\| \in \mathcal{L}^\infty$ and M is finite, g^c is integrable. From 10.9,

$$n^c(\mathcal{R}^{c1}) = \int \mathcal{R}^{c1} g^c dm$$

is relatively norm-compact in F^c. Therefore $(\xi_k) = (n^c(\varphi_k))$ has a norm-convergent subsequence.

We know now that if M is finite and $n\colon L^1 \to F$ weakly M-compact, then the scalarly locally integrable derivative $g\colon X \to F$ of 37.3 is actually (Bochner) integrable. For arbitrary $(\mathcal{R},M)$, let P denote a

T-invariant adequate partition (34B, T as in 37.1). For each $K \in P$, Kg is a $(K\mathcal{R}, KM)$-integrable derivative of Kn with respect to Km. Hence g is measurable on each $K \in P$, and so g is measurable also in this case (18.9).

From $\|n\| = \sup\{\|\int g\varphi dm\|: \varphi \in \mathcal{R}, M(\varphi) \leq 1\}$ and 21.3, $\|n\| = M_\infty(g)$.

(ii): From 10.6, the uniform limit of functions in M^∞_{FW} (measurable functions with relatively weakly compact image) belongs to $\mathcal{M}^\infty_{FW}$. Hence M^∞_{FW} is closed in L^∞_F. The rest is clear from (i) and 10.5.#

<u>37.6. Corollary</u>. Let $g: X \to F$ be a scalarly measurable map that is almost weakly compact valued (in the sense that the family $WC(g)$ of sets $K \subset X$ for which $g(K)$ is relatively weakly compact in F is dense).

Then g is scalarly equivalent to a measurable map h, and

$$\|h\| \leq \|g\| \quad \text{a.e.}$$

<u>Proof</u>. Let $K \in WC(g)$ and consider the measure n defined by

$$\langle n(\varphi);\eta\rangle = \int K\langle g;\eta\rangle\varphi dm \qquad , \varphi \in \mathcal{R}, \eta \in F'.$$

Since g is bounded on K, this integral exists and defines a bounded linear map $n: \mathcal{R} \to F''$. Let $C \subset F$ denote the (weakly compact) weakly closed convex equilibrated hull of $g(K)$. If η is in the polar C^o of C and $\varphi \in \mathcal{R}_1[M]$, then $\langle n(\varphi);\eta\rangle \leq 1$ (10.5). Therefore $n(\varphi) \subset C^{oo} = C \subset F$, and n is weakly compact. There exists a derivative $h_K = Kh_K = dn/dm \in \mathcal{L}^\infty_F(\mathcal{R}, M)$. If $P \subset WC(g)$ is an adequate partition,

$$h = \sum \{h_K: K \in P\}$$

is evidently measurable (18.9) and scalarly equivalent to g.

Let $K \in WC(g)$ and $a = \sup\{\|g(x)\|: x \in K\}$. From

$$|\langle n(\varphi);\eta\rangle| \le a\|\eta\| \int K|\varphi|dm,$$

the measure aKm majorizes $n = Khm$, and hence $Kh \lesssim aK$ (22.3). Since $WC(g)$ is dense $h \dot{\lesssim} g$.#

<u>37.7. Theorem</u>. Let $n: \mathcal{R} \to F$ be a measure of finite variation, absolutely continuous with respect to m; and let $(\mathcal{R},M)$ be a strictly localizable upper integral for m.

Then there is a locally integrable derivative $dn/dm \in \mathcal{L}_F^{\ell}(\mathcal{R},M)$ if and only if either of the following conditions holds.

(i) n is almost (norm,M)-compact.

(ii) n is almost weakly M-compact.

(iii) n is almost $|n|$-compact.

(iv) n is almost weakly $|n|$-compact.

Consequently, $L_F^{\ell}(\mathcal{R},M)$ is isomorphic with the vector space of almost (weakly) M-compact measures in $(m)_F$; and $L_F^{\infty}(\mathcal{R},M)$ is isometrically isomorphic with the space $AL_0(L^1,F)$ of almost (weakly or norm) compact linear maps from L^1 to F.

Conditions (iii) and (iv) show that having a locally integrable derivative dn/dm is a property of n alone, and is independent of the (strictly localizable) measure m with respect to which n is absolutely continuous.

<u>Proof of 37.7</u>. The implications (i) $\Leftrightarrow$ (iii) $\Rightarrow$ (ii) $\Leftrightarrow$ (iv) are clear from 37.2, and (i) is necessary for the existence of a locally integrable derivative. If (iv) holds, then the F-valued, almost weakly compact valued,

scalarly locally integrable derivative g is scalarly equivalent to a measurable function h with $\|h\| \dot{\leq} \|g\| \in \mathcal{L}^{\ell}(\mathcal{R},M)$. Hence h is locally integrable, and from

$$\langle \int \varphi h dm;\eta \rangle = \int \langle g;\eta \rangle \varphi dm = n(\varphi) \qquad , \varphi \in \mathcal{R},\ \eta \in F',$$

h is a locally integrable derivative of n with respect to m.#

37.8. Corollary. Suppose $n: \mathcal{R} \to F$ is a measure of finite variation, with values in the reflexive space F, and absolutely continuous with respect to m. If the upper integral $(\mathcal{R},M)$ for m is strictly localizable, then there is an F-valued locally $(\mathcal{R},M)$-integrable derivative $g = dn/dm$, and

$$\int f dn = \int \langle g;f \rangle dm \qquad , \text{ for } f \in \mathcal{L}^1_{F'}(\mathcal{R}, \|g\| M).$$

Proof. Since the unit ball of a reflexive Banach space is weakly compact, n is weakly $|n|$-compact.#

37.9. Corollary. Let $U: \mathcal{R} \otimes E \to G$ be a linear map whose variation $|U|$ is finite and absolutely continuous with respect to m. If the upper integral $(\mathcal{R},M)$ for m is strictly localizable then there is a function g on X with values in $L(G',E')$, and having the following properties:

(i) for every $\eta \in G'$ and $\varphi \in \overline{\mathcal{R}}_E$, the function $\langle \varphi;g\eta \rangle$ is integrable, and

$$\langle U(\varphi);\eta \rangle = \int \langle \varphi;g\eta \rangle dm;$$

(ii) $$\|g\| \leq d|U|/dm.$$

Moreover, if E is reflexive, then g can be chosen so that $g\eta: X \to E'$ is locally integrable for all $\eta \in G'$.

<u>Proof</u>. Put $|U| = \gamma m$ with $\gamma = d|U|/dm \in \mathcal{L}^{\ell}(\mathcal{R},M)$. Choose an $(\mathcal{R},M)$-adequate partition P such that γ is bounded on each set of P. Discarding the trivial case (16.9), assume that P is infinite and fix a lifting T for $(\mathcal{R},M)$ which leaves P invariant (35.2). The E'-valued measures $U\eta$, defined for each $\eta \in G'$ by

$$\langle (U\eta)(\varphi);\xi \rangle = \langle U(\varphi\xi);\eta \rangle \qquad , \xi \in E, \varphi \in \mathcal{R},$$

all have finite variation (4.14) absolutely continuous with respect to m; and since the unit ball of E' is E-weakly compact, there exists a scalarly E-measurable function $\tilde{g}_\eta : X \to E'$ such that

$$\langle U(\varphi\xi);\eta \rangle = \langle (U\eta)(\varphi);\xi \rangle = \int \langle \varphi\xi;\tilde{g}_\eta \rangle dm$$

for all $\varphi \in \mathcal{R}$ and $\xi \in E$. Clearly $\|\tilde{g}_\eta\| \lesssim \|\eta\|\gamma$. Replacing each $\tilde{g}_\eta$ by $g_\eta = T\,\tilde{g}_\eta$ (34.6) we obtain E-scalarly locally integrable derivatives g_η which are T-invariant and majorized by $\|\eta\|T\gamma$. For every x with $(T\gamma)(x) < \infty$, the map $\eta \to g_\eta(x) \in E'$ is linear and bounded and defines a continuous linear map $g(x) \in L(G',E')$ by

$$g(x)\eta = g_\eta(x) \qquad , \eta \in G'.$$

By its very definition, g satisfies (i); and (ii) follows by linearity and continuity in (γM)-mean from

$$\langle U(\varphi\xi);\eta \rangle = \langle (U\eta)(\varphi);\xi \rangle = \int \langle \varphi\xi; g\eta \rangle dm \qquad (\xi \in E, \eta \in G).$$

If E is reflexive, the $g_\eta = g\eta$ so defined are automatically strongly measurable.#

C. Supplements

37.10. $L_F^\infty(\mathcal{R},M)$ is isometrically isomorphic with the space of bounded almost weakly compact maps from $L^1(\mathcal{R},M)$ to F (F a Banach space, $(\mathcal{R},M)$ a strictly localizable upper integral).

37.11. Any S-measure $n\colon \mathcal{R} \to F$ of finite variation is almost weakly $|n|$-compact if and only if it is almost (norm-) $|n|$-compact (use the Gelfand-Bauer transform). Moreover, if n is $*$-continuous, it enjoys these properties if and only if

$$|n| = \sup\{K|n| : K \in \mathcal{K}^*(\mathcal{R}),\ Kn \text{ is (weakly) } |n|\text{-compact}\}.$$

LITERATURE: [22], [11].

38. Linear maps on L_E^p

Let $1 \leq p < \infty$ be a number and p' its conjugate, defined by $1/p + 1/p' = 1$. Furthermore, $(\mathfrak{R},M)$ is an essential and strictly localizable upper integral for the positive measure m.

A. The dual of L_E^p

We consider a linear map U from $L_E^p = L_E^p(\mathfrak{R},M)$ to the Banach space G and try to represent it in integral form

$$(1) \qquad U(f) = \int fgdm \qquad , \ f \in \mathcal{L}_E^p,$$

where g is an appropriate function from X to $L(E,G)$. Such g hardly ever exists, and we need some assumptions to come even close:

It will henceforth be assumed that U is continuous in the p-norm. Adjusting by taking multiples, it is assumed that the operator norm of U is one. The semivariation $\|U\|$ is then majorized by M_p; and if $p = 1$ then U has actually finite variation. Each of the measures $U\eta$, $\eta \in G'$ (4.14) has finite variation on $\mathcal{L}^p$, and

$$(2) \qquad |U\eta|(f) \leq \|\eta\| \cdot \|U\|(f) \leq \|\eta\| M_p(f) \qquad , \text{ for } f \in \mathcal{L}_+^p.$$

Since the M_p-negligible sets and the M-negligible sets coincide, $|U\eta|$ is absolutely continuous with respect to m on $\mathfrak{R}^S$ (6.7), and so in particular is S-continuous (3.10). Since the unit ball E_1' of E' is E-weakly compact, there is an E-scalarly measurable and E'-valued derivative g_η such that (37.3)

$$(3) \qquad \|g_\eta\| = d|U\eta|/dm \in \mathcal{L}^\ell(\mathfrak{R},M) \qquad \text{and}$$

(4) $$\langle U(f);\eta\rangle = \int \langle g_\eta ; f\rangle dm \qquad , \text{ for } f \in \mathcal{L}^1_E(\mathcal{R}, \|g_\eta\|M).$$

From (2), $\int \|g_\eta\| f dm \leq \|\eta\| M_p(f)$, for $f \in \mathcal{L}^p_+$, and so

$$M_{p'}(g_\eta) \leq \|\eta\| \qquad (21.3).$$

From Minkowski's inequality (21.1),

$$\|g_\eta\| M \leq \|\eta\| M_p \text{ , and so}$$

(5) $$\langle U(f);\eta\rangle = \int \langle g_\eta ; f\rangle dm \qquad , \text{ for } f \in \mathcal{L}^p_E(\mathcal{R},M).\#$$

Definition. For any essential upper integral $(\mathcal{R},M)$, any q with $1 < q \leq \infty$, and any Banach space F, let $\mathcal{L}^q_{F'W}(\mathcal{R},M)$ denote the vector space of F'-valued, scalarly F-measurable functions g satisfying

$$M_q(g) < \infty.$$

$\mathcal{L}^q_{F'W}$ consists of the F-scalarly q-integrable functions. Two such functions g,h, are called scalarly equivalent, if

$$\langle g;\xi\rangle \doteq \langle g';\xi\rangle \ (M) \qquad , \text{ for all } \xi \in F;$$

and $L^q_{F'W}$ denotes the Banach space of equivalence classes modulo scalar equivalence. #

38.1. Theorem. Let $(\mathcal{R},M)$ be an essential and strictly localizable upper integral with restriction m to $\mathcal{R}$; let E be a Banach space and $1 \leq p < \infty$.

Then the dual of L^p_E is isometrically isomorphic with $L^{p'}_{E'W}$ under the duality

$$(g,f) \to \int \langle g;f\rangle dm\,, \qquad (f \in L^{p'}_{E'W},\ f \in L^{p}_{E},\ 1/p + 1/p' = 1.)$$

Moreover, if E is reflexive, then

$$(L^p_E)' = L^{p'}_{E'}\ .$$

<u>Proof</u>. We apply the foregoing to $G = \mathbb{R}$, with $\eta = 1$; it shows that $(L^p_E)' \subset L^{p'}_{E'W}$. Conversely, every function $g \in \mathcal{L}^{p'}_{E'W}$ defines a linear functional U by (5).

If E is reflexive, $E = E''$, then a scalarly E-measurable function $g: X \to E'$ is scalarly measurable and by (37.6), is scalarly equivalent to a strongly measurable function. Hence $L^{p'}_{E'W} = L^{p'}_{E'}$ in this case.#

B. Linear maps

<u>38.2. Theorem</u>. Let $(\mathcal{R},M)$, m, p as before, and let $U: L^p_E \to G$ be a continuous map of norm one. Suppose E is reflexive.

If U has finite variation, then $|U|$ is absolutely continuous with respect to m and there exists a function $g: X \to L(G',E')$ with the following properties:

(i) $g\eta:\ x \to g(x)\eta$ is $\mathcal{L}^{p'}_{E'W}$ for all $\eta \in G'$;

(ii) $\langle f;g\eta\rangle$ is integrable for all $f \in \mathcal{L}^p_E$, $\eta \in G'$ and
$\langle U(f);\eta\rangle = \int \langle f;g\eta\rangle dm$;

(iii) $\|g\| \le d|U|/dm$.

<u>Proof</u>. From 4D,

$$\begin{aligned}|U|(\varphi) &= \sup \Sigma \|U(\varphi_i)\| = \sup \Sigma \langle U(\varphi_i);\eta_i\rangle \\ &= \sup \Sigma (U\eta_i)(\varphi_i) = \sup \Sigma |U\eta_i|(\|\varphi_i\|) \\ &\le \sup \Sigma |U\eta_i|(\varphi)\end{aligned}$$

where $\varphi_i \in \mathcal{R} \otimes E$, $\sum \|\varphi_i\| \leq \varphi \in \mathcal{R}_+$, $\eta_i \in G'_1$, and where the suprema are taken over all finite sums with such ingredients. Since the $|U\eta_i|$ are absolutely continuous with respect to m, so is $|U|$. From 37.5, there exists a map $g: X \to L(G',E')$ such that $\langle g\eta;\varphi\rangle$ is integrable for all $\eta \in G'$ and $\varphi \in \overline{\mathcal{R}}_E$, and

$$\langle U(\varphi);\eta\rangle = \int \langle g\eta;\varphi\rangle dm.$$

Comparison with (5) shows that $g\eta = g_\eta$ scalarly; hence $\langle g\eta;f\rangle$ is integrable for all $f \in \mathcal{L}^p_E$ and

$$\langle U(f);\eta\rangle = \int \langle g\eta;f\rangle dm,$$

and since g_η belongs to $\mathcal{L}^{p'}_{E'W}$ so does $g\eta$. Equation (iii) is clear.#

LITERATURE: [11], [14], [22].

SUBJECT INDEX

10A refers to subsection 10A; 10.2 refers to the result with this number in section 10; and (15) refers to page number 15. If the first letter of a word is underlined then more information can be found where this word is listed.

INDEX OF NOTATIONS

Sets are identified with their characteristic functions (5).
Concerning the symbol * see (32).
Restriction of A to B: A|B.

LITERATURE

[1] BAUER, H. Probability Theory and Elements of Measure Theory. Holt, Rinehart and Winston, N.Y., 1972.

[2] ________. Sur l'equivalence des theories de l'integration selon N. Bourbaki et selon M. H. Stone. Bull. Soc. Math. France, 85 (1957) 51-75.

[3] BERBERIAN. S. Measure and Integration. Macmillan, N.Y., 1965.

[4] BILLINGSLEY, P. Ergodic Theory and Information. John Wiley Inc., N.Y., 1965.

[5] BOURBAKI, N. Integration, Ch. I-IX. Hermann, Paris, 1965 (2nd ed.).

[6] CHATTERJI, S. D. Comments on the Martingale Convergence Theorem. In Symposium on Probability Methods in Analysis. Lecture Notes in Mathematics #31, Springer, N.Y., 1967.

[7] ________. Martingales of Banach-Valued Random Variables. Bull. Am. Math. Soc., 66 (1960) 395-398.

[8] ________. A Note on the Convergence of Banach-Space Valued Martingales. Math. Annalen, 153 (1964) 142-149.

[9] CHOQUET, G. Lectures on Analysis I, II, III. Benjamin, N.Y., 1969.

[10] DANIELL, D. J. A General Form of Integral. Annals of Mathematics, 19 (1917) 279-294.

[11] DINCULEANU, N. Vector Measures. VEB, Berlin, 1967.

[12] DIXMIER, J. Les C*-algebres et leurs representations. Gauthiers-Villars, Paris, 1964.

[13] DOOB, J. L. Stochastic Processes. Wiley, N.Y., 1953.

[14] DUNFORD, N. and SCHWARTZ, J. Linear Operators I and II. Interscience, N.Y., 1967 (4th ed.).

[15] EDWARDS, D. A. Functional Analysis. Holt, Rinehart and Winston, N.Y., 1965.

[16] FELLER, W. An Introduction to Probability Theory and its Applications, Vol. I, II. Wiley, N.Y., 1968.

[17] GILLMAN, L. and JERISON, M. Rings of Continuous Functions. van Nostrand, Princeton, 1960.

[18] HALMOS, P. R. Lectures on Ergodic Theory. Chelsea, N.Y., 1956.

[19] ________. Measure Theory. van Nostrand, Princeton, 1950.

[20] HAYES, C. A. and PAUC, C. Y. Derivation and Martingales. Springer, N.Y., 1970.

[21] HEWITT, E. and STROMBERG. K. Real and Abstract Analysis. Springer, N.Y., 1965.

[22] IONESCU-TULCEA, A. and C. Topics in the Theory of Lifting. Springer, N.Y., 1969.

[23] IONESCU-TULCEA, A. and C. On the Existence of a Lifting Commuting with the Left Translations of an Arbitrary Locally Compact Group. Proc. Fifth Berkeley Symposium on Math. Stat. and Probability, p. 63-97, Univ. of California Press (1967).

[24] ISBELL, J. Uniform Spaces. Am. Math. Soc., Providence, R.I., 1964.

[25] KAKUTANI, S. Concrete Representations of Abstract (M)-Spaces. Ann. of Math.(2), 42 (1941) 994-1024.

[26] KELLEY, J. L. General Topology. van Nostrand, Princeton, 1955.

[27] KOELZOW, D. Characterisierung der Masse welche zu einem Integral im Sinne von Stone oder von Bourbaki gehoeren. Arch. Math. 16 (1965) 200-207.

[28] _______. Differentiation von Massen. Lecture Notes, #65, Springer, 1968.

[29] KURATOWSKI, C. Topologie I and II, 2nd ed. Warsaw, 1948.

[30] LEWIS, D. R. Integration with Respect to Vector Measures. Pacific J. Math. 33/1 (1970) 157-167.

[31] LOEVE, M. Probability Theory. van Nostrand, Princeton, 1962.

[32] McSHANE, E. J. Integration. Princeton University Press, Princeton, 1944.

[33] MEYER, P. A. Probability and Potentials. Blaisdell, 1966.

[34] RICKART, C. E. Decomposition of Additive Set Functions. Duke Math. J., 10 (1943) 653-665.

[35] ROYDEN, H. L. Real Analysis. Macmillan, N.Y., 1963.

[36] SAKS, S. Theory of the Integral. Dover, N.Y., 1964.

[37] SCALORA, F. S. Abstract Martingale Convergence Theorems. Pacific J. Math., 2 (1961) 347-374.

[38] SCHAEFER, H. H. Topological Vector Spaces. Macmillan, N.Y., 1966.

[39] SCHWARTZ, L. Les mesures de Radon dans les espaces topologiques arbitraires. Lecture Notes, Paris, 1964.

[40] SEGAL, I. E. and KUNZE, R. A. Integrals and Operators. McGraw-Hill, N.Y., 1968.

[41] SIERPINSKI, W. General Topology, 2nd ed. Toronto, 1952.

[42] SIMMONS, G. F. Introduction to Topology and Modern Analysis. McGraw-Hill, N.Y., 1963.

[43] SION, M. On Analytical Sets in Topological Spaces. Trans. Amer. Math. Soc., 96 (1960) 341-354.

[44] _______. On Capacibility and Measurability. Ann. Inst. Fourier Grenoble, 13 (1963).

[45] SONDERMANN, D. Masse auf lokalbeschränkten Räumen. Ann. Inst. Fourier Grenoble, XIX/2 (1970) 35-113.

[46] STONE, M. H. Notes on Integration I-IV. Proc. Nat. Acad. Sci. USA, 34 (1948) 336-342, 447-455, 483-490; 35 (1949) 50-58.

[47] VARADARAJAN. Measures on Topological Spaces. Transl. Amer. Math. Soc. (2), 78 (1965 161-228.

[48] YOSIDA, K. Functional Analysis. Springer, Berlin, 1966.

[49] ZAANEN, Adriaan C. An Introduction to the Theory of Integration. Interscience, N.Y., 1958.

.ecture Notes in Mathematics

omprehensive leaflet on request

ol. 146: A. B. Altman and S. Kleiman, Introduction to Grothendieck uality Theory. II, 192 pages. 1970. DM 18,-

ol. 147: D. E. Dobbs, Cech Cohomological Dimensions for Com- utative Rings. VI, 176 pages. 1970. DM 16,-

ol. 148: R. Azencott, Espaces de Poisson des Groupes Localement ompacts. IX, 141 pages. 1970. DM 16,-

ol. 149: R. G. Swan and E. G. Evans, K-Theory of Finite Groups and rders. IV, 237 pages. 1970. DM 20,-

ol. 150: Heyer, Dualität lokalkompakter Gruppen. XIII, 372 Seiten. 970. DM 20,-

ol. 151: M. Demazure et A. Grothendieck, Schémas en Groupes I. SGA 3). XV, 562 pages. 1970. DM 24,-

ol. 152: M. Demazure et A. Grothendieck, Schémas en Groupes II. SGA 3). IX, 654 pages. 1970. DM 24,-

ol. 153: M. Demazure et A. Grothendieck, Schémas en Groupes III. SGA 3). VIII, 529 pages. 1970. DM 24,-

ol. 154: A. Lascoux et M. Berger, Variétés Kähleriennes Compactes. II, 83 pages. 1970. DM 16,-

ol. 155: Several Complex Variables I, Maryland 1970. Edited by Horváth. IV, 214 pages. 1970. DM 18,-

ol. 156: R. Hartshorne, Ample Subvarieties of Algebraic Varieties. V, 256 pages. 1970. DM 20,-

ol. 157: T. tom Dieck, K. H. Kamps und D. Puppe, Homotopietheorie. , 265 Seiten. 1970. DM 20,-

ol. 158: T. G. Ostrom, Finite Translation Planes. IV. 112 pages. 1970. M 16,-

ol. 159: R. Ansorge und R. Hass. Konvergenz von Differenzenver- hren für lineare und nichtlineare Anfangswertaufgaben. VIII, 145 eiten. 1970. DM 16,-

ol. 160: L. Sucheston, Constributions to Ergodic Theory and Proba- lity. VII, 277 pages. 1970. DM 20,-

ol. 161: J. Stasheff, H-Spaces from a Homotopy Point of View. , 95 pages. 1970. DM 16,-

ol. 162: Harish-Chandra and van Dijk, Harmonic Analysis on Reduc- ve p-adic Groups. IV, 125 pages. 1970. DM 16,-

ol. 163: P. Deligne, Equations Différentielles à Points Singuliers eguliers. III, 133 pages. 1970. DM 16,-

ol. 164: J. P. Ferrier, Seminaire sur les Algebres Complétes. II, 69 pa- es. 1970. DM 16,-

ol. 165: J. M. Cohen, Stable Homotopy. V, 194 pages. 1970. DM 16,-

ol. 166: A. J. Silberger, PGL_2 over the p-adics: its Representations, pherical Functions, and Fourier Analysis. VII, 202 pages. 1970. M 18,-

ol. 167: Lavrentiev, Romanov and Vasiliev, Multidimensional Inverse oblems for Differential Equations. V, 59 pages. 1970. DM 16,-

ol. 168: F. P. Peterson, The Steenrod Algebra and its Applications: conference to Celebrate N. E. Steenrod's Sixtieth Birthday. VII, 7 pages. 1970. DM 22,-

ol. 169: M. Raynaud, Anneaux Locaux Henséliens. V, 129 pages. 1970. M 16,-

ol. 170: Lectures in Modern Analysis and Applications III. Edited by T. Taam. VI, 213 pages. 1970. DM 18,-

l. 171: Set-Valued Mappings, Selections and Topological Properties 2^X. Edited by W. M. Fleischman. X, 110 pages. 1970. DM 16,-

ol. 172: Y.-T. Siu and G. Trautmann, Gap-Sheaves and Extension Coherent Analytic Subsheaves. V, 172 pages. 1971. DM 16,-

ol. 173: J. N. Mordeson and B. Vinograde, Structure of Arbitrary urely Inseparable Extension Fields. IV, 138 pages. 1970. DM 16,-

ol. 174: B. Iversen, Linear Determinants with Applications to the card Scheme of a Family of Algebraic Curves. VI, 69 pages. 1970. M 16,-

ol. 175: M. Brelot, On Topologies and Boundaries in Potential Theory. , 176 pages. 1971. DM 18,-

ol. 176: H. Popp, Fundamentalgruppen algebraischer Mannigfaltig- eiten. IV, 154 Seiten. 1970. DM 16,-

ol. 177: J. Lambek, Torsion Theories, Additive Semantics and Rings Quotients. VI, 94 pages. 1971. DM 16,-

Vol. 178: Th. Bröcker und T. tom Dieck, Kobordismentheorie. XVI, 191 Seiten. 1970. DM 18,-

Vol. 179: Seminaire Bourbaki – vol. 1968/69. Exposés 347-363. IV. 295 pages. 1971. DM 22,-

Vol. 180: Séminaire Bourbaki – vol. 1969/70. Exposés 364-381. IV, 310 pages. 1971. DM 22,-

Vol. 181: F. DeMeyer and E. Ingraham, Separable Algebras over Commutative Rings. V, 157 pages. 1971. DM 16,-

Vol. 182: L. D. Baumert. Cyclic Difference Sets. VI, 166 pages. 1971. DM 16,-

Vol. 183: Analytic Theory of Differential Equations. Edited by P. F. Hsieh and A. W. J. Stoddart. VI, 225 pages. 1971. DM 20,-

Vol. 184: Symposium on Several Complex Variables, Park City, Utah, 1970. Edited by R. M. Brooks. V, 234 pages. 1971. DM 20,-

Vol. 185: Several Complex Variables II, Maryland 1970. Edited by J. Horváth. III, 287 pages. 1971. DM 24,-

Vol. 186: Recent Trends in Graph Theory. Edited by M. Capobianco/ J. B. Frechen/M. Krolik. VI, 219 pages. 1971. DM 18,-

Vol. 187: H. S. Shapiro, Topics in Approximation Theory. VIII, 275 pages. 1971. DM 22,-

Vol. 188: Symposium on Semantics of Algorithmic Languages. Edited by E. Engeler. VI, 372 pages. 1971. DM 26,-

Vol. 189: A. Weil, Dirichlet Series and Automorphic Forms. V, 164 pages. 1971. DM 16,-

Vol. 190: Martingales. A Report on a Meeting at Oberwolfach, May 17-23, 1970. Edited by H. Dinges. V, 75 pages. 1971. DM 16,-

Vol. 191: Séminaire de Probabilités V. Edited by P. A. Meyer. IV, 372 pages. 1971. DM 26,-

Vol. 192: Proceedings of Liverpool Singularities – Symposium I. Edited by C. T. C. Wall. V, 319 pages. 1971. DM 24,-

Vol. 193: Symposium on the Theory of Numerical Analysis. Edited by J. Ll. Morris. VI, 152 pages. 1971. DM 16,-

Vol. 194: M. Berger, P. Gauduchon et E. Mazet. Le Spectre d'une Variété Riemannienne. VII, 251 pages. 1971. DM 22,-

Vol. 195: Reports of the Midwest Category Seminar V. Edited by J.W. Gray and S. Mac Lane.III, 255 pages. 1971. DM 22,-

Vol. 196: H-spaces – Neuchâtel (Suisse)- Août 1970. Edited by F. Sigrist, V, 156 pages. 1971. DM 16,-

Vol. 197: Manifolds – Amsterdam 1970. Edited by N. H. Kuiper. V, 231 pages. 1971. DM 20,-

Vol. 198: M. Hervé, Analytic and Plurisubharmonic Functions in Finite and Infinite Dimensional Spaces. VI, 90 pages. 1971. DM 16,-

Vol. 199: Ch. J. Mozzochi, On the Pointwise Convergence of Fourier Series. VII, 87 pages. 1971. DM 16,-

Vol. 200: U. Neri, Singular Integrals. VII, 272 pages. 1971. DM 22,-

Vol. 201: J. H. van Lint, Coding Theory. VII, 136 pages. 1971. DM 16,-

Vol. 202: J. Benedetto, Harmonic Analysis on Totally Disconnected Sets. VIII, 261 pages. 1971. DM 22,-

Vol. 203: D. Knutson, Algebraic Spaces. VI, 261 pages. 1971. DM 22,-

Vol. 204: A. Zygmund, Intégrales Singulières. IV, 53 pages. 1971. DM 16,-

Vol. 205: Séminaire Pierre Lelong (Analyse) Année 1970. VI, 243 pages. 1971. DM 20,-

Vol. 206: Symposium on Differential Equations and Dynamical Systems. Edited by D. Chillingworth. XI, 173 pages. 1971. DM 16,-

Vol. 207: L. Bernstein, The Jacobi-Perron Algorithm – Its Theory and Application. IV, 161 pages. 1971. DM 16,-

Vol. 208: A. Grothendieck and J. P. Murre, The Tame Fundamental Group of a Formal Neighbourhood of a Divisor with Normal Crossings on a Scheme. VIII, 133 pages. 1971. DM 16,-

Vol. 209: Proceedings of Liverpool Singularities Symposium II. Edited by C. T. C. Wall. V, 280 pages. 1971. DM 22,-

Vol. 210: M. Eichler, Projective Varieties and Modular Forms. III, 118 pages. 1971. DM 16,-

Vol. 211: Théorie des Matroïdes. Edité par C. P. Bruter. III, 108 pages. 1971. DM 16,-

Please turn over

Vol. 212: B. Scarpellini, Proof Theory and Intuitionistic Systems. VII, 291 pages. 1971. DM 24,-

Vol. 213: H. Hogbe-Nlend, Théorie des Bornologies et Applications. V, 168 pages. 1971. DM 18,-

Vol. 214: M. Smorodinsky, Ergodic Theory, Entropy. V, 64 pages. 1971. DM 16,-

Vol. 215: P. Antonelli, D. Burghelea and P. J. Kahn, The Concordance-Homotopy Groups of Geometric Automorphism Groups. X, 140 pages. 1971. DM 16,-

Vol. 216: H. Maaß, Siegel's Modular Forms and Dirichlet Series. VII, 328 pages. 1971. DM 20,-

Vol. 217: T. J. Jech, Lectures in Set Theory with Particular Emphasis on the Method of Forcing. V, 137 pages. 1971. DM 16,-

Vol. 218: C. P. Schnorr, Zufälligkeit und Wahrscheinlichkeit. IV, 212 Seiten 1971. DM 20,-

Vol. 219: N. L. Alling and N. Greenleaf, Foundations of the Theory of Klein Surfaces. IX, 117 pages. 1971. DM 16,-

Vol. 220: W. A. Coppel, Disconjugacy. V, 148 pages. 1971. DM 16,-

Vol. 221: P. Gabriel und F. Ulmer, Lokal präsentierbare Kategorien. V, 200 Seiten. 1971. DM 18,-

Vol. 222: C. Meghea, Compactification des Espaces Harmoniques. III, 108 pages. 1971. DM 16,-

Vol. 223: U. Felgner, Models of ZF-Set Theory. VI, 173 pages. 1971. DM 16,-

Vol. 224: Revètements Etales et Groupe Fondamental. (SGA 1). Dirigé par A. Grothendieck XXII, 447 pages. 1971. DM 30,-

Vol. 225: Théorie des Intersections et Théorème de Riemann-Roch. (SGA 6). Dirigé par P. Berthelot, A. Grothendieck et L. Illusie. XII, 700 pages. 1971. DM 40,-

Vol. 226: Seminar on Potential Theory, II. Edited by H. Bauer. IV, 170 pages. 1971. DM 18,-

Vol. 227: H. L. Montgomery, Topics in Multiplicative Number Theory. IX, 178 pages. 1971. DM 18,-

Vol. 228: Conference on Applications of Numerical Analysis. Edited by J. Ll. Morris. X, 358 pages. 1971. DM 26,-

Vol. 229: J. Väisälä, Lectures on n-Dimensional Quasiconformal Mappings. XIV, 144 pages. 1971. DM 16,-

Vol. 230: L. Waelbroeck, Topological Vector Spaces and Algebras. VII, 158 pages. 1971. DM 16,-

Vol. 231: H. Reiter, L^1-Algebras and Segal Algebras. XI, 113 pages. 1971. DM 16,-

Vol. 232: T. H. Ganelius, Tauberian Remainder Theorems. VI, 75 pages. 1971. DM 16,-

Vol. 233: C. P. Tsokos and W. J. Padgett. Random Integral Equations with Applications to Stochastic Systems. VII, 174 pages. 1971. DM 18,-

Vol. 234: A. Andreotti and W. Stoll. Analytic and Algebraic Dependence of Meromorphic Functions. III, 390 pages. 1971. DM 26,-

Vol. 235: Global Differentiable Dynamics. Edited by O. Hájek, A. J. Lohwater, and R. McCann. X, 140 pages. 1971. DM 16,-

Vol. 236: M. Barr, P. A. Grillet, and D. H. van Osdol. Exact Categories and Categories of Sheaves. VII, 239 pages. 1971, DM 20,-

Vol. 237: B. Stenström. Rings and Modules of Quotients. VII, 136 pages. 1971. DM 16,-

Vol. 238: Der kanonische Modul eines Cohen-Macaulay-Rings. Herausgegeben von Jürgen Herzog und Ernst Kunz. VI, 103 Seiten. 1971. DM 16,-

Vol. 239: L. Illusie, Complexe Cotangent et Déformations I. XV, 355 pages. 1971. DM 26,-

Vol. 240: A. Kerber, Representations of Permutation Groups I. VII, 192 pages. 1971. DM 18,-

Vol. 241: S. Kaneyuki, Homogeneous Bounded Domains and Siegel Domains. V, 89 pages. 1971. DM 16,-

Vol. 242: R. R. Coifman et G. Weiss, Analyse Harmonique Non-Commutative sur Certains Espaces. V, 160 pages. 1971. DM 16,-

Vol. 243: Japan-United States Seminar on Ordinary Differential and Functional Equations. Edited by M. Urabe. VIII, 332 pages. 1971. DM 26,-

Vol. 244: Séminaire Bourbaki - vol. 1970/71. Exposés 382-399. IV, 356 pages. 1971. DM 26,-

Vol. 245: D. E. Cohen, Groups of Cohomological Dimension One. V, 99 pages. 1972. DM 16,-

Vol. 246: Lectures on Rings and Modules. Tulane University Ring and Operator Theory Year, 1970-1971. Volume I. X, 661 pages. 1972. DM 40,-

Vol. 247: Lectures on Operator Algebras. Tulane University Ring and Operator Theory Year, 1970-1971. Volume II. XI, 786 pages. 1972. DM 40,-

Vol. 248: Lectures on the Applications of Sheaves to Ring Theory. Tulane University Ring and Operator Theory Year, 1970-1971. Volume III. VIII, 315 pages. 1971. DM 26,-

Vol. 249: Symposium on Algebraic Topology. Edited by P. J. Hilton. VII, 111 pages. 1971. DM 16,-

Vol. 250: B. Jónsson, Topics in Universal Algebra. VI, 220 pages. 1972. DM 20,-

Vol. 251: The Theory of Arithmetic Functions. Edited by A. A. Gioia and D. L. Goldsmith VI, 287 pages. 1972. DM 24,-

Vol. 252: D. A. Stone, Stratified Polyhedra. IX, 193 pages. 1972. DM 18,-

Vol. 253: V. Komkov, Optimal Control Theory for the Damping of Vibrations of Simple Elastic Systems. V, 240 pages. 1972. DM 20,-

Vol. 254: C. U. Jensen, Les Foncteurs Dérivés de lim et leurs Applications en Théorie des Modules. V, 103 pages. 1972. DM 16,-

Vol. 255: Conference in Mathematical Logic - London '70. Edited by W. Hodges. VIII, 351 pages. 1972. DM 26,-

Vol. 256: C. A. Berenstein and M. A. Dostal, Analytically Uniform Spaces and their Applications to Convolution Equations. VII, 130 pages. 1972. DM 16,-

Vol. 257: R. B. Holmes, A Course on Optimization and Best Approximation. VIII, 233 pages. 1972. DM 20,-

Vol. 258: Séminaire de Probabilités VI. Edited by P. A. Meyer. VI, 253 pages. 1972. DM 22,-

Vol. 259: N. Moulis, Structures de Fredholm sur les Variétés Hilbertiennes. V, 123 pages. 1972. DM 16,-

Vol. 260: R. Godement and H. Jacquet, Zeta Functions of Simple Algebras. IX, 188 pages. 1972. DM 18,-

Vol. 261: A. Guichardet, Symmetric Hilbert Spaces and Related Topics. V, 197 pages. 1972. DM 18,-

Vol. 262: H. G. Zimmer, Computational Problems, Methods, and Results in Algebraic Number Theory. V, 103 pages. 1972. DM 16,-

Vol. 263: T. Parthasarathy, Selection Theorems and their Applications. VII, 101 pages. 1972. DM 16,-

Vol. 264: W. Messing, The Crystals Associated to Barsotti-Tate Groups: with Applications to Abelian Schemes. III, 190 pages. 1972. DM 18,-

Vol. 265: N. Saavedra Rivano, Catégories Tannakiennes. II, 418 pages. 1972. DM 26,-

Vol. 266: Conference on Harmonic Analysis. Edited by D. Gulick and R. L. Lipsman. VI, 323 pages. 1972. DM 24,-

Vol. 267: Numerische Lösung nichtlinearer partieller Differential- und Integro-Differentialgleichungen. Herausgegeben von R. Ansorge und W. Törnig. VI, 339 Seiten. 1972. DM 26,-

Vol. 268: C. G. Simader, On Dirichlet's Boundary Value Problem. IV, 238 pages. 1972. DM 20,-

Vol. 269: Théorie des Topos et Cohomologie Etale des Schémas. (SGA 4). Dirigé par M. Artin, A. Grothendieck et J. L. Verdier. XIX, 525 pages. 1972. DM 50,-

Vol. 270: Théorie des Topos et Cohomologie Etle des Schémas. Tome 2. (SGA 4). Dirige par M. Artin, A. Grothendieck et J. L. Verdier. V, 418 pages. 1972. DM 50,-

Vol. 271: J. P. May, The Geometry of Iterated Loop Spaces. IX, 175 pages. 1972. DM 18,-

Vol. 272: K. R. Parthasarathy and K. Schmidt, Positive Definite Kernels, Continuous Tensor Products, and Central Limit Theorems of Probability Theory. VI, 107 pages. 1972. DM 16,-

Vol. 273: U. Seip, Kompakt erzeugte Vektorräume und Analysis. IX, 119 Seiten. 1972. DM 16,-

Vol. 274: Toposes, Algebraic Geometry and Logic. Edited by. F. W. Lawvere. VI, 189 pages. 1972. DM 18,-

Vol. 275: Séminaire Pierre Lelong (Analyse) Année 1970-1971. VI, 181 pages. 1972. DM 18,-

Vol. 276: A. Borel, Représentations de Groupes Localement Compacts. V, 98 pages. 1972. DM 16,-

Vol. 277: Séminaire Banach. Edité par C. Houzel. VII, 229 pages. 1972. DM 20,-